32

£ 15.95
m

Experimental Researches in Chemistry and Physics

Experimental Researches
in
Chemistry and Physics

MICHAEL FARADAY

Taylor & Francis

London • New York • Philadelphia

1991

UK   Taylor & Francis Ltd, 4 John St., London WC1N 2ET

USA   Taylor & Francis Inc., 1900 Frost Road, Suite 101, Bristol PA 19007

Foreword © J. M. Thomas 1991

British Library Cataloguing in Publication Data

Faraday, Michael *1791-1867*
   Experimental researches in chemistry and physics.
   1. Physical sciences
   I.Title
   500.2

   ISBN 0-85066-841-7

Library of Congress Cataloging-in-Publication Data

Faraday, Michael, 1791-1867.
   Experimental researches in chemistry and physics/Michael Faraday;
   with a foreword by J.M. Thomas.
      p. cm.
   Originally published in 1859.
   Includes index.
   ISBN 0-85066-841-7
   1. Physics—Experiments—History. 2. Chemistry—Experiments
   —History. I. Title.
   QC33.F23   1991
   530'.72—dc20   90-41828
                                   CIP

Printed in Great Britain by Burgess Science Press, Basingstoke

Foreword
to Bicentennial (1991) Edition
by
Professor J.M. Thomas, F.R.S.,

Fullerian Professor of Chemistry and
Director of The Royal Institution

Michael Faraday was incontestably one of the greatest experimental philosophers of all time, a figure that still bestrides the world of chemistry and physics like a Colossus. But in addition to his genius as a scientist, he also possessed numerous other qualities which have endeared him to succeeding generations: warm humanity, genuine humility and piety and a poetic ability to express, in beautifully evocative terms, his love of natural phenomena. 'Nothing is too wonderful to be true, if it be consistent with the laws of nature and, in such things as these, experiment is the best test of such consistency'. 'Let us now consider how wonderfully we stand upon this world. Here it is we are born, bred, and live, and yet we view these things with an almost entire absence of wonder.'

From his early days as an assistant bookbinder to his final years, when he had become one of the world's most famous scientists, he never ceased to be struck by a sense of awe in contemplating the inner quality of things. Faraday was endowed both with an irrepressible urge to write and a memorable way of communicating his thoughts. In his youth he kept a 'common place book' or notebook in which he jotted down ideas, facts, quotations and questions as they occurred to him. The earliest one has the following description on the first page: 'A collection of Notices, Occurrences, Events, etc., relating to the Arts and Sciences collected from the Public Papers, Reviews, Magazines, and other miscellaneous works. Intended to promote both Amusement and Instruction and also to corroborate or invalidate those theories which are continually starting in the world of science'. Posterity is indeed fortunate that this unique scholar and discoverer has bequeathed a gigantic store of recorded contributions which, more than a century after his death, continues to reward the historian of science, to kindle the hearts of the young and to inspire aspiring and mature research scientists alike.

Even if the corpus of Faraday's legacy of written material were only a fraction of what it is, the story of his achievements, and the manner in which they were accomplished, bears repetition. His rise from the village blacksmith's son to his role, first as a chemical assistant to Humphry Davy at the Royal Institution — with his remarkable assimilation of skills as an experimentalist that was soon to make him the premier analytical chemist of his age — through to his election as Fellow of the Royal Society and his appointment as the Fullerian Professor of

Chemistry at the Royal Institution, is one of the most romantic in the annals of science. He commenced his duties with Davy in March 1813. In October of that year Faraday accompanied Davy and his wife on a prolonged tour of Europe that brought him into contact with Ampère, Cuvier, Humboldt, Gay Lussac, Arago and Volta. When he returned to England in April 1815 his education as an experimental philosopher had advanced enormously and he had also become versed in French and Italian.

Largely because of their immediacy in the world of technology, and also because of their unification and extension by Clerk Maxwell and Einstein, Faraday's discoveries in electricity and magnetism, encompassing as they do[1] the suggestion of the ideas of a field and lines of force and demonstrations of electromagnetic rotation (1821), electromagnetic induction (1831), the identities of electricities from various sources (1832), electrostatics and dielectrics (1835), discharge of gases (1835), light, electricity and magnetism (1845) and diamagnetism (1845), tend to dwarf, at least in the minds of physicists, cosmologists and engineers, his contributions to other branches of science. Yet he is also one of the founders of chemistry. His pioneering work includes: the discovery and characterization of the compounds of chlorine and carbon (1820), the introduction of photochemistry as a preparative procedure (1821), the discovery of benzene (1825), researches on alloys and steel (1818), the manufacture of optical quality glass (1825), the liquefaction of gases (1823 and 1845), the laws of electrolysis (1833) and the introduction (with William Whewell, the philosopher and polymathic Master of Trinity College Cambridge) of the terms anode, cathode, ion etc. (1834).

In 1858 Faraday assembled his collected researches in chemistry and physics that had appeared over the preceding forty years or so in *Philosophical Transactions of the Royal Society, The Journal of the Royal Institution, Philosophical Magazine* and in other publications. This collection is revealing both in what it does and does not include. His first ever paper published in 1816, 'On the Native Caustic Lime of Tuscany' is reproduced primarily for touching, personal reasons: Faraday records (p.1), in a footnote added in 1858, 'It was the beginning of my communications to the public'; and it represents his first attempt in chemistry 'at a time when my fear was greater than my confidence, and both far greater than my knowledge'. More important is his choice of the whole of his (first) Bakerian lecture to the Royal Society — delivered in three instalments in 1829 — 'On the Manufacture of Glass for Optical Purpose'. Some historians of science hold that Faraday's researches on glass constitute a relatively barren period. True, he spent a long time on them. But Faraday's added footnote on p. 231 shows him to have been proud of the crucial role his pioneering work on the preparation of glass played in his own subsequent researches into diamagnetism and magneto-optics. The optical fibres that now dominate the world of communications are generated by spinning molten glass through small orifices in platinum-rich bushes, the very material which Faraday had so painstakingly established as the most appropriate for the production of defect-free glass.

The reader is struck by Faraday's versatility, originality, intellectual energy, the enormous sweep of his brush and his limpid style. For the practitioner, the book is a gold-mine strewn with glittering nuggets. Some of the experiments are ideally suited as class demonstrations for contemporary primary schoolchildren; others may be used, without modification, as practical exercises for pre-university or college students; whilst yet others identify themes that are currently at the frontier of modern chemical physics.

For example, Faraday's account (p. 57) 'On the Vapour of Mercury at common Temperatures' would appeal to enquiring young children even now. His summary of the 'Composition of Crystals of Sulphate of Soda' (p. 153) could be adopted, without modification, for children in their mid-teens. For those interested in lecture demonstrations, the description of the 'Apparatus for the combustion of the Diamond' (pp. 11–13) is archetypal; and details of the roaring flame given on p. 24 could readily be adapted for a Christmas Lecture aimed at young audiences. The observation (p. 212) that sulphur may remain liquid-like at room temperatures — well below its freezing point — ties up with work recently reported[2] at a Discussion meeting of the Royal Society of Chemistry on the nature of the liquid–vapour interface of gallium monitored by synchrotron radiation. The behaviour of the crystals formed when 'a strong solution of chloride of silver in ammonia was left for some weeks in a bottle stopped only by a piece of paper' (p. 19) is in line with modern theories of topochemistry[3,4]. And when Faraday writes (p. 393) '. . .at one time I hoped that I had altered one coloured ray (of light) into another by means of gold . . .' one is tempted to believe that Faraday, long before the discovery of second harmonic or sum-frequency generation, was seeking, with his unerring intuition, to devise a means of altering the intrinsic quality of a ray by a device that would nowadays be called a frequency doubler.

This book is suffused with the very essence of the scientific method that Karl Popper advocated a hundred years later in his book *Conjectures and Refutations*[5]. It also reveals many aspects of Faraday's character, in a manner that reinforces the long-recognized example he provides on how properly to pursue science; it reflects his basic philosophy and his generosity if spirit. Moreover, it provides fascinating insights into his psychology, the social context of his work, his skills as an experimentalist, his prodigious capacity for hard work and his intuitive grasp of the unity of science. Some of the articles spill over into fields of a more general nature like 'turning the tables' (see below) on spiritualists and their followers, and his exposition on mental education. We know from *Faraday's Diary*[6] that he described each experiment, in full and careful detail, on the day on which it was made. (What an example for modern researchers!) Many of his diary entries discussed the conclusions drawn from his observations. This is turn led him to philosophical reflections which appear — still redolent of their freshly minted character — in many of the articles herein reproduced. It has often been remarked that, for Faraday, life and science were indissolubly linked. This view is reinforced on reading *Experimental Researches in Chemistry and Physics*.

Present day research chemists who wish to know how best to attack the problem of identifying and characterizing a new compound or material would benefit from reading Faraday's account 'On two new Compounds of Chlorine and Carbon, and on a new Compound of Iodine, Carbon and Hydrogen' (pp. 33–53). Schoolteachers who wish to improve their pupils' writing up of their experiments have model presentations on p. 18 of this book ('Exposed to the air, these crytals . . .'). And if they wish to instil their pupils with enthusiasm and the joy of science, they could do worse than quote Faraday's words (p. 53): '. . . as yet, I have not succeeded in procuring an iodide of carbon, but intend to pursue these experiments in a brighter season of the year, and expect to obtain this compound'. Consider, again, the magisterial terms and broad perspectives that characterize the opening (p. 391) of his 1857 Bakerian Lecture on 'Experimental Relations of Gold (and other metals) to Light', which delves largely into the preparation and properties and colloidal dispersions of metals, a topic of growing importance today[7]:

> That wonderful production of the human mind, the undulatory theory of light, with the phenomena for which it strives to account, seems to me, who am only an experimentalist, to stand midway between what we may conceive to be the coarser mechanical actions of matter, with their explanatory philosophy, and that other branch, which includes, or should include, the physical idea of forces acting at a distance; . . . I have often struggled to perceive how far that medium might account for or mingle in with such actions generally; and to what extent experimental trials might be devised, which, with their results and consequences, might contradict, confirm, enlarge, or modify the idea we form of it, always with the hope that the corrected or instructed idea would approach more and more to the truth of nature, and in the fullness of time, concide with it.

His reference to 'contradict, confirm, enlarge etc. . . .' echoes the remarks of Hooke in the Preface of whose *Micrographia*[8] we read: '. . . we have the power of considering, comparing, altering, assisting and improving . . . the works of nature'. And we are aware that Faraday had in his possession a copy of the *Micrographia*.

In 1853, Faraday was bombarded with requests and enquiries pertaining to the supposed phenomenon of table-turning. This allegedly arose because of the consequences of the spiritualist movement which began in 1848 in Hydesville, New York with the rappings and knockings of the Fox girls[9]. It spread rapidly to other parts of the United States, this country and continental Europe. The spiritual phenomena encompassed the tilting and levitation of tables and chairs and the movement of objects in the dark and became the subjects of several books published in England and France in 1853. The rather rapid and uncritical acceptance of spiritualistic forces greatly disturbed Faraday because of what they revealed about the general level of intelligence. One may gauge the measure of his exasperation from the following excerpt of a letter that he wrote to Professor C.F. Schonbein, the German–Swiss scientist who discovered ozone. Only July 25, 1853 he wrote:

> I have not been at work except in turning the tables upon the table-turners, nor should I have done that, but that I thought it better to stop the inpouring flood by letting all know at once what my views and thoughts were. What a weak, credulous, incredulous, unbelieving, superstitious, bold, frightened, what a ridiculous world ours is, as far as concerns the mind of man. How full of inconsistencies, contradictions, and absurdities it is . . .

It is chastening to reflect that superstition and credulity are still widespread: many people have no idea that one cannot do things that violate the laws of physics.

This is the background to the two articles that appear on pp. 382–91. The first being a letter to the editor of *The Times,* the second a fuller and ingenious account of his experiments repudiating the claims of the table-movers published originally in *The Athenaeum* which continues today as the *New Statesman.*

The final article in this book contains his 'Observations on Mental Education' and summarizes the views which he aired at the Royal Institution on May 6, 1854 before His Royal Highness The Prince Consort, with whom Faraday had considerable rapport. This often reprinted article is replete with words of wisdom relevant to the present age. As Faraday makes clear at its outset, a discourse on mental education is tantamount to a discourse on judgement. Who can argue with his assertion (p. 465): '. . . but if we desire to know how far education is required, we do not consider the few who need it not, but the many who have it not; and in respect of judgement, the number of the latter is almost infinite.'? Throughout this article, Faraday's luminous intelligence and self-critical approach shine through. So does his essential humility. In another context, he once remarked 'The man who is certain he is right is almost sure to be wrong; and he has the additional misfortune of inevitably remaining so'.

I am indebted to Irena McCabe and Max Perutz for helpful comments.

References

1. R. King, *Michael Faraday of the Royal Institution* (The Royal Institution of Great Britain, London, 1973).
2. S.A. Rice, *Faraday Discussion No. 89* (Royal Society of Chemistry, London, in press).
3. H.W. Kohlschutter, *Naturwissenschaften* **11** (1923), 865.
4. J.M. Thomas, *Philosophical Transactions of the Royal Society* **277** (1974), 251.
5. K.R. Popper, *Conjectures and Refutations* (Routledge and Kegan Paul, London, 1963).
6. T. Martin (Ed.), *Faraday's Diary (1820-1862)* Vol i-vi (G. Bell and Sons, London, 1932).

7. *Faraday Discussion No. 92* (to be held at The Royal Institution, September 1991) on 'The chemistry and physics of small particles'.
8. R. Hooke, *Micrographia* (James Allestry, London, 1667).
9. *The Scientific Monthly* (September 1956), 145.

J.M. Thomas
The Royal Institution, May 1990

EXPERIMENTAL RESEARCHES

IN

CHEMISTRY AND PHYSICS.

The Royal Institution from the author Feb. 7. 1859

EXPERIMENTAL RESEARCHES

IN

CHEMISTRY AND PHYSICS.

BY

MICHAEL FARADAY, D.C.L., F.R.S.,

FULLERIAN PROFESSOR OF CHEMISTRY IN THE ROYAL INSTITUTION OF GREAT BRITAIN.

HON. MEM. R.S.ED., CAMB. PHIL., AND MED. CHIRURG. SOCC., F.G.S., ORD. BORUSSI "POUR LE MÉRITE" EQ., COMMANDER OF THE LEGION OF HONOUR, INSTIT. IMP. (ACAD. SC.) PARIS. SOCIUS, ACADD. IMP. SC. VINDOB. ET PETROP., REG. SC. BEROL., TAURIN., HOLM., MONAC., NEAPOL., AMSTELOD., BRUXELL., BONON., ITAL. MUT., SOCC. REG. GOTTING., ET HAFN., UPSAL., HARLEM. ACAD. AMER. BOST., ET SOC. AMER. PHILAD. SOCIUS, ACAD. PANORM., SOCC. GEORG. FLORENT., ET PHILOM. PARIS., INSTIT. WASHINGTON., ET ACAD. IMP. MED. PARIS. CORRESP., ETC.

REPRINTED FROM THE PHILOSOPHICAL TRANSACTIONS OF 1821—1857; THE JOURNAL OF THE ROYAL INSTITUTION; THE PHILOSOPHICAL MAGAZINE, AND OTHER PUBLICATIONS.

LONDON:

RICHARD TAYLOR AND WILLIAM FRANCIS,

PRINTERS AND PUBLISHERS TO THE UNIVERSITY OF LONDON,

RED LION COURT, FLEET STREET.

1859.

PRINTED BY TAYLOR AND FRANCIS,
RED LION COURT, FLEET STREET.

ALERE FLAMMAM.

PREFACE.

THE reasons which induce me to gather together in this Volume the various physical and chemical papers scattered in the Philosophical Transactions and elsewhere, are the same as those which caused the 'Experimental Researches in Electricity ' to be collected into one Series. As investigations, several of them are very imperfect; but it was thought a duty to print them just as they were, that they might be referred to as safely for facts, opinions, and dates, as the original papers. The correction of certain phrases and typographical errors, and the addition of some matter here and there with its proper date, is not considered as interfering with this intention.

MICHAEL FARADAY.

October, 1858.

CONTENTS.

viii CONTENTS.

EXPERIMENTAL RESEARCHES

IN

CHEMISTRY AND PHYSICS.

Analysis of Native Caustic Lime.*

ON THE NATIVE CAUSTIC LIME OF TUSCANY.
BY THE MARQUIS RIDOLFI.

THE interesting communication of Dr. Giovacchino Taddei respecting his discovery of caustic lime in the water of the ancient bath of Santa Gonda, in August 1815, induced me to visit the spot. The following is the result of my researches:—

The bath is situated in a laguna in the corner of a field near the high road to Pisa, which divides the plain called La Catena from the mountains of Cigoli and San. Miniato. The soil is a mixture of clay, calcareous earth, siliceous sand, and vegetable matter. There are two sources of water; one issues from the bottom of the laguna, and the other from the side. The first is hot, raising the thermometer of Reaumur to $35\frac{1}{4}°$. It is so saturated with lime, that upon cooling the water, it

---

* Quarterly Journal of Science, i. 260.

I reprint this paper at full length. It was the beginning of my communications to the public, and in its results very important to me. Sir Humphry Davy gave me the analysis to make as a first attempt in chemistry at a time when my fear was greater than my confidence, and both far greater than my knowledge; at a time also when I had no thought of ever writing an original paper on science. The addition of his own comments and the publication of the paper encouraged me to go on making, from time to time, other slight communications, some of which appear in this volume. Their transference from the 'Quarterly' into other Journals increased my boldness; and now that forty years have elapsed and I can look back on what the successive communications have led to, I still hope, much as their character has changed, that I have not, either now or forty years ago, been too bold.—M. F.

deposits a considerable quantity. It contains also muriate of lime and muriate of soda. The upper spring contains a little carbonic acid gas, some sulphuretted hydrogen, and some sulphate of soda. The following is the manner in which the caustic lime is formed in this bath. The lower spring yields a quantity of lime, but as this spring does not rise freely, but oozes through the bottom of the bath, the lime forms a stratum at the bottom of the lagune; which stratum, absorbing the carbonic acid gas of the water above, passes to the state of a carbonate, and thus forms a defence to the lime, which is continually depositing itself underneath, and prevents it losing its causticity. In fact, the caustic lime is found enclosed between the stratum of the carbonate of lime and the clayey bottom of the laguna.

Signor Taddei found the masses of caustic lime so large, that he could not get them out but by breaking them into pieces. He, however, succeeded in removing the whole of it: and I, having visited the spot two months after, found small incrustations of the same substance newly formed.

ANALYSIS OF THE NATIVE CAUSTIC LIME. By MR. FARADAY, ASSIST-
ANT IN THE LABORATORY OF THE ROYAL INSTITUTION.

THIS substance came to England in a bottle filled up with water, the atmospherical air being perfectly excluded.

It is almost entirely soluble in muriatic acid without effervescence, leaving nothing but a few light flocculi. The solution, when tested, was found to contain lime and iron.

A clean uniform piece of the substance was dried, as much as could be, by bibulous paper. A fragment of it being heated red, lost 62·26 per cent. of water.

The remainder of the original substance, weighing 188 grains, was dissolved in muriatic acid, and evaporated at a high heat on the sand-bath, acid was again added, and the evaporation repeated. Water was poured on it, and the silica separated: when well washed, dried, and heated red, it weighed 7·5 grains.

The filtered solution was precipitated by carbonate of potash, and the precipitate boiled in solution of pure potash. The solution was separated from the solid matter, neutralized by sulphuric acid, and precipitated by carbonate of ammonia.

The precipitate, when well washed and dried, weighed 0·95 of a grain. It was soluble in sulphuric acid, and possessed the properties of alumina.

Diluted sulphuric acid was added to the solid matter not acted upon by the potash; the whole boiled for some time, and then filtered. The sulphate of lime obtained weighed, after being heated red, 136 grains, which, estimating the lime at 43 per cent., is equivalent to 58·48 grains of lime.

The sulphuric solution was precipitated by ammonia, and two grains of oxide of iron were obtained.

Supposing the quantity of water in every part of the piece first taken to be uniform, it would follow that the 188 grains contained 117·05 of water; so that 70·95 was the quantity of dry matter acted upon. The results were—

	Grains.
Silica	7·50
Alumina	0·95
Lime	58·48
Oxide of iron	2·00
	68·93

The loss is therefore rather more than two grains, which may, perhaps, actually have taken place, and the difference may have been derived from the unequal diffusion of water throughout the piece.

Supposing 100 parts of the specimen to have been taken, the analysis will stand thus :—

	Grains.
Lime	82·424
Silex	10·570
Iron	2·820
Alumina	1·340
Loss	2·846
	100·000

It is perhaps worthy of observation, that during the solution of the substance in muriatic acid, a part only of the silica separated; the greater part remained in solution until heat was applied, when it gelatinized, as in the case where it is separated by an acid and heat from its combination with alkali.

OBSERVATIONS ON THE PRECEDING PAPER.
By Sir H. DAVY, *V.P.R.I., F.R.S.*

THE Duchess of Montrose was so good as to send me the
caustic lime which is the subject of the preceding analysis:
her Grace received it immediately from Tuscany. It was in a
bottle, carefully sealed and full of water. Some of the exterior
portions had become combined with carbonic acid before they
were collected, and from the colour, it appeared that there
were different portions of protoxide of iron in different parts
of the substance.

On examining the water, it was found to be a saturated
solution of lime, and it contained fixed alkali, but in quantities so minute, that after the lime was separated, it could be
made evident only by coloured tests.

It appears from Mr. Faraday's analysis, that the menstruum
which deposits the solid substance must be a solution of silica
in lime-water, and heat is evidently the agent by which the
large quantity of lime deposited is made soluble and is enabled
to act on silica; and the fact offers a new point of analogy
between the alkalies and the alkaline earths.

Vestiges of extinct volcanoes exist in all the low countries
on the western side of the Apennines; and the number of
warm springs in the Tuscan, Roman, and Neapolitan States,
prove that a source of subterraneous heat is still in activity
beneath a great part of the surface in these districts.

Carbonic acid is disengaged in considerable quantities in
several of the springs at the foot of the Apennines; and some
of the waters that deposit calcareous matter are saturated
solutions of this substance. Calcareous tufas of recent formation are to be found in every part of Italy. The well-
known Travertine marble, *Marmor Tiburtinum*, is a production of this kind; and the Lago di Solfaterra near Tivoli,
of which I shall give a particular account on a future occasion,
annually deposits masses of this stone of several inches in
thickness.

It is scarcely possible to avoid the conclusion, that the
carbonic acid, which by its geological agency has so modified
the surface of Italy, is disengaged in consequence of the
action of volcanic fires on the limestone, of which the Apennines are principally composed, and liberated at their feet,

where the pressure is comparatively small; but the Tuscan laguna offers the only instance in which the action of these fires extends, or has extended, to the surface at which the water collected in the mountains finds its way to the sea, so as to enable it to dissolve caustic calcareous matter.

On the Escape of Gases through Capillary Tubes*.

As the mobility of a body, or the ease with which its particles move among themselves, depends entirely upon its physical properties, little delay would arise in the mind, on a consideration of the probable comparative mobilities of the different gases. These bodies being nearly similar in all the physical properties, except specific gravity, which can interfere with internal motions generated in them, would be supposed to have those motions retarded in proportion as this latter character increased; but as this supposition has not been distinctly verified, the following experiments, though possessed of no peculiar claim to attention, may deserve to be recorded.

The apparatus was a copper vessel of the capacity of 100 cubic inches nearly, to which a condensing gauge was attached. Four atmospheres of the gas to be tried were thrown into it, and then a fine thermometer tube, 20 inches in length, was fixed on by adjusting pieces: the gas was suffered to escape until reduced to an atmosphere and a quarter, and the time noticed by a seconds' pendulum. In this way,—

> Carbonic acid gas required 156·5 minutes to escape.
> Olefiant gas   ,,   135·5   ,,   ,,
> Carbonic oxide   ,,   133   ,,   ,,
> Common air   ,,   128   ,,   ,,
> Coal-gas   ,,   100   ,,   ,,
> Hydrogen   ,,   57   ,,   ,,

These experiments tend to show, that the mobility of the gases tried decreases as their specific gravity increases, and they are corroborated by others made with vanes. A wheel, having small planes attached to it, as radii perpendicular to the plane of motion, was made to rotate by a constant force in atmospheres of different gases, and the times which the motion

---

* Quarterly Journal of Science, iii. 354.

continued, after the force was removed, diminished as the specific gravity increased; as for instance, in

Carbonic acid it continued 6 seconds.		
Common air	,,	8	,,
Coal-gas	,,	10	,,
Hydrogen gas	,,	17	,,

There is therefore every reason to believe, that the actual relative mobilities of the gases are inversely as their specific gravities.

These experiments have been carried much further, in consequence of some peculiar results obtained at low pressures; but as I have not been able to satisfy myself respecting the causes, and have probably taken a wrong view of the phenomena, I shall refrain from detailing them, and merely observe, that there is no apparent connexion between the passage of gases through small tubes and their densities at low pressures. Olefiant gas then passes as readily as hydrogen, and twice as rapidly as either carbonic oxide or common air, and carbonic acid escapes far more readily than much lighter gases. Similar results are also obtained by diminishing the bore of the tube, and then even at considerable pressures, the effect produced by mobility alone is interfered with by other causes, and different times are obtained. These anomalies depend, probably, upon some peculiar loss or compensation of forces in the tube, and offer interesting matter of discussion to mathematicians.

Experimental Observations on the Passage of Gases through Tubes.*

IN a previous communication I have noticed briefly some curious effects which take place when gases are passed through tubes at low pressures. They consist in an apparent inversion of the velocities; those which traverse quickest when the pressure is high, moving more tardily as it is diminished, until they are among those which require the longest time in passing the tube; thus with equal high pressures equal volumes of hydrogen gas and olefiant gas passed through the same tube in the following times :—

* Quarterly Journal of Science, vii. 106.

Hydrogen in   57''
Olefiant gas in 135''·5 ;
but equal volumes of each passed through the same tube at
equal low pressures in the following times :—
Hydrogen   8' 15''
Olefiant gas 8' 11''.

Again, equal volumes of carbonic oxide and carbonic acid
gases passed at equal high pressures through the same tube,
occupied, the   Carbonic oxide 133''
Carbonic acid   156''·5 ;
but at low pressures, Carbonic oxide 11' 34''
Carbonic acid   9' 56''.

I have lately had my attention again called to the subject,
but have not yet been able to satisfy myself of the cause of
this curious effect; nevertheless, as experiments do not always
owe their value to the hypothesis which accompanies them,
a few short observations on some made on this subject may be
acceptable.

The effect is always produced by fine tubes at low pressures,
but does not appear to belong to the mere obstruction by the
tube to the passage of the gas, nor have I been able to produce it without a tube. A very fine needle-hole was made in
a piece of platinum foil, and so arranged on a mercurial gasometer, that the pressure of a small column of mercury sent
seven cubical inches of the following gases through in the
times mentioned, namely—
Hydrogen   3'·8 nearly, taking a mean,
Olefiant gas 9'·2 ;
and when the pressure was increased, the same proportions in
the times was observed. Other similar experiments gave
similar results.

Slits, cut in platinum foil by the edge of a penknife, did not
give so great a superiority to hydrogen as that mentioned
above, and the proportion varied with different slits; still the
hydrogen passed most rapidly, and a difference of pressure
caused no difference in the relative times.

Three diaphragms were placed in different parts of the
same tube, each being perforated with a small hole, but the
effects produced in tubes were not observable here. Hydrogen
passed in 3·8 minutes, and olefiant gas in 9·1 minutes.

The gases were passed through discs of paper, and the number of discs was increased so as to increase the obstruction, the pressure and quantity of gas remaining the same. With one disc of drawing-paper 6·5 cubical inches of hydrogen passed in 7$'$

     6·5 ,,   ,, of olefiant gas in 18$'$;

with two discs the hydrogen passed in 15$'$·4

  ,, ,, olefiant gas ,, 38$'$;

with three discs the hydrogen ,, 22$'$·5

  ,, ,, olefiant gas ,, 57$'$·75.

Lastly, for the effect of obstruction, I used a tube filled with pounded glass. This was uncertain, because on moving the tube it was impossible, almost, not to move some of the particles within, and then, of course, circumstances were changed ; but by sending the gases through one after the other, results were obtained, the mean of which gave for hydrogen 3$'$·4

       ,,   ,, for olefiant gas 4$'$·7.

It would seem from these experiments that mere obstruction is not the cause of the effect observed in tubes, for when the tubes are removed, and obstructions which retard much more placed for them, the effect is lost; and as the same aperture produces no difference of effect at high or low pressures, the variations between different apertures should probably be referred to some other cause.

I then endeavoured to ascertain some of the circumstances attending on tubes. Both glass and metal tubes produce the effect, and a metal tube, down which a wire had been thrust, did not seem to have this influence on the passage of gases through it altered. The effect is heightened as the gas is made to pass more slowly through the tube; and this, whether the increased time be caused by diminished pressure, increased length of tube, or diminished diameter. This may be well illustrated by putting several very fine tubes together, for the particular effect is thus increased whilst the time is shortened. Two brass planes were ground together, and a few scratches made down one of them so as to form very fine tubes ; through these olefiant gas passed in 26$'$·2, and hydrogen in 32$'$·5.

Three glass tubes were taken of different diameter, and cut into such lengths that they passed nearly equal quantities of hydrogen gas in equal times by the same pressure ; their lengths were 42, 10·5, and 1·6 inches. The longer tube passed the

hydrogen in　.　.　.　$3'·7$, the olefiant gas in $2'·75$
the second in .　.　.　$3'·5$　　,,　　,,　　$2'·5$
the smaller in .　.　.　$3'·45$　　,,　　,,　　$2'·8$;

and in several other experiments there seemed to be nearly an equal effect, when the quantity of gas passed in the same time was the same.

I imagined that the specific gravity of the gases might have some constant influence, but this does not seem to be the case; carbonic oxide and olefiant gas are nearly of the same density; and if the effect depended upon their weight, it should be nearly the same for both of them; but this is not so; seven cubical inches of carbonic oxide required $4'·6$ minutes to pass through a tube which was traversed by the same quantity of olefiant gas under the same pressure in $3·3$ minutes, each gas having been placed over caustic lime for some time previously; and oxygen required to pass through the same tube $5·45$ minutes of time.

I placed three gauges in different parts of a tube, of such a size that it passed olefiant and hydrogen gas in nearly equal times; the gauges were very obedient to the pressure of the gas in the different parts of the tube, but I could not perceive any difference between the effect of the different gases.

Such are some of the circumstances which affect and produce this curious effect: that the velocity of gases in passing through tubes should be in some proportion to the pressure on them is nothing particular; but the singularity is, that the ratio for the same gas varies with the pressure, and that this variation differs in different gases; thus the one which passes with the greatest facility at low pressure, passes with the least at high pressure.

It may be deduced from the experiments at high pressures and on obstructions, that the fluidity of the gas has little or nothing to do in this case, for where it alone can have an influence, the indications are the same at all pressures, and the gas of least density passes in the shortest time; thus comparing hydrogen with olefiant gas, and considering its time 1, the time of the latter will be in the experiments already mentioned, as $2·38, 2·42, 2·4, 2·57, 2·46, 2·57$ ratios, which do not differ much from each other, though the times, pressure, obstructions, and quantities of gas used vary very considerably.

Neither is the variation among the different gases between the ratio of the velocity and pressure, connected with specific gravity, at least I have not been able to observe such a connexion. I have quoted an experiment, or rather the general result of several, on carbonic oxide and olefiant gases, and it is adverse to the supposition; and in others, made on sulphurous acid gas and ammoniacal gas, still further departures from the order of the densities were observed.

If a tube sufficiently fine and long be connected with a portion of gas under high pressure, so that the time occupied in its passage through it be considerable, the effect will be produced, *i. e.* the times of different gases will vary from each other, but not according to their specific gravities; if the tube, however, be cut off so that the gases pass quickly, then the times will be as the specific gravity. Now, in the long tube, the pressure and velocity will vary throughout its length, the pressure being greatest at the internal or connected end, and least at the other extremity, whilst the velocity is least at the end towards the reservoir, and greatest at the other. But the ratio by which the pressure and velocity decrease and increase, appears different for and peculiar to each gas. At the end of the long tube the olefiant gas issues more rapidly than hydrogen, though the pressure at the reservoir is the same; but shorten the tube, and let that part in which high pressures only exist confine the gases in their passage, and the hydrogen gas will surpass the olefiant gas in velocity as far as 4 or 5 does 2. It would seem, therefore, that in the long tube the pressure or elasticity of the olefiant gas diminishes less rapidly than that of the hydrogen, or that its velocity increases more rapidly.

Perhaps these effects may be accounted for by the supposition of some power of expansion peculiar to each gas, which, if existing, a tube would for many reasons be well calculated to exhibit. The experiment requires numerous repetitions and much time, and I have not yet had sufficient to satisfy myself on the subject. I will therefore refrain from mixing up crude notions with facts, and at some more convenient opportunity endeavour to supply what is wanting in this paper.

Combustion of the Diamond *.

Sir H. Davy was the first to show that the diamond was capable of supporting its own combustion in oxygen without the continued application of extraneous heat, and he thus obviated one of the anomalies exhibited by this body when compared with charcoal. This phenomenon, though rarely observed, is easily exhibited. If the diamond, supported in the perforated cup, be fixed at the end of a jet, so that a stream of hydrogen can be thrown on to it, it is easy, by inflaming the jet, to heat the gem, and whilst in that state to introduce it into a globe or flask containing oxygen. On turning off the hydrogen the diamond enters into combustion, and will remain burning until nearly consumed. The loss of weight in the diamond, the formation of carbonic acid, and the actual combustion are thus very easily shown.

Description of a New Apparatus for the Combustion of the Diamond†.

In the course of the experiments which Sir Humphry Davy made at Florence on the combustion of the diamond, he discovered that when the gem began to burn in an atmosphere of pure oxygen, having free access to it on all sides, it would continue burning, though the original source of heat were removed, until the particles were rendered so small as to be too readily cooled by the little platinum tray which supported them. (Philosophical Transactions, 1814, p. 557.) In consequence of this observation, an idea arose, that if the diamond were well heated, and then introduced into oxygen, it would go on burning, and afford an easy method of exhibiting its combustibility. Upon trial this was found to be the case, and a notice to that effect put in this Journal (see above). Since then, an apparatus of this kind has been perfected, and is now represented in Plate I. fig. 1.

It consists of a glass globe, of the capacity of about 140 cubic inches, furnished with a cap, having a large aperture; the stopcock, which screws into this cap, has a jet **A** rising

---

* Quarterly Journal of Science, iv. 155.   † Ibid. ix. 264.

from it, nearly into the centre of the globe; this is destined to convey a small stream of hydrogen or other inflammable gas. Two wires, *c c*, terminate at a very little distance from each other, just above this jet, and are intended to light the stream of hydrogen by electrical sparks; one of them commences from the side of the jet, the other is enclosed and insulated nearly in its whole length in a glass tube: the tube and wire pass through the upper part of the stopcock, and the wire terminates on the outside in a ball or ring, D, at which sparks are to be taken from the machine, either directly or by a chain. On the end of the jet is fixed, by a little socket, a small capsule, B, made of platinum foil. This capsule is pierced full of small holes, and serves as a grate to hold the diamonds. Its distance is about three-quarters of an inch from the end of the jet; and the arm, by which it is supported, is bent round, so that the stream of hydrogen shall not play against it. The stopcock screws, by its lower termination, on to a small pillar fixed on a stand, and at the side of this pillar is an aperture by which a bladder filled with gas may be connected with the apparatus.

On using the apparatus, the diamond is to be placed in the capsule, and then the globe being screwed on to the stopcock, the latter is to be removed from the pillar and placed on the air-pump; the globe is then to be exhausted and afterwards filled with pure oxygen: or, lest the stream of oxygen in entering should blow away the diamond, the globe may be filled with the gas first, and then, dexterously taking out the stopcock for a short time, the diamonds may be introduced and the stopcock replaced. The apparatus is then to be fixed on the pillar, and a bladder of hydrogen gas attached to the aperture. Now, passing a current of sparks between the wires, a small stream of hydrogen is to be thrown in, which inflaming, immediately heats the capsule and diamonds white-hot; the diamonds will then enter into combustion, and the hydrogen may be immediately turned off and the bladder detached. The diamonds will continue to burn, producing a strong white heat, until so far reduced in size as to be cooled too low by the platinum with which they lie in contact.

When the flame of hydrogen is used to heat the diamonds, it is evident a little water will be found in the globe; but this is of no consequence, except in attempts to detect hydrogen in the diamond; the inconvenience may be obviated, if required, by using the flame of carbonic oxide. As, however, no hydrogen has at any time been detected in the diamond, it is better to use that gas as the heating agent; for then the carbonic acid, produced by the combustion, is unmixed with that from any other source, and may be collected, and its quantity ascertained.

On the Solution of Silver in Ammonia*.

THE ease with which the compounds of silver are dissolved by ammonia, and the frequent formation of powerfully detonating and dangerous substances in these solutions, are well known. I have been induced to examine some of the phenomena presented by these bodies, and perhaps an account of what is, I believe, original, may not be unacceptable as an addition to the scanty stock of information published on this subject.

When the oxide of silver, precipitated either by the alkalies or alkaline earths, is put into solution of ammonia, it is entirely dissolved, producing a pale brownish solution. If this solution be exposed in an open vessel, a brilliant pellicle forms on its surface, which, when removed, is succeeded by another and another, until most of the metal is separated.

This, which is an oxide of silver, was noticed long ago by Berthollet in the 'Annales de Chimie,' tome i., and he there states its production to be dependent on the abstraction of ammonia by the atmosphere.

From some difference which exists between this solution of silver and that of the nitrate when treated by precipitants, and from other circumstances, I was induced to collect and analyse some of the oxide, to ascertain its identity with the common oxide, or that previously dissolved. 20 grains that had been dried for some hours on the sand-bath, were put into a small glass retort, they were decomposed by heat, and the gas liberated received over water; it equalled 2·75 cubical inches. 18 grains of silver remained in the retort, and the 2·75 of oxygen being equivalent to ·935 grain, we have those numbers as the proportions of the elements in the oxide, the

* Quarterly Journal of Science, iv. 268.

loss being supposed to be water, some of which had condensed in the neck of the retort.   Now—

Oxygen.	Silver.	Oxygen.	Silver.
·935	18	7·5	1·444

The same method of analysis was applied to oxide of silver, precipitated by potash from nitrate of silver, it having been well washed and dried : 40 grains gave 7·9 cubical inches, and 36·4 grains of silver remained.   The 7·9 cubical inches = 2·686 grains, and

Oxygen.	Silver.	Oxygen.	Silver.
2·686	36·4	7·5	101·6,

the number for silver very nearly as given in the most correct elementary treatises on chemistry.   There appears, therefore, to be no error in the mode of analysis, and the oxide by ammonia seems to contain less oxygen than that precipitated by alkalies.   Again,—

30 grains of the oxide of silver were put into a retort, and decomposed with every precaution as before ; the silver left weighed 27·4 grains, and the quantity of gas given off was 4·125 cubical inches.   I suspected that a small portion of carbonate of silver had been mixed with the oxide, for when the ammoniacal solution has been long exposed to the air, much carbonate of ammonia and of silver is formed in it ; the gases were therefore placed over solution of potash, and were reduced in bulk to 3·625, which was pure oxygen.   This volume is equivalent to 1·2325 gr., and 1·2325 : 27·4 : : 7·5 : 166·7, a proportion of silver still higher than in the first experiment, but which may be accounted for by the purification of gas and the small quantity of oxygen that remained in the retort.

In a third experiment, 24 grains of silver were left; 4·25 cubical inches of gas were given off, which decreased over potash to 4.   In order to estimate the proportion of azote arising from the air in the retort, the 4 were treated with nitrous gas of known purity, and gave results equal to 3·475 of pure oxygen.   This is equal to 1·1815 gr., and 1·1815 : 24 : : 7·5 : 152·3.

One or two other experiments varied considerably from this, giving a greater proportion of silver, but the mean of many gave the oxygen to the silver as 7·5 to 157·4.

There is every reason, therefore, to believe this a protoxide
of silver, containing about two-thirds the quantity of oxygen
found in the common oxide, or that obtained by precipitation
from the nitrate; and there are also other circumstances
observable in its solution and during its formation which
favour this notion.

When this oxide forms on the surface of an ammoniacal
solution by slow spontaneous evaporation, it takes a crystalline form, which, however, is quickly lost by its covering the
whole surface of the liquor. It is of a grey colour by reflected light, and highly resplendent; the light transmitted
through thin films is of a bright yellow colour. When heated
gradually it is reduced, giving off oxygen without change of
form; but heated suddenly, it fuses first, and leaves a solid
button of silver: under pressure, it perhaps might be fused
without decomposition.

Potash precipitates the solution of oxide of silver in ammonia white; carbonate, or subcarbonate more abundantly,
and white; alcohol and æther throw down precipitates, at first
white, but rapidly changing colour; when dry, they detonate
by heat or friction. Chromate of ammonia does not precipitate until nitric acid be added. Tincture of galls gives
a very copious black precipitate, different in appearance to
that obtained from the nitrate of silver by adding ammonia
after the tincture. Solution of iodine in water gives a brown
curdy precipitate, but with nitrate of silver a yellow turbidity.
Muriatic acid or muriates always form chloride of silver.

It is probable, from these circumstances, that part of the
silver exists in the solution in the state of protoxide, and as no
gas is given off during the solution of the original oxide, that
a portion of nitric acid and water have been formed.

M. Berthollet has in the paper before referred to, described
a fulminating compound of silver and ammonia, obtained from
solutions similar to those from which the above oxide had been
obtained, and has stated it to be his opinion that it is a compound of protoxide of silver and ammonia. As it is frequently
left in the form of a black powder when oxide of silver is dissolved in ammonia, I imagined it might be a compound of the
peroxide with the alkali, as protoxide was formed and held in
solution; and that the circumstance of the liberation of azote,

which gave rise to the idea of its being a combined protoxide, might be explained by the further formation of a portion of oxide similar to that already described.

The method of obtaining this compound has been to pre-cipitate oxide of silver from the nitrate by alkalies, or better, by lime-water, to wash and dry it well, and then to leave it in contact with liquid ammonia for ten or twelve hours; the greater part is dissolved, but a black powder remains, which is fulminating silver; if the solution be heated, azote is given off, and a further quantity of fulminating silver is obtained (Annales de Chimie, tome i.).

I find that fulminating silver may be formed from any precipitated oxide of silver, whether moist or dry, recent or old. Boil the oxide carefully in a tube with a mixed solution of potash and ammonia for a few moments; the potash absorbs all the carbonic acid that may have been united to the oxide, and to a certain degree prevents its solution in the ammonia; a black powder, similar to that procured by the other process, results.

In order to gain some evidence respecting the nature of the oxide combined with the ammonia in fulminating silver, I endeavoured to ascertain the mode of formation of that com-pound. It appears to be formed in every case where common oxide of silver is dissolved in ammonia, and the entire solution of all solid matter is no evidence of its non-existence, for the compound is itself soluble in ammonia, though not so much so as the oxide. When there is an excess of oxide, unless it predominate in a great degree, the undissolved portion will be found to contain fulminating silver, and when the whole is dis-solved, by heating the solution, it is thrown down.

To ascertain whether the liberation of azote depended upon the formation of the fulminating compound, I boiled, for a few moments, a solution of the oxide in ammonia; the solu-tion became highly coloured, azote was given off, and a black curdy precipitate formed, which left the liquid colourless; separated by a filter, the precipitate proved to be fulminating silver. The solution was again heated, it again blackened, gave off azote, and again a precipitate formed; this was not fulminating silver, but merely oxide: filtered and again heated, it gave off azote, and more of the oxide was formed; and this occurred with the same solution a fourth and fifth time. The liberation of the azote, therefore, does not belong exclusively to the formation of fulminating silver, but seems rather to depend on the production of protoxide.

I endeavoured to form fulminating silver by using the protoxide described in the first part of this paper, but could not succeed: I got nothing but a black powder from it, which appeared to be the same oxide in another form. I endeavoured also to form fulminating silver from those portions of oxide given off by the further boiling of solutions which had previously yielded the detonating compound, but failed; I presume from its being also a protoxide. When the fulminating compound is dissolved in the acids, it gives off a gas which I believe to be oxygen, but I could not work with quantities sufficient to ascertain this point. Perhaps to these reasons for supposing fulminating silver to be a compound rather of the peroxide than the protoxide, may be added the easy solubility of the protoxide in ammonia, and the difficult solubility of the detonating compound.

The oxide which is obtained by boiling solution of silver in ammonia, I have supposed to be a protoxide similar to the one obtained by spontaneous evaporation. This opinion is founded on the liberation of azote during its formation in consequence of the decomposition of ammonia by oxygen, and on its apparent incapability of forming fulminating compounds: the idea is supported by the following circumstance. A tube, in which solutions of silver in ammonia had been repeatedly boiled, became coated on the inside with the oxide, so as to be perfectly opake; on pouring dilute nitric acid into it to remove the oxide, the tube became lined with brilliant metallic silver, which, however, was soon dissolved by the continued action of the acid. I attribute this phenomenon to the reduction of one part of the oxide by another, which was thus rendered soluble in the acid.

When a portion of the ammoniacal solution is evaporated to dryness in a platinum capsule, it leaves a film of oxide, which, when decomposed by heat, gives a perfectly continuous and smooth coat of silver to the vessel. I have also covered other metals, as iron and copper, with silver in the same way, and found that the burnisher might be applied without any injury

c

to the coating. It is probable that a solution of silver of this kind might be applied in some cases in the arts, to the purposes of ornament and utility.

Combinations of Ammonia with Chlorides*.

It has been already shown, particularly by Sir H. Davy, that several of the binary compounds of chlorine, as those of phosphorus, tin, &c., exert a strong affinity for ammonia, condensing it when in the gaseous state, and neutralizing its alkaline properties. The combinations which will here be offered to notice are of a different kind, and if they deserve any attention, it will be in consequence of the weakness of the power which is exerted in their formation, and the slight change of properties induced on the substances by union.

It has been frequently observed by chemists, that if well-fused chloride of lime be placed in ammoniacal gas, there is a rapid absorption of the gas, and the chloride becomes covered with a white powder. If ammonia be repeatedly added until the absorption ceases, the mass of chloride swells, cracks, splits in all directions, and at last forms a white pulverulent substance.

Exposed to the atmosphere, it deliquesces, but not so rapidly as chloride of lime. Thrown into water it dissolves, forming a strong alkaline solution. Heated, it gives off ammonia, and the chloride remains unchanged. Placed in chlorine it inflames spontaneously, and burns with a pale yellow flame.

The fused chlorides of barium and strontium suffer a very slight change in ammoniacal gas in many days; after more than a fortnight, the chloride of strontium, weighing about 30 grains, had absorbed only a cubical inch of gas, and a slight efflorescent appearance was seen on the broken edge.

A piece of fused chloride of silver, weighing about 30 grains, placed in ammoniacal gas, gradually absorbed more than 40 cubical inches. The action took place over the whole surface of the mass, but most speedily at the fractured edges. The chloride swelled considerably, and crumbled into powder. The substance formed was at first white, but it

* Quarterly Journal of Science, v. 74.

blackened by exposure to light, though without liberating any gas. Thrown into water the ammonia was separated, forming a solution, and the chloride remained unchanged. Heated, the whole of the ammonia was given off. Placed in chlorine it inflamed spontaneously, and the ammonia was decomposed.

Chloride of silver that had been well dried, but not fused, gave the same compounds with ammonia, but in a much shorter time.

A strong solution of chloride of silver in ammonia was left for some weeks in a bottle stopped only by a piece of paper. At the end of that time several perfectly colourless and transparent crystals had formed in it; some of them being as much as a quarter of an inch in width. Their general form was that of a flat rhomboid, but sometimes two acute angles of the rhomboïd were wanting, and then the crystals looked like hexahedral prisms with oblique bases.

Exposed to the air, these crystals became opake, gradually losing the whole of the ammonia, and were then so friable as to fall into powder by a slight touch; the substance remaining was a dry chloride of silver. Placed in water, the same change occurred, but more readily; the water separated the ammonia, and they instantly became opake. Heated, they gave off much ammoniacal gas, and the chloride remained unaltered. Exposed to light, they gradually blackened, though covered by the solution from which they were deposited.

If the ammoniacal solution be weak, other crystals are formed which are pure chloride of silver.

Dry corrosive sublimate placed in ammoniacal gas had suffered no change in fourteen days, nor had any action been exerted on the ammonia; there was a diminution of a quarter of a cubical inch of gas, probably owing to a little water being present. The corrosive sublimate heated gave out no ammonia, and the whole of the gas remaining was absorbed by water.

The precipitate obtained by adding ammonia to a solution of corrosive sublimate appears to be a compound of the two bodies, but the alkali is neutralized in this case, and it is therefore more analogous to the combination of ammonia with the chloride of tin. When the precipitate is distilled, it gives off ammoniacal gas and also some azote, and the corrosive sublimate is converted into calomel in consequence of the action of the ammonia at high temperatures. Heated with potash, the ammonia is driven off, the chlorine is removed from the mercury, and red oxide results.

Some crystals of calomel were introduced into ammoniacal gas; they immediately blackened on the surface, and gas was absorbed. The action appeared to be exactly similar to that exerted when calomel is thrown into solution of ammonia. A black substance is produced, which though repeatedly washed in distilled water, gives off ammonia by heat, and calomel with a little mercury sublimes.

A piece of fused chloride of lead exerted but little action in a fortnight; a small quantity of gas was absorbed, and a very superficial combination had been formed.

Chloride of bismuth absorbed a small quantity of ammoniacal gas, which was again given out by heat; there was no remarkable change in appearance.

A small piece of chloride of nickel being placed in ammoniacal gas, absorbed it, and in twenty-four hours was converted into a bulky powder of a pale rose tint. The ammonia was separated by exposure to air, to water, or to heat.

Chloride of copper fused was powerfully acted upon by ammonia. It immediately burst open upon being placed in the gas, and absorbing great quantities fell into a blue powder. The compound placed in water was decomposed, and an ammoniacal solution of copper produced. Heated, it fused, boiled, the ammonia flew off, and the chloride remained.

The protochloride of iron introduced immediately after fusion into ammoniacal gas, exerted an instantaneous action; great quantities of gas were absorbed, and a very light adhesive white powder was formed. Exposed to the air, it immediately changed colour, became yellow, brown, then green, and ultimately black: this effect resulted from the presence of water in the atmosphere, and the separation of oxide by the ammonia; and the substance offers a test, if one should be wanted, for the presence of aqueous vapour. A portion of it thrown up into a small receiver of common air over mercury, immediately changed colour, and became brown. When the powder was heated out of the contact of air it gave off ammonia, and the chloride remained.

I have not examined the action of ammonia upon the other chlorides; with some of them it would probably form neutral compounds, with other combinations similar to those described. Nearly all those mentioned are formed by the exertion of an affinity so weak that it is overcome by the attraction of water for the ammonia, and yet in one instance it is capable of giving a definite crystalline form.

The facility with which many of them afford dry ammoniacal gas at low temperatures in considerable quantities, may perhaps in some cases make them convenient sources of that substance *; 19 grains of the compound with chloride of lime which had been made many days, gave 19·4 cubical inches of gas. They also offer a convenient means of ascertaining the specific gravity of ammonia, by the quantity of gas given off, and the loss of weight in the substance.

On the Sounds produced by Flame in Tubes, &c. †

THERE is an experiment usually made in illustration of the properties of hydrogen gas, which was first described by Dr. Higgins in the year 1777 ‡, and in which tones are produced by burning a jet of hydrogen within a glass jar or tube. These tones vary with the diameter, the thickness, the length, and the substance of the tube or jar; and also with changes in the jet. They have frequently attracted attention, and some attempts to explain their origin have been made.

After Dr. Higgins, Brugnatelli in Italy, and M. Pictet at Geneva, described the experiment, and the effects produced by varying the position and other circumstances of the jet and tube; and M. De la Rive read a paper at Geneva (published in the 'Journal de Physique,' lv. 165), in which he accounted for the phenomenon by the alternate expansion and contraction of aqueous vapour. That they are not owing to aqueous vapour, will be evident from some experiments to be described. I have no doubt they are caused by vibrations, similar to those described by M. De la Rive; but the vibrations are produced

---

* See 'Condensation of Gases.'
† Quarterly Journal of Science, v. 274.
‡ Nicholson's Journal, vol. i. p. 130.

in a different manner, and may result from the action of any flame.

I was induced to make a few experiments on this subject, in consequence of the request of Mr. J. Stodart that it should be introduced at one of the evening meetings of the Members and friends of the Royal Institution; and was soon satisfied that no correct explanation had been given. That the sounds were not owing to any action of aqueous vapour, was shown by heating the whole tube above 212°; and still more evidently by an experiment, in which I succeeded in producing them from a jet of *carbonic oxide*. That they do not originate in vibrations of the tube, caused by the current of air passing through it, was shown by using cracked glass tubes, tubes wrapped up in a cloth; and I have obtained very fine sounds by using a tube formed at the moment by rolling up half a sheet of cartridge-paper, and keeping it in form by grasping it in the hand. The sounds have been accounted for, as well as their supposed peculiarity of production by hydrogen, by the supposition of a rapid current of air through the tube; but that this is not essential, is shown by using tubes closed at one end, and bell-glasses, as described by Mr. Higgins in his first experiment.

I was surprised to find, on my first trials with other gases, that I could produce those sounds from them which had been supposed to be generated exclusively by hydrogen; and this, with the insufficiency of the explanations that had yet been given, induced me to search after the cause of an effect which appeared to be produced generally by all flame.

In examining attentively the appearance of a flame when introduced into a tube, it will commonly be found, that, on coming within its aperture, a current of air is established through the tube, which compresses the flame into a much smaller space; it is slightly lengthened, but its diameter is considerably diminished: on being introduced a little further, and as the tube becomes warm, this effect is increased, and the flame is gradually compressed a little above its commencement at the orifice of the jet, more than at any other part; a very faint sound begins to be heard, and as it increases, vibrations may be perceived in the flame, which are most evident in the upper part, but frequently also perceptible in the lower and smaller portion; these increase with the sound, which at last becomes very loud, and if the flame be further introduced into the tube, it is generally blown out. Such are the general appearances with hydrogen. If a jet of olefiant or coal-gas, both of which I have ascertained may be used successfully, be substituted, then, in addition to those appearances, it will be perceived, that as the bright flame of the gas enters the tube, its splendour is diminished, and it burns with less light.

By substituting other gases and inflammable vapours for hydrogen, and using other vessels than tubes, I was enabled so to magnify the effects, as to perceive more distinctly what took place in the flame at these times, and soon concluded that the sound was nothing more than the report of a continued explosion.

Sir H. Davy has explained the nature of flame perfectly; and has shown that it is always a combination of the elements of explosive atmospheres. In continued flame, as of a jet of gas, the combination takes place successively, and without noise, as the explosive mixture is made. In what is properly called an explosion, the combination takes place at once throughout a considerable quantity of mixture, and sound results from the mechanical forces thus suddenly brought into action; and a roaring flame presents some of the characters of both. If a strong flame be blown on by the mouth, a pair of bellows, the draught of a chimney, or other means, the air and the gaseous inflammable matter are made to mix in explosive proportions in considerable quantities at once, and these being fired by the contiguous flame, combine at once throughout their whole extent, and produce sound: the effect is rapidly repeated in various parts of the flame as long as the air is mixed thus forcibly with it, and a repetition of noise is produced, which constitutes the roar.

Now this I believe to be exactly analogous to that which takes place in what have been called the singing tubes; but in them the explosions are generally more minute and more rapid. By placing the flame in the tube, a strong current of air is determined up it, which envelopes the flame on every side. The current is stronger in the axis of the tube than in any other part, in consequence of the friction at the sides and the position of the flame in the middle; and just at the entrance of the tube an additional effect of the same kind is produced, by the edge obstructing the air which passes near it; the air is therefore propelled on to the flame, and mingling with the inflammable matter existing there, forms portions of exploding mixtures, which are fired by the contiguous burning parts, and produce sound, in the manner already described, with a roaring flame; only, the impelled current being more uniform, and the detonations taking place more rapidly and regularly, and in smaller quantities, the sound becomes continuous and musical, and is rendered still more so by the effect of the tube in forming an echo.

That the roaring flame gives sound in consequence of explosion, can hardly be doubted; and the progress from a roar to a musical tone is easily shown in the following manner:—Take a lamp with a common cotton wick, and trim it with æther or alcohol; light it, and hold a tube over the flame (that which I have used is a thin tube of glass about an inch in diameter, and nearly 30 inches long); in a few seconds after introducing the flame, the draught will be sufficiently strong to blow it out, but if the current be obstructed by applying the fingers round the lamp at the bottom of the tube, combustion will go on, though irregularly; then, by a little management in admitting the air on one side or the other, and in greater or lesser quantity, it may be impelled on to the flame in various degrees, so as to produce a rough roaring sound, or one more continued and uniform, of a higher note, and more musical; and these may be made to pass into each other at pleasure: then, by substituting a stream of æthereal vapour for the wick, which may be easily done from a small flask through a tube, the tones may be brought out more and more clearly, until they exactly resemble those of hydrogen.

A similar experiment may be made with coal-gas: light a small Argand burner with a low flame, and bring a glass tube, which is very little larger than the diameter of the flame, down upon it so as nearly to include it: the current of air will be impelled on the external part of the flame, it will remove the limit of combustion a little way up from the burner, that part of the flame will vibrate rapidly, burning with continued explosions, and an irregular tone will be obtained. Remove the burner, and fix on a long slender pipe to the gas-tube, so as to afford a candle-flame that may be introduced into the tube; light it, and introduce it about 5 or 6 inches, and a clear musical tone will be obtained.

During the experiments that were made in consequence of this view of the subject, many appearances occurred which might be added to the above account, to support the opinion that the vibration of the flame, in consequence of rapid successive explosions, is the cause of the sound; but they are neglected, because they are supposed unnecessary.

If the explanation given be true, then the only requisite to the production of these sounds is the successive sudden inflammation of portions of gaseous explosive mixtures. These mixtures are most easily made by propelling a stream of air on to a stream of inflammable gaseous matter; but it is also possible to make them in other ways, and the same phenomenon may be produced in a different manner.

That the tube is not essentially necessary, is shown by making it swell out into a cylinder of 3 or 4 inches diameter, except above and below; or part of it may be extended into a globe. I took two air-jars that were open above, but with contracted apertures; one of these was inverted over an inflamed jet of hydrogen, so as to form a lamp or bell-glass about it: there was no effect of sound, because the downward currents from above interfered with the stream of air issuing up from beneath, and made it irregular; but placing the second receiver on the first, applying them edge to edge, so as to preserve the current of air upwards from disturbing forces, the sounds were immediately produced; and lastly, I succeeded in obtaining the tones by the draught of a common chimney; for, by attaching a large inverted air-jar to the end of a funnel-pipe that came from the flue, closing the other lower opening into it, and introducing an inflamed jet of hydrogen within the lower contracted orifice of the glass, the sounds were produced.

That the same sounds may be obtained by means different to those above described, though depending on the same cause, is shown by some experiments made by Sir H. Davy, in his first researches on the miners' safety-lamp. Small wire-gauze safety-lamps being introduced into air-jars filled with explosive atmospheres, the gases burnt on the inside of the cylinder, and produced sounds similar to those obtained from a jet of flame in a tube.

Having thus endeavoured to account for the phenomenon of sounds produced by jets of flame, in tubes and other vessels, I shall notice shortly the combustible bodies I have tried. Carbonic oxide, olefiant gas, light hydrocarbonate, coal-gas, sulphuretted hydrogen, and arseniuretted hydrogen, were burned at the end of a long narrow brass tube rising up from a transferring jar placed under pressure in a pneumatic trough. Æther was burned from the end of a tube fixed in a flask containing a small quantity which was heated; but a better method, and one I afterwards adopted, is to pour a little æther into a bladder, and then force common air in; so much æther rises in vapour as to prevent the mixture being detonating, and it may be pressed out and burnt at the end of a tube. All these were very successful. Alcohol was more difficult to manage from being less volatile; but it succeeded when raised in vapour from a flask and burnt at a tube. In trials made with a wax taper, no distinct tone could be produced; but when the tube was made very hot, so as to assist the current through it, something like the commencement of a sound was heard at the moment the taper was blown out by the current.

Hydrogen is by far the best substance by which to produce these tones; and its superiority depends upon the low temperature at which it inflames, the intense heat it produces in combustion, and the small quantity of oxygen that a given bulk of it requires. It is in consequence less easily extinguished by the current than other gases, the current formed is more powerful and rapid, and an explosive mixture is sooner made. With gases producing little heat by combustion, and therefore occasioning but a feeble current, the effect is increased by first heating the tube at a fire, and when not heated previously, the tone is perceived to improve as the tube becomes hot from the flame playing in it.

Some variations of the form of the vessel enclosing the flame, and the material used, have been mentioned. Globes from 7 to 2 inches in diameter, with short necks, give very low tones: bottles, Florence flasks, and phials have always succeeded: air-jars from 4 inches diameter to a very small size may be used. I constructed some angular tubes of long narrow slips of glass and wood, placing three or four together, so as to form a triangular or square tube, tying them round with packthread, and easily obtained tones from hydrogen by means of them; and it is evident that variations of the channel, the use of which is to form and direct the current of air, may be made without end.—*May* 11, 1818.

Boracic Acid, action on Turmeric.

It may be observed, in connexion with the changes of colour produced by acids, that boracic acid reddens turmeric paper in all states of dilution. When a very weak solution is used, it requires a few minutes to produce the effect; but when produced, it exactly resembles that of alkali. It has been said that strong solutions of alkaline borates, which have been made purposely acid, have become alkaline on being diluted. This has probably arisen from a careless observance of the effect above noticed, and a want of corroboration by the effect on litmus paper of the diluted solution. I find that solutions once made acid redden litmus paper, however diluted; though at the same time they also redden turmeric paper. Paper coloured by rhubarb is not affected in this way.

Boracic Acid†.

I mentioned above the property possessed by boracic acid in all states of dilution, of reddening turmeric paper in the manner of an alkali. Since then the attention of M. Desfosses has been drawn to the action of boracic acid on this colouring matter (Annales de Chimie, xvi. p. 75), apparently without a knowledge of the previous remark; and he has shown that a mixture of boracic with other acids, reddens turmeric very deeply, and that turmeric, when acted on by this mixture of acids, has its nature altered, for it approaches somewhat to turnsole, and is rendered blue by alkalies.

There is something so curious in this action of boracic acid on turmeric, that I am tempted to offer a few more results on the subject.

* Quarterly Journal of Science, vi. 152.   † Ibid. xi. 403.

Turmeric paper dipped into a solution of pure boracic acid very speedily receives a slight tint of brownish-red, which, when the paper is dry, is very marked, and resembles that produced by a weak alkali. In this state the properties of the colouring matter are entirely different to what they were before: sulphuric, nitric, muriatic, and phosphoric acids, even when very dilute, produce a bright red colour on this paper, and a strong solution of oxalic acid also reddens it. Alkalies, on the contrary, make it blue, gradually passing to shades of purplish-blue, yellowish-red, &c. As long as the acids or alkalies remain on the paper, if not so strong as to destroy the colouring matter, the new colour remains, but a slight washing removes them, and then the boracic acid tint returns, and the paper has its first peculiar properties. When altered by muriatic acid or ammonia, the mere volatilization restores the paper to its first state; with the ammonia the restoration is very ready and perfect; with the acid it is longer and not so complete. If the paper reddened by boracic acid be heated, the yellow of the turmeric is almost restored, and then it takes from acids a weaker red tinge, and from alkalies a more purplish colour than before.

Turmeric, thus altered by boracic acid, is readily restored to its original state by washing; altered turmeric paper when put in water for two or three hours resumes its original properties, and acts, as at first, in testing the alkalies.

When the altered paper is placed in sunlight a few days, the colour is soon destroyed as with turmeric alone, and then neither acid nor alkali will affect it.

When turmeric paper is dipped into neutral or slightly alkaline borate of ammonia, it soon becomes of the red tint produced by boracic acid, and is in every respect as if altered by boracic acid alone; when this paper is made blue by ammonia, the ammonia easily washes out and the blue tint disappears, and afterwards the boracic acid or borates will wash out and leave the paper as at first.

Borax itself at first reddens turmeric paper because of the excess of alkali, but as the colouring matter becomes altered by the presence of the boracic acid, the tint becomes of a dirty bluish colour, and then the paper is changed by acids or alkalies, just as if it had been altered by boracic acid.

Hence it is probable that the neutral borates have the same

power as the boracic acid, of altering the colouring matter of turmeric, for it is not probable there should be an actual separation of the elements of the salts by it, especially as they both wash out from it and leave it unaltered.

Hence also both acid and alkaline borates redden turmeric.

M. Desfosses proposes this effect of boracic acid as a test for its presence; for a very small quantity of it mixed with other acid has the power of reddening turmeric paper in consequence of these changes.

On the Changing of Vegetable Colours as an Alkaline Property, and on some Bodies possessing it*.

THE changes produced by acids and alkalies on vegetable colours have long been considered as very distinctive and peculiar effects, and even sufficient of themselves to indicate the presence of these bodies. Since the introduction of substances before excluded, as of silica, various oxides, and vegetable substances, into the class of acids, and of oxide of lead, morphia, &c., into the class of alkalies, it becomes more important to substantiate any particular property as peculiar to those classes, or show its fallacy, by pointing out to what substances excluded from them it also belongs.

At present I shall detail the results of a few experiments made on the colouring matter of turmeric and rhubarb, comparing the changes produced on them by alkalies to those occasioned by some other bodies. Formerly†, I mentioned the property possessed by muriatic acid gas and strong acids in general, of reddening or browning turmeric paper. I find that in general they have the same effect on rhubarb paper, and a very weak nitric acid gives a brown tint to it, exactly like that of an alkali: strong solution of muriatic acid does not affect it much, but sulphuric acid does.

At pp. 27, 28, &c. I have shown the effect of boracic acid in reddening turmeric paper. Mr. South, I believe, first showed that the subacetate of lead reddened turmeric, and this has been considered sufficient evidence by many, that the oxide of lead merited, in some degree, the name of an alkali. I find on trial,

---

* Quarterly Journal of Science, xiii. 315.   † Ibid. v. 125.

however, so many substances possessing this property, that it must either be limited more exactly than has yet been done, or else given up as a distinguishing property.

All the soluble salts of iron that I have tried, except the acetate, brown turmeric paper. Weak solutions of the green and red muriate seem very alkaline indeed, and even common green vitriol strongly so. They do not, however, produce the same effect on rhubarb paper, but the persalts give it an olive-green tint, whilst the protosalts produce no change at first, but gradually give green tints, apparently from becoming persalts.

If a strong solution of muriate of zinc be boiled on zinc, it gradually oxidizes the metal and dissolves it; and a concentrated solution is obtained, which, when diluted, deposits either an oxide or a submuriate of zinc. This strong solution is apparently alkaline to turmeric paper. If it be diluted with about its bulk of water, and filtered, it will still redden turmeric paper, though it proves slightly acid by litmus paper. If further diluted, more precipitate will fall, and the solution will appear alkaline or not, according as the dilution has been small or great. This substance has the same effect on rhubarb paper, and the tint is very like that of a true alkali.

The acid nitrate of bismuth appears alkaline to turmeric; if diluted till a little oxide becomes deposited, it is more so:—the common solution of chloride of antimony in muriatic acid added to water till a precipitate falls, appears alkaline:—permuriate of tin produces a strong change; protomuriate of tin a very decided reddening; sulphate of tin, slight only. When the acid nitrate of bismuth, and the three salts of tin mentioned, are applied to rhubarb paper, they produce but little effect at first, but if dried by the fire, it becomes quite brown.

A strong solution of chloride of manganese seems feebly but evidently alkaline to turmeric paper.

It may be supposed, that with many of these substances the effect is produced principally by the acid present, inasmuch as they were all more or less acid to litmus, and it has been shown that these bodies, especially nitric acid, have this power. But the very slight excess in some of them, as in the salts of iron, zinc, manganese, &c., and the greater effect produced by further dilution, as with bismuth and antimony, show that the whole substance must act; for if it be considered due to the acid alone, it is not, in the first case, of sufficient quantity, and in the second, it would be diminished by dilution.

It is probable that if the tints produced be compared exactly with those occasioned by alkalies, slight differences might be perceived among some of them; or that if the properties of the altered colouring matter were examined, they would be found different with the different substances, as M. Desfosses has shown with regard to boracic acid and other acids; but my object has not been to trace these changes as far as possible, but merely to show their general appearance; to guard against any deceptive conclusion with respect to solutions tested by turmeric; and to call attention to the distinguishing characters of acids and alkalies.

Action of Salts on Turmeric Paper *.

AMONG the salts not alkaline, which have the power of affecting turmeric paper like alkalies (see pp. 29, 30), those of uranium are perhaps most powerful. The muriate, sulphate and acetate affect turmeric paper strongly even when considerably diluted; but the nitrate is the most powerful. A strong solution scarcely seems to have its power diminished by dilution with ten or twelve times its weight of water, and even when the solution contains only $\frac{1}{600}$th of dry nitrate of uranium, it sensibly browns turmeric paper.

The muriate of zirconia also possesses this same property to a considerable degree.

On the Decomposition of Chloride of Silver by Hydrogen and by Zinc †.

M. ARFVEDSON some time ago communicated to me a mode of reducing chloride of silver by hydrogen ‡. In a few experiments made some time since, in consequence of this communication, I found myself unable to decompose the chloride by a stream of pure hydrogen gas, or by allowing an atmosphere of the gas to remain for a long time in contact with it; I supposed, therefore, that the effect was produced by the hydrogen in its

* Quarterly Journal of Science, xiv. 234.
† Ibid. viii. 374.   ‡ Ibid. v. 360.

nascent state. But lately resuming the experiment, with the intention of ascertaining why the nascent state was more favourable for the combination of the elements than that of development, I found reason to suppose that the hydrogen was not at all concerned in liberating the chlorine from the silver.

When zinc is thrown into chloride of silver, diffused through dilute sulphuric or muriatic acid, hydrogen is liberated, and the chloride suffers decomposition. But the same effect takes place if zinc be thrown into chloride of silver, diffused through pure water, so that the hydrogen which escapes in the state of gas, cannot, in its nascent state, have been the decomposing agent. It may, however, be supposed that water is decomposed even when no acid is present, and that thus hydrogen is still the agent. But I find that zinc decomposes chloride of silver even more rapidly when unembarrassed by water, than when water is present. Thus, if a little fused chloride of silver and a small portion of zinc be heated in a glass tube, a violent action takes place; chloride of zinc is formed and silver liberated, and the heat rises so high as generally to fuse the silver; or if dry chloride of silver in powder be triturated in a mortar with zinc filings, the two bodies immediately act, and a heat above that of boiling water is produced.

Zinc is not the only common metal which thus rapidly decomposes chloride of silver, in the dry way. Tin acts even more powerfully when triturated with it; and copper and iron have both of them affinities for chlorine strong enough to produce the same effect.

There is therefore no occasion to assume hydrogen as the decomposing agent, when chloride of silver is reduced in contact with zinc or iron (iron acts as zinc does in all these experiments, though not so powerfully); for the metals, by their attraction for chlorine, are sufficiently energetic to produce the effect. Yet, as I had supposed, from general opinion, that hydrogen could, by its attraction for chlorine, separate that element from silver, I endeavoured to ascertain in what circumstances it had the power of doing so. If a stream of hydrogen, rapidly generated from iron or zinc, be sent against moist chloride of silver, in a dark place or by candlelight, it appears to alter it; but this effect must be due to metals or impurities held in solution, for when purified it has no power of changing it out of daylight; nor have I been able, even in the sunshine of this month, to make hydrogen act on chloride of silver in several hours.

Still, however, hydrogen may be allowed in certain circumstances to have the power of decomposing chloride of silver, but the circumstances are not such as were, I believe, generally supposed to have place in the experiment first referred to. When zinc, iron, tin, &c. are thrown into moist chloride of silver, the first decomposition is occasioned by the action of the zinc on the chloride, afterwards a voltaic circle is formed by the zinc, the reduced silver and the water; water is decomposed, the zinc takes oxygen, the hydrogen liberated at the surface of the silver takes the chlorine from the chloride in its immediate neighbourhood, and thus the reduction will go on to the distance of an inch or more from the piece of zinc, and the consequent products are silver and solution of muriate of zinc. But as this is a case of decomposition entirely different to the supposed one of the reduction of chloride of silver by hydrogen, any denial of the latter is not at all invalidated by the truth of the former.

On two new Compounds of Chlorine and Carbon, and on a new Compound of Iodine, Carbon, and Hydrogen*.

[Read Dec. 21, 1820.]

ONE of the first circumstances that induced Sir H. Davy to doubt the compound nature of what was formerly called oxymuriatic acid gas, was the want of action of heated charcoal upon it; and considerable use of the same agent, and of the phenomena exhibited by it in different circumstances with chlorine, was afterwards made in establishing the simple nature of that body.

The true nature of chlorine being ascertained, it became of importance to form all the possible compounds of it with other elementary substances, and to examine them in the new view had of their nature. This investigation has been pursued with such success at different times, that very few elements remain uncombined with it; but with respect to carbon, the very cir-

* Phil. Trans. 1821, p. 47, and Phil. Mag. lix. p. 337.

D

cumstance which first tended to correct the erroneous opinions which, after Scheele's time, and before the year 1810, had gone abroad respecting its nature, proved an obstacle to the formation of its compounds; and up to the present time the chlorides of carbon have escaped the researches of chemists.

That the difficulty met with in forming a compound of chlorine and carbon was probably not owing to any want or weakness of affinity between the two bodies, was pointed out by Sir H. Davy; who, reasoning on the triple compound of chlorine, carbon and hydrogen, concluded that the attraction of the two bodies for each other was by no means feeble; and the discovery of phosgene gas by Dr. Davy, in which chlorine and carbon are combined with oxygen, was another circumstance strongly in favour of this opinion.

I was induced last summer to take up this subject, and have been so fortunate as to discover two chlorides of carbon, and a compound of iodine, carbon and hydrogen, analogous in its nature to the triple compound of chlorine, carbon and hydrogen, sometimes called chloric æther. I shall endeavour in the following pages to describe these substances, and give the experimental proofs of their nature.

If chlorine and olefiant gas be mixed together, it is well known that condensation takes place, and a colourless limpid volatile fluid is produced, containing chlorine, carbon and hydrogen. If the volumes of the two gases are equal, the condensation is perfect. If the olefiant gas is in excess, that excess is left unchanged. But if the chlorine is in excess, the fluid becomes of a yellow tint, and acid fumes are produced. This circumstance alone proves that chlorine can take hydrogen from the fluid; and on examination I found it was without the liberation of any carbon or chlorine.

That the action thus begun might be carried to its utmost extent, some of the pure fluid (chloric æther) was put into a retort with chlorine, and exposed to sunshine. At the first instant of contact between the chlorine and the fluid, the latter became yellow; but when in the sun's rays, a few moments sufficed to destroy the colour both of the fluid and the chlorine, heat being at the same time evolved. On opening the retort, there was no absorption, but it was found full of muriatic acid gas. This was expelled and more chlorine introduced, and the whole again exposed to sunlight: the colour again disappeared, and a few moist crystals were formed round the edge of the fluid. Chlorine being a third time introduced, and treated as before, it still removed more hydrogen; and now a sublimate of crystals lined the retort. Proceeding in this way until the chlorine exerted no further action, the fluid entirely disappeared, and the results were, the dry crystalline substance and muriatic acid gas.

A portion of olefiant gas was then mixed in a retort with eight or nine times its bulk of chlorine, and exposed to sunlight. At first the fluid formed, but this instantly disappeared; the retort became lined with crystals, and the colour of the chlorine very much diminished.

On examining these crystals, I found they were the compound I was in search of; but before I give the proofs of their nature, I will describe the process by which this chloride of carbon can be obtained pure.

Perchloride of Carbon.—A glass vessel was made in the form of an alembic head, but without the beak; the neck was considerably contracted, and had a brass cap with a stopcock cemented on; at the top was a small aperture, into which a ground stopper fitted air-tight. The capacity of the vessel was about 200 cubic inches. Being exhausted by the air-pump, it was nearly filled with chlorine; and being then placed over olefiant gas, and as much as could enter having passed in, the stopcocks were shut, and the whole left for a short time. When the fluid compound of chlorine and olefiant gas had formed and condensed on the sides of the vessel, it was again placed over olefiant gas, and in consequence of the condensation of a large portion of the gases, a considerable quantity more entered. This was left, as before, to combine with part of the remaining chlorine, to condense, and to form a partial vacuum; which was again filled with olefiant gas, and the process repeated until all the chlorine had united to form the fluid, and the vessel remained full of olefiant gas. Chlorine was then admitted in repeated portions as before; consequently more of the fluid formed; and ultimately a large portion was obtained in the bottom of the vessel, and an atmosphere of chlorine above it. It was now exposed to sunlight. The chlorine immediately disappeared, and the vessel became filled with muriatic acid gas. Having ascertained that water did not interfere with the action of the substances, a small portion was admitted into the vessel, which absorbed the muriatic acid gas, and then another atmosphere of chlorine was introduced. Again exposed to the light, this was partly combined with the carbon, and partly converted into muriatic acid gas; which being, as before, absorbed by the water, left space for more chlorine. Repeating this action, the fluid gradually became thick and opake from the formation of crystals in it, which at last adhered to the sides of the glass as it was turned round ; and ultimately the vessel only contained chlorine with the accumulated gaseous impurities of the successive portions, a strong solution of muriatic acid coloured blue from the solution of a little brass, and the solid substance.

I have frequently carried the process thus far in retorts; and it is evident that any conveniently formed glass vessel will answer the purpose. The admission of water during the process prevents the necessity of repeated exhaustion by the air-pump, which cannot be done without injury to the latter; but to have the full advantage of this part of the process, the gases should be as pure as possible, that no atmosphere foreign to the experiment may collect in the vessel.

In order to cleanse the substance, the remaining chlorine and muriatic acid were blown out of the vessel by a pair of bellows, introduced at the stoppered aperture, and the vessel afterwards filled with water, to wash away the muriatic acid and other soluble matters. Considerable care is then requisite in the further purification of the chloride. It retains water, muriatic acid, and a substance which I find to be a triple compound of chlorine, carbon and hydrogen, formed from the cement of the cap ; and as all these contain hydrogen, a small quantity of any one remaining with the chloride would, in analysis, give erroneous results. Various methods of purification may be devised, founded on the properties of the substance, but I have found the following the most convenient:—The substance is to be washed from off the glass, and poured with the water into a jar; a little alcohol will remove the last portions which adhere to the glass; and this, when poured into the water, will precipitate the chloride, and the whole will fall to the bottom of the vessel. Then having decanted the water, the chloride is to be collected on a filter, and dried as much as may be, by pressure between folds of bibulous paper. It should next be introduced into a glass tube, and sublimed by a spirit-lamp: the pure substance with water will rise at first, but the last portions will be partially decomposed, muriatic acid will be liberated, and charcoal left. The sublimed portion is then to be dissolved in alcohol, and poured into a weak solution of potash, by which the substance is thrown down, and the muriatic acid neutralized and separated; then wash away the potash and muriate by repeated affusions of water, until the substance remains pure; collect it on a filter, and dry it, first between folds of paper, and afterwards by sulphuric acid in the exhausted receiver of the air-pump.

It will now appear as a white pulverulent substance; and if perfectly pure, will not, when a little of it is sublimed in a tube, leave the slightest trace of carbon, or liberate any muriatic acid. A small portion of it dissolved in æther should give no precipitate with nitrate of silver. If it be not quite pure, it must be resublimed, washed, and dried until it is pure.

This substance does not require the direct rays of the sun for its formation. Several tubes were filled with a mixture of one part of olefiant gas with five or six parts of chlorine, and placed over water in the light of a dull day; in two or three hours there was very considerable absorption, and crystals of the substance were deposited on the inside of the tubes. I have also often observed the formation of the crystals in retorts in common daylight.

A retort being exhausted had 12 cubic inches of olefiant gas introduced, and 24·75 cubic inches of chlorine: as soon as the condensation occasioned by the formation of the fluid had taken place, 21·5 cubic inches more of chlorine were passed in, and the retort set aside in a dark place for two days. At the end of that time muriatic acid gas and the solid chloride had formed, but the greater part of the fluid remained unchanged. Hence it will form even in the dark by length of time.

I tried to produce the chloride by exposure of the two gases in tubes over water to strong lamp-light for two or three hours, but could not succeed.

The perchloride of carbon, when pure, is, immediately after fusion, or sublimation, a transparent colourless substance. It has scarcely any taste. Its odour is aromatic, and approaching to that of camphor. Its specific gravity is as nearly as possible 2. Its refractive power is high, being above that of flint-glass (1·5767). It is very friable, easily breaking down under pressure; and when scratched has much of the feel and appearance of white sugar. It does not conduct electricity.

The crystals obtained by sublimation and from solutions of the substance in alcohol and æther, are dendritical, prismatic, or in plates; the varieties of form, which are very interesting, are easily ascertained, and result from a primitive octahedron.

It volatilizes slowly at common temperatures, and passes, in the manner of camphor, towards the light. If warmed, it rises more rapidly, and then forms fine crystals: when the temperature is further raised, it fuses at 320° Fahr. and boils at 360° under atmospheric pressure. When condensed again from these rapid sublimations, it concretes in the upper part of the tube or vessel containing it, in so transparent and colourless a state, that it is difficult, except from its high refractive power, to perceive where it is lodged. As the crust it forms becomes thicker, it splits, and cracks like sublimed camphor; and in a few minutes after it is cold, is white, and nearly opake. If the heat be raised still higher, as when the substance is passed through a red-hot tube, it is decomposed, chlorine is evolved, and another chloride of carbon, which condenses into a fluid, is obtained. This shall be described presently.

It is not readily combustible; when held in the flame of a spirit-lamp, it burns with a red flame, emitting much smoke and acid fumes; but when removed from the lamp, combustion ceases. In the combustion that does take place in the lamp, the hydrogen of the alcohol, by combining with the chlorine of the compound, performs the most important part; nevertheless, when the substance is heated red in an atmosphere of pure oxygen, it sometimes burns with a brilliant light.

It is not soluble in water at common temperatures, or only in very small quantity. When a drop or two of the alcoholic solution is poured into a large quantity of water, it renders it turbid from the deposition of the substance. It does not appear that hot water dissolves more of it than cold water.

It dissolves in alcohol with facility, and in much greater quantity with heat than without. A saturated hot solution crystallizes as it cools, and the cold solution also gives crystals by spontaneous evaporation. When poured into water, the chloride is precipitated, and falls to the bottom in flakes. If burnt, the flame of the alcohol is brightened by the presence of the substance, and fumes of muriatic acid are liberated. Solution of nitrate of silver does not produce any turbidness in it, unless it be in such quantity that the water throws down the substance; but no chloride of silver is formed.

It is much more soluble in æther than in alcohol, and more so in hot than in cold æther. The hot solution deposits crystals as it cools; and the crystallization of a cold solution, when evaporated on a glass plate, is very beautiful. This solution is not precipitated by water, unless the æther has previously been dried, and then water occasions a turbidness. Nitrate of silver does not precipitate it. When burned, muriatic acid fumes are liberated, but the greater part of the chloride remains in the capsule.

It is soluble in the volatile oils, and on evaporation is again obtained in crystals. It is also readily soluble in fixed oils. The solutions when heated liberate muriatic acid gas, and the oil becomes of a dark colour, as if charred.

Solutions of the acids and alkalies do not act with any energy on the substance. When boiled with solutions of pure potash and soda, it rises and condenses in the upper part of the vessel; and though it be brought down to the alkali many times and reboiled, still the alkali, when examined, is not found to contain any chlorine, nor is any change produced. Ammonia in solution is also without action upon it. These solutions do not appear to dissolve more of it than pure water.

Muriatic acid in solution does not act at all upon it. Strong nitric acid boiled upon it dissolves a portion, but does not decompose it: as it cools, part of the chloride is deposited unaltered, and the concentrated acid, when diluted, lets more fall down. The diluted portion being filtered, and tested with nitrate of silver, gives no precipitate. It does not appear to be either soluble in, or acted upon by, concentrated sulphuric acid. It sinks slowly in the acid, and, when heated, is converted into vapour, which, rising through the acid, condenses in the upper part of the tube.

It is not acted upon by oxygen at temperatures under a red

heat. A mixture of oxygen and the vapour of the substance would not inflame by a strong electric spark, though the temperature was raised by a spirit-lamp to about 400°. When oxygen mixed with the vapour of the substance is passed through a red-hot tube, there is decomposition; and mixtures of chlorine, carbonic oxide, carbonic acid, and phosgene gases are produced. A portion of the chloride was heated with peroxide of mercury, in a glass tube over mercury; as soon as the oxide had given off oxygen, and the heat had risen so high as to soften the glass considerably, the vapour suddenly detonated with the oxygen with bright inflammation. The substances remaining were oxygen, carbonic acid, and calomel; and I believe there was no decomposition or action, until so much mercury had risen in vapour as to aid the oxygen by a kind of double affinity in decomposing the chloride of carbon.

Chlorine produces no change on the substance, either by exposure to light or heat.

When iodine is heated with it at low temperatures, the two substances melt and unite, and there is no further action. When heated more strongly in vapour, the iodine separates chlorine, reducing the perchloride to the fluid protochloride of carbon, and chloriodine is produced. This dissolves, and if no excess of iodine be present, the whole remains fluid at common temperatures. When water is added, it generally liberates a little iodine; and on heating the solution, so as to drive off all free iodine, and testing by nitrate of silver, chloride and iodide of silver are obtained.

Hydrogen and the vapour of the substance would not inflame at the temperature of 400° Fahr. by strong electrical sparks; but when the mixture was sent through a red-hot tube, the chloride was decomposed, and muriatic acid gas and charcoal produced.

The vapour of the perchloride of carbon readily detonates by the electric spark with a mixture of oxygen and hydrogen gases; but the gaseous results are very mixed and uncertain, from the near equipoise of affinities that exists among the elements.

Sulphur readily unites to it when melted with it, and the mixture crystallizes on cooling into a yellowish mass. When heated more strongly, the substance rises unchanged, and leaves the sulphur unaltered ; but when the mixed vapours are raised to a still higher temperature, chloride of sulphur and protochloride of carbon are formed. Sometimes there are appearances as if a carburet of sulphur were formed, but of this I have not satisfied myself.

Phosphorus at low temperatures melts and unites with the substance without any decomposition. If heated in the vapour of the substance, but not too highly, it takes away chlorine, and forms the protochlorides of phosphorus and carbon. If heated more highly, it frequently inflames in the vapour with a brilliant combustion, and abundance of charcoal is deposited. Sometimes I have had the charcoal left in films stretching across the tubes, and occupying the space where the flame passed. The appearance is then very beautiful.

When phosphorus is heated with the vapour of the substance over mercury, so as not to inflame in it, there is generally a small portion of muriatic acid gas formed. If great care be taken, this is in very minute quantity; and its variable proportion sufficiently shows, that the hydrogen which forms it does not come from the substance. I am induced to believe that it is derived from moisture adhering to the phosphorus. The action of iodine on phosphorus shows that it is very difficult to dry the latter substance perfectly.

A stick of phosphorus put into the alcoholic or æthereal solution of the perchloride did not exert any action upon it.

Charcoal heated in the vapour of the substance appears to have no action upon it.

Most of the metals decompose it at high temperatures. Potassium burns brilliantly in the vapour, depositing charcoal and forming chloride of potassium. Iron, zinc, tin, copper and mercury act on it at a red heat, forming chlorides of those metals and depositing charcoal; and when the experiments are made with pure substances, and very carefully, no other results are obtained. Some of the substance was passed over iron turnings heated in a glass tube. At the commencement of the sublimation of the chloride through the hot iron, the common air of the vessels was expelled, and received in different tubes; but before one-third of the substance had been passed, all liberation of gas ceased, and the remainder was decomposed by the iron, without the production of any gaseous matters. The different portions of air that were thrown out being examined, the first proved to be common air, and the last carbonic oxide. This had resulted, probably, from the action of the chlorine on the lead of the glass tube. An evident action had taken place, and the oxygen evolved, meeting with the liberated carbon, would produce the carbonic oxide. This experiment has been repeated several times with the same results.

When the perchloride of carbon is heated with metallic oxides, different results are produced according to the proportions of oxygen in the oxides. The peroxides, as of mercury, copper, lead and tin, produce chlorides of those metals, and carbonic acid; and the protoxides, as those of zinc, lead, &c., produce also chlorides, but the gaseous products are mixtures of carbonic acid and carbonic oxide. I have frequently perceived the smell of phosgene gas, on passing the chloride over oxide of zinc; and as the substance easily liberates chlorine at high temperatures, it will be readily seen how a small portion of that gas may be formed. It also happens, sometimes, that the protoxides become blackened from the deposition of charcoal.

When the vapour of the chloride is passed over lime, baryta or strontia, heated red-hot, a very vivid combustion is produced. The oxygen and the chlorine change places, and both the metals and the carbon are burnt. Chlorides are produced, carbonic acid is formed and absorbed by the undecomposed parts of the earths, and carbon is deposited. In these experiments no carbonic oxide is produced. When passed over magnesia, there is no action on the earth, but the perchloride of carbon is converted by the heat into protochloride.

In these experiments with the oxides no trace of water could be perceived.

Having thus far described the properties of the substance, I shall now give the reasons which induce me to consider it a true chloride of carbon, and shall endeavour to assign its composition. My first object was to ascertain whether hydrogen existed in it or not. When phosphorus is heated in it, a small quantity of muriatic acid is generally formed; but doubt arises as to the cause of its production, from the circumstance that the phosphorus, as already mentioned, may be the source of the hydrogen.   When potassium is heated in the vapour of the substance, there is generally a small expansion of volume, and inflammable gas produced; but it is very difficult to cleanse potassium both from naphtha and an adhering crust of moist potash; and either of these, though in extremely minute quantities, would give fallacious results.

A more unexceptionable experiment made with iron has been already described; and the inferences from it are against the presence of hydrogen in the compound.

Some of the substance in vapour was electrized over mercury by having many hundred sparks passed through it.   Calomel was formed and carbon deposited.   A very minute bubble of gas was produced, but it was much too small to interfere with the conclusions drawn respecting the binary nature of the compound; and was probably caused by air that had adhered to the sides of the tube when the mercury was poured in.

The most perfect demonstration that the body contains no hydrogen, and indeed of its nature altogether, is obtained from the circumstances which attend its formation.   When the fluid compound of chlorine and olefiant gas is acted on by chlorine and solar light in close vessels, although the whole of the chlorine disappears, yet there is no change of volume, its place being occupied by muriatic acid gas.   Hence, as muriatic acid gas is known to consist of equal volumes of chlorine and hydrogen, combined without condensation, it is evident that half the chlorine introduced into the vessel has combined with the elements of the fluid, and liberated an equal volume of hydrogen; and as, when the chloride is perfectly formed, it condenses no muriatic acid gas, a method, apparently free from all fallacy, is thus afforded of ascertaining its nature.

I have made many experiments on given volumes of chlorine and olefiant gases.   A clean dry retort was fitted with a cap and stopcock.   Its capacity was 25·25 cubic inches.   Being exhausted by the air-pump, it was filled with nitrogen (24·25 cubic inches being required), and being again exhausted, 5 cubic inches of olefiant gas, and 10 cubic inches of chlorine, were introduced.   It was then set aside for half an hour, that the fluid compound might form, and afterwards being placed again over a jar of chlorine, 19·25 cubic inches entered; so that the condensation had been as nearly as possible 10 cubic inches, or twice the volume of the olefiant gas (barometer 29·1 inches). It was now placed for the day (Oct. 18) in the rays of the sun; but the weather was not very fine. In the evening the solid crystalline substance had formed in abundance, and very little fluid remained. When placed over chlorine, not the slightest change in volume had been produced. The stopcock was now opened under mercury, and a small portion of the metal having entered, it was agitated in the retort, to absorb the chlorine; the neck of the retort was left open under the mercury all night, and the whole agitated from time to time. Next morning (barometer 29·6) the mercury which had entered, being passed into the neck of the retort, stood at a certain mark 6 inches above the level of the mercury in the trough, occupying 1·25 cubic inch, and leaving 24 cubic inches filled by the expanded muriatic acid gas and nitrogen. These volumes, corrected to the pressure of 29·1 inches, give 5·78 cubic inches for the chlorine absorbed, and 19·47 cubic inches for the muriatic acid gas, &c. These absorbed by water left 1·2 cubic inch of nitrogen; so that the gases in the retort, after the action of solar light, were,—

	Cubic inches.
Muriatic acid gas	18·27
Chlorine	5·78
Nitrogen, &c.	1·2

and before that action,—

Chlorine	29·25
Olefiant gas	5·0
Nitrogen	1·0

Hence 23·47 cubic inches of chlorine had disappeared, and 9·13 of these had entered into combination with an equal volume of 9·13 cubic inches of hydrogen liberated from the 5 cubic inches of olefiant gas, to form muriatic acid; and consequently 14·34 cubic inches of chlorine remained combined with the carbon of the 5 cubic inches of olefiant gas. Here the volume of chlorine actually employed is not quite five times that of the olefiant gas, nor the volume of muriatic acid gas produced equal to four times that of the olefiant gas; but they approximate; and when it is remembered that the conversion was not quite perfect, and that the gases used would inevitably contain a slight portion of impurity, the causes of the deficiency can easily be understood.

In other experiments made in the same way, but with smaller quantities, more accurate results were obtained: 1 cubic inch of olefiant gas with 12·25 cubic inches of chlorine, produced by the action of light 3·67 cubic inches of muriatic acid gas, 4·963 of the chlorine having been used. 1·4 cubic inch of olefiant gas with 12·5 cubic inches of chlorine produced 5·06 cubic inches of muriatic acid gas, 6·7 cubic inches of chlorine having been used. Other experiments gave very nearly the same results; and I have deduced from them, that one volume of olefiant gas requires five volumes of chlorine for its conversion into muriatic acid and chloride of carbon; that four volumes of muriatic acid gas are formed; that three volumes of chlorine combine with the two volumes of carbon in the olefiant gas to form the solid crystalline chloride; and that, when chlorine acts on the fluid compound of chlorine and olefiant gas, for every volume of chlorine that combines, an equal volume of hydrogen is separated.

I have endeavoured to verify these proportions by analytical experiments. The mode I adopted was, to send the substance in vapour over metals and metallic oxides at high temperatures. Considerable care is requisite in such experiments; for if the process be carried on quickly, a portion of fluid chloride of carbon is formed, and escapes decomposition. The following are two results from a number of experiments agreeing well with each other.

Five grains were passed over peroxide of copper in an iron tube, and the gas collected over mercury; it amounted to 3·9 cubic inches; barometer 29·85; thermometer 54° Fahr. Of these nearly 3·8 cubic inches were carbonic acid, and rather more than ·1 of a cubic inch was carbonic oxide. These are nearly equal to ·5004 of a grain of carbon. Hence 100 of the chloride would give 10 of carbon nearly, but by calculation 100 should give 10·19. The difference is so small as to come within the limits of errors in experiment.

Five grains were passed over peroxide of copper in a tube made of green phial glass, and the chlorine estimated in the same manner as before. 17·7 grains of chloride of silver were obtained, equal to 4·36 grains of chlorine. This result approaches much nearer to the calculated result than the former; but there had still been action on the tube, and a minute portion of the substance had passed undecomposed, and condensed at the opposite end of the tube in crystals.

Experiments made by passing the perchloride over hot lime or barytes, promise to be more accurate and easy of performance. In the mean time, the above analytical results will, perhaps, be considered as strong corroboration of the opinion of the nature of the compound, deduced from the synthetical experiments; and the composition of the perchloride of carbon will be—

Three proportions of chlorine . . . . $= 100 \cdot 5$
Two   ,,   carbon . . . . $= \ 11 \cdot 4$

                                            $111 \cdot 9$

Protochloride of Carbon.—Having said so much on the nature of the perchloride of carbon, I shall have less occasion to dwell on the proofs that the compound I am about to describe, is also a binary combination of carbon and chlorine.

When the vapour of the perchloride of carbon is heated to dull redness, chlorine is liberated, and a new compound of that element and carbon is produced. This is readily shown by heating the bottom of a small glass tube, containing some of the perchloride in a spirit-lamp. The substance at first sublimes; but as the vapour becomes heated below, it is gradually converted into protochloride, and chlorine is evolved.

It is not without considerable precaution that the protochloride of carbon can be obtained pure; for though passed through a great length of heated tube, part of the perchloride frequently escapes decomposition. The process I have adopted is the following:—Some of the perchloride is introduced into the closed end of a tube, and the space above it, for 10 or 12 inches, filled with small fragments of rock-crystal; the part of the tube beyond this is then bent up and down two or three times, so that the angles may form receivers for the new compound; then heating the tube and crystal to bright redness, and dipping the angles in water, the perchloride is slowly sublimed by a spirit-lamp, and, on passing into the hot part of the tube, is decomposed; a fluid passes over, which is condensed in the angles of the tube, and chlorine is evolved; part of the gas escapes, but the greater portion is retained in solution by the fluid, and renders it yellow. Having proceeded thus far, by the careful application of a lamp and blowpipe, the bent part of the tube may be separated from that within the furnace, and the end closed, so as to form a small retort; and on distilling the fluid four or five times from one angle to the other, all the chlorine may be driven off without any loss of the substance, and it becomes limpid and colourless. It still, however, always contains some perchloride, which has escaped decomposition; and, to separate this, I have boiled the fluid until the tube was nearly full of its vapour, and then closing the end that still remained open, by a lamp and blowpipe, have afterwards left the whole to cool. It is then easy, by collecting all the fluid into one end of the tube, and introducing that end through a cork into a receiver, under which a very small flame is burning, to distil the whole of the fluid at a temperature very little above that of the atmosphere. The solid chloride being less volatile does not rise so soon, and the pure protochloride collects at the external end of the tube. To ascertain its purity, a drop may be placed on a glass plate; it will immediately evaporate, and if it contains perchloride, that substance will be left behind; otherwise, no trace will remain on the glass. The presence or absence of free chlorine may be ascertained by dissolving a little of the fluid in alcohol or æther, and testing by nitrate of silver.

The pure protochloride of carbon is a highly limpid fluid, and perfectly colourless. Its specific gravity is 1·5526. It is a non-conductor of electricity. I am indebted to Dr. Wollaston for the determination of the refractive power of this chloride, and for the approximation to the refractive power given of the perchloride. In the present case it is 1·4875, being very nearly that of camphor. It is not combustible except when held in a flame, as of a spirit-lamp, and then it burns with a bright yellow light, much smoke, and fumes of muriatic acid.

It does not become solid at the zero of Fahrenheit's scale. When its temperature is raised under the surface of water to between 160° and 170°, it is converted into vapour, and remains in that state until the temperature is lowered. When heated more highly, as by being passed over red-hot rock-crystal in a glass tube, a small portion is always decomposed;

nearly all the fluid may, however, be condensed again; but it passes slightly coloured, and the tube and crystal are blackened on the surface by charcoal. I am uncertain whether this decomposition ought not to be attributed rather to the action of the glass at this high temperature than to the heat alone.

It is not soluble in water, but remains at the bottom of it in drops, for many weeks, without any action.

It is soluble in alcohol and æther, and the solutions burn with a greenish flame, evolving fumes of muriatic acid.

It is soluble in the volatile and fixed oils. The volatile oils containing it burn with the emission of fumes of muriatic acid. When the solutions of it in the fixed oils are heated, they do not blacken or evolve fumes of muriatic acid. It is therefore probable, that when this happens with the solution of the perchloride in fixed oils, it is from its conversion by the heat into protochloride and the liberation of chlorine.

It is not soluble in alkaline solutions, nor is any action apparent after several days. Neither is it at all soluble in, or affected by, strong nitric, muriatic, or sulphuric acids.

Solutions of silver do not act on it.

Oxygen decomposes it at high temperatures, forming carbonic oxide or acid, and liberating chlorine.

Chlorine dissolves in it in considerable quantity, but has no further action, or only a very slow one, in common daylight; on exposure to solar light, a different result takes place. I have only had two days, and those in the middle of November, on which I could expose the protochloride of carbon in atmospheres of chlorine to solar light; and hence the conversion of the whole of the protochloride was not perfect; but at the end of those two days the retorts containing the substances were lined with crystals, which, on examination under the microscope, proved to be quadrangular plates, resembling those of the perchloride of carbon. There were also some rhomboidal crystals here and there. After the formation of these crystals, there was considerable absorption in the retort; hence chlorine had combined; and the gas which remained was chlorine unmixed with anything else, except a slight impurity. The solid body, on examination, was found to be volatile, soluble in alcohol, precipitable by water, and had the smell and other properties of perchloride of carbon. Hence, though heat in separating chlorine from the perchloride of carbon produces its decomposition, light occasions its reproduction.

It dissolves iodine very readily, and forms a brilliant red solution, similar in colour to that made by putting iodine into sulphuret of carbon or chloric æther. It does not exert any further action on iodine at common temperatures.

An electric spark passed through a mixture of the vapour of the chloride with hydrogen, does not cause any detonation; but when many are passed, the decomposition is gradually effected, and muriatic acid is formed. When hydrogen and the vapour of the protochloride are passed through a red-hot tube, there is a complete decomposition effected, muriatic acid gas being formed, and charcoal deposited. The mixed vapour and gas burn with flame as they arrive in the hot part of the tube. The vapour of the protochloride detonates readily by the electric spark with a mixture of oxygen and hydrogen gases, and a complete decomposition is effected. It will not detonate with the vapour of water.

Sulphur and phosphorus both dissolve in it, but exert no decomposing action at temperatures at or below the boiling-point of the chloride. The hot solution of sulphur becomes a solid crystalline mass by cooling. Phosphorus decomposes it at a red heat.

Its action on metals is very similar to that of the perchloride. When passed over them at a red heat, it forms chlorides, and liberates charcoal. Potassium does not act on it immediately at common temperatures; but, when heated in its vapour, burns brilliantly, and deposits charcoal.

When passed over heated metallic oxides, chlorides of the metals are formed, and carbonic oxide or carbonic acid, according to the state of oxidation of the metal. When its vapour is transmitted over heated lime, baryta, or strontia, the same brilliant combustion is produced as with the perchloride.

While engaged in analysing this chloride of carbon for the purpose of ascertaining the proportions of its elements, I endeavoured at first to find how much chlorine was liberated from a certain weight of perchloride during its conversion into protochloride, and for this purpose distilled the perchloride through red-hot tubes into solution of nitrate of silver, receiving the gas into tubes filled with and immersed in the same solu-

E

tion; but I could never get accurate results in this way, from the difficulty of producing a complete decomposition, and also from the formation of chloric acid. Five grains of perchloride distilled in this manner gave 4·3 grains of chloride of silver, which are equivalent to 1·06 grain of chlorine; but some of the chloride evidently passed undecomposed, and crystallized in the tube.

2·7 grains of the pure protochloride were passed over red-hot pure baryta in a glass tube: a very brilliant combustion with flame took place, chloride of barium and carbonic acid were produced, and a little charcoal deposited. When the tube was cold, the barytes was dissolved in nitric acid, and the chlorine precipitated by nitrate of silver. 9·4 grains of dry chloride of silver were obtained =2·32 grains of chlorine.

Other experiments were made with lime, which gave results very near to this, the quantity of chloride being rather less.

Three grains of pure protochloride were passed over peroxide of copper heated red-hot in an iron tube, and the gas received over mercury. 3·5 cubic inches of carbonic acid gas came over, mixed with ·1 of a cubic inch of common air. These 3·5 cubic inches are nearly equal to ·449 of a grain of carbon.

These experiments indicate the composition of the fluid chloride of carbon to be one proportion of chlorine and one of carbon, or 33·5 of the former, and 5·7 of the latter. The difference between these theoretical numbers, and the results of the experiments, is not too great to have arisen from errors in working on such small quantities of the substance.

A mixture of equal volumes of oxygen and hydrogen was made, and two volumes of it detonated with the vapour of the protochloride in excess over mercury by the electric spark. The expansion was very nearly to four volumes; of these, two were muriatic acid, and the rest pure carbonic oxide: and calomel had been formed, its presence being ascertained by potash. Hence it appears, that one volume of hydrogen and half a volume of oxygen had decomposed one proportion of the protochloride, forming the two volumes of muriatic acid gas and one volume of carbonic oxide; and that at the intense temperature produced within the tube by the inflammation, the rest of the oxygen and the mercury had decomposed a further portion of the substance, giving rise to the second volume of the carbonic oxide, and to the calomel.

A mixture of two volumes of hydrogen and one volume of oxygen was made, and three volumes of it detonated with the vapour, as before. After cooling, the expansion was to six volumes, four of which were muriatic acid, and two carbonic oxide. There was no action on the mercury in this experiment. Again, five volumes of the same mixture being detonated with the vapour of the substance expanded to 9·75 volumes, of which 6·25 were absorbed by water and were muriatic acid, and 3·5 were carbonic oxide mixed with a very small portion of air introduced along with the fluid chloride. These experiments, I think, establish the composition of the protochloride of carbon, and prove that it contains one proportion of each of its elements.

From a consideration of the proportions of these two chlorides of carbon, it seems extremely probable that another may exist, composed of two proportions of chlorine combined with one of carbon. I have searched assiduously for such a compound, but am undecided respecting its production. When the fluid protochloride was exposed with chlorine to solar light, crystals were formed, as before described. The greater number of these were certainly the perchloride first mentioned in this paper; but when the retort was examined by a microscope, some rhomboidal crystals were observed here and there among those of the usual dendritic and square forms. These may perhaps be the real perchloride; but I had not time, before the season of bright sunshine passed away, to examine minutely what happens in these circumstances; and must defer this, with many other points, till the next year brings more favourable weather.

Compound of Iodine, Carbon, and Hydrogen.—The analogy which exists between chlorine and iodine, naturally suggested the possible existence of an iodide of carbon, and the means which had succeeded with the one element offered the best promise of success with the other.

Iodine and olefiant gas were put in various proportions into retorts, and exposed to the sun's rays. After awhile, colourless crystals formed in the vessels, and a partial vacuum was produced. The gas in the vessels being then examined, was

E 2

found to contain no hydriodic acid, but only pure olefiant gas. Hence the effect had been simply to produce a compound of the iodine with the olefiant gas.

The new body formed was obtained pure by introducing a solution of potash into the retort, which dissolved all the free iodine; the substance was then collected together and dried. It is a solid white crystalline body, having a sweet taste and aromatic smell. It sinks readily in sulphuric acid of specific gravity 1·85. It is friable; is not a conductor of electricity. When heated, it first fuses, and then sublimes without any change. Its vapour condenses into crystals, which are either prismatic or in plates. On becoming solid after fusion, it also crystallizes in needles. The crystals are transparent. When highly heated it is decomposed, and iodine evolved. It is not readily combustible; but when held in the flame of a spirit-lamp, burns, diminishing the flame, and giving off abundance of iodine and some fumes of hydriodic acid. It is insoluble in water, or in acid and alkaline solutions. It is soluble in alcohol and æther, and may be obtained in crystals from these solutions. The alcoholic solution is of a very sweet taste, but leaves a peculiarly sharp biting sensation on the tongue.

Sulphuric acid does not dissolve it. When heated in the acid to between 300° and 400°, the compound is decomposed, apparently by the heat alone; and iodine and a gas, probably olefiant gas, are liberated. Solution of potash acts on it very slowly, even at the boiling-point, but does gradually decompose it.

This substance is evidently analogous to the compound of olefiant gas and chlorine, and remarkably resembles it in the sweetness of its taste, though it differs from it in form, &c. It will, with that body, form a new class of compounds, and they will require names to distinguish them. The term chloric æther, applied to the compound of olefiant gas and chlorine, did not at any time convey a very definite idea, and the analogous name of iodic æther would evidently be very improper for a solid crystalline body heavier than sulphuric acid. Mr. Brande has suggested the names of hydriodide of carbon and hydrochloride of carbon for these two bodies. Perhaps, as their general properties range with those of the combustibles, while the specific nature of the compound is decided by the supporter of

combustion which is in combination, the terms of hydrocarburet of chlorine and hydrocarburet of iodine may be considered as appropriate for them.

As yet, I have not succeeded in procuring an iodide of carbon,
but I intend to pursue these experiments in a brighter season
of the year, and expect to obtain this compound.

On a new Compound of Chlorine and Carbon.
By Phillips *and* Faraday*.

[Read July 12, 1821.]

M. Julin, of Abo in Finland, is proprietor of a manufactory,
in which nitric acid is prepared by distilling calcined sulphate
of iron with crude nitre in iron retorts, and collecting the products in receivers connected by glass tubes, in the manner of
Woulfe's apparatus. In this process he observed, that when a
peculiar kind of calcined vitriol, obtained from the waters of
the mine of Fahlun, and containing a small portion of pyrites,
known in Sweden by the name of calcined aquafortis vitriol
No. 3, was used, the first tube was lined with sulphur, and the
second with fine white feathery crystals. These were in very
small quantity, amounting only to a few grains from each distillation; but M. Julin, by degrees, collected a portion of it, and,
having brought it to this country, inserted a short account of
its properties in the 'Annals of Philosophy,' vol. i. p. 216, to
which a few observations were added by ourselves.

The following are the properties of this substance, as described by M. Julin. It is white; consists of small soft adhesive fibres; sinks slowly in water; is insoluble in it whether
hot or cold; is tasteless; has a peculiar smell, somewhat
resembling spermaceti; is not acted on by sulphuric, muriatic,
or nitric acid, except that the latter by boiling on it gives traces
of sulphuric acid; boiled with caustic potash, has a small portion of sulphur dissolved from it; dissolves in hot oil of turpentine, but most of it crystallizes in needles from the solution on
cooling; dissolves in boiling alcohol of ·816, but by far the
greater part crystallizes on cooling; burns in the flame of a
lamp with a greenish-blue flame, giving a slight smell of chlorine

* Phil. Trans. 1821, p. 392, and Phil. Mag. lix. p. 33.

gas; when heated, melting, boiling, and subliming at a temperature between 350° and 400°, and subliming slowly without
melting at a heat of about 250°, forming long needles. Potassium burned with a vivid flame in its vapour in an open tube,
and carbon was deposited; a solution made of the residuum,
and saturated with nitric acid, gave a copious precipitate with
nitrate of silver. M. Julin then remarks, that the small quantity he possessed, with want of leisure, prevented him from
making any further experiments on it, and concludes by comparing it with the chlorides of carbon that have lately been
formed.

The small quantity of the substance which, by the kindness
of M. Julin, we had at our disposal at that time, was insufficient
to enable us satisfactorily to ascertain its nature. We found it
mixed with free sulphur, and sulphate and muriate of ammonia.
When purified, our first object, in consequence of M. Julin's
suggestion, was to compare it with the perchloride of carbon,
but it was found entirely distinct from it in its properties.

Since M. Julin's return from the continent, he has very
kindly placed some further portions of this substance at our
disposal. We have therefore been enabled to continue our
experiments, and have come to the very unexpected conclusion
of its being another chloride of carbon, in addition to the two,
an account of which has been published in the Transactions of
the Royal Society for this year.

The substance, after being boiled in solution of potash,
washed in water, dried and sublimed, formed beautiful acicular
crystals, which appeared to Mr. W. Phillips to be four-sided
prisms. They contained no sulphur, and, when dissolved in
alcohol or æther, gave no traces of chlorine or muriates by
nitrate of silver. They burned in the air with a strong bright
flame at a heat below redness, and agreed with the description
given by M. Julin of the properties of the substance.

When heated moderately, it sublimed unaltered; but on
passing a portion over rock-crystal, heated to bright redness,
in a green glass tube, it was decomposed, charcoal was deposited, and the gas, passed into solution of nitrate of silver, precipitated it, and proved to be chlorine.

A portion was repeatedly sublimed in a small retort filled
with chlorine, which was made red-hot in several places; it

however underwent no change, but on cooling crystallized as at first. It was also exposed in the same gas to sunlight for many days, but no change took place.

When raised in vapour over hot mercury, and detonated with excess of oxygen, a quantity of carbonic acid gas and chloride of mercury were produced. There was no change in the volume of gas used; and lime-water being passed into it absorbed the carbonic gas, became turbid, and left a residuum of pure oxygen. Acetic acid being then added, to dissolve the carbonate of lime, the solution was tested for chlorine, which was readily found in it. When detonated with oxygen, the substance being in excess, there was expansion of volume, carbonic oxide, carbonic acid and chloride of mercury being formed.

When phosphorus, iron, tin, &c. were heated to redness in its vapour over mercury, it was decomposed, chlorides of those substances being formed, and charcoal deposited; and M. Julin has shown that the same effect is produced by potassium.

Three grains of this substance were passed in vapour over pure peroxide of copper, heated to redness in a green glass tube: a very small portion passed undecomposed. The gas received over mercury equalled 5·7 cubic inches; it was carbonic acid gas. A small part of the oxide of copper was reduced, and portions of a crystalline body appeared within the tube, which, on examination, proved to be chloride of copper. Some of this was used in making experiments on its nature; but when that was ascertained, the remaining contents of the tube were dissolved in nitric acid, and precipitated by nitrate of silver: 6·1 grains of chloride of silver were obtained.

Two grains were passed over pure quick-lime, raised to a red heat in a green glass tube. The moment the vapour came in contact with the hot lime, ignition took place, and the earth burned as long as the vapour passed over it. When cold, the tube was examined, and much charcoal found deposited at the spot where the ignition occurred. The contents of the tube were dissolved in nitric acid, and the filtered solution precipitated by nitrate of silver: 5·9 grains of chloride of silver were obtained.

These results afford us sufficient data from which to deduce the nature and composition of this body. All the experiments of decomposition indicate it to contain chlorine and carbon, and those with oxygen and the metals sufficiently prove the absence of hydrogen and oxygen. With regard to the proportions of the elements, three grains of the substance gave 5·7 cubic inches of carbonic acid gas, therefore two grains will give 3·8 cubic inches. One hundred cubic inches of carbonic acid gas weigh 46·47 grains, and contain 12·72 grains of carbon; and 3·8 cubic inches will therefore contain 0·483 grain of carbon. The two grains of the substance decomposed by heated lime gave 5·9 grains of chloride of silver, which, according to Dr. Wollaston's scale, equal 1·45 of chlorine; hence the two grains gave—

$$\begin{array}{ll} \text{Chlorine} & 1\text{·}45 \\ \text{Carbon} & \cdot483 \\ \hline & 1\text{·}933 \end{array}$$

The loss here is 0·067, which is by no means important, when the small quantity of the substance and the nature of the experiments are considered.

As to the proportion of these two bodies to each other, if we consider chlorine as represented by 33·5 and carbon by 5·7, or with Dr. Wollaston by 44·1 and 7·5, then the 1·45 of chlorine would be equivalent to 0·2466 of carbon. This is the constitution of the fluid or protochloride of carbon; and if we double the 0·2466, the product 0·4932 approaches so near to the experimental result 0·483, that we do not hesitate to regard this compound as consisting of one portion of chlorine and two portions of carbon, or

$$\begin{array}{lll} \text{Chlorine} & 44\text{·}1 & 33\text{·}5 \\ \text{Carbon} & 15 & 11\text{·}4 \end{array}$$

It is remarkable that another of these compounds should be found so soon after the discovery of the two former chlorides of carbon. Its physical properties and its chemical energies are in every respect analogous to those of the former compounds; and its constitution increases the probability that another chloride of carbon may be found, consisting of two portions of chlorine and one of carbon.

All the endeavours we have yet made to form the chloride of carbon now described, or to convert it into either of the other chlorides, have been unsuccessful. We expected that when decomposed by heat, it would produce the protochloride with the liberation of carbon, as the perchloride does with the liberation of chlorine, but we have not yet been able to ascertain that point. We have only to offer as an apology for this and other imperfections in the present paper, the smallness of the quantity of this substance that we possessed.

On the Vapour of Mercury at common Temperatures[*].

It has long been admitted, that in the upper part of the barometer and thermometer an atmosphere of mercury exists, even at common temperatures, but having a very small degree of tension. The following experiment renders it easy to show this atmosphere even when the air has not, as in the instruments above mentioned, been removed. A small portion of mercury was put through a funnel into a clean dry bottle, capable of holding about six ounces, and formed a stratum at the bottom not one-eighth of an inch in thickness : particular care was taken that none of the mercury should adhere to the upper part of the inside of the bottle. A small piece of leaf-gold was then attached to the under part of the stopper of the bottle, so that when the stopper was put into its place, the leaf-gold was enclosed in the bottle. It was then set aside in a safe place, which happened to be both dark and cool, and left for between six weeks and two months. At the end of that time it was examined, and the leaf-gold was found whitened by a quantity of mercury, though every part of the bottle and mercury remained apparently just as before.

This experiment has been repeated several times, and always with success. The utmost care was taken that mercury should not get to the gold, except by passing through the atmosphere of the bottle. I think therefore it proves, that at common temperatures, and even when the air is present, mercury is always surrounded by an atmosphere of the same substance.

Experiments on the Alloys of Steel, made with a View to its Improvement. By Stodart and Faraday [†].

In proposing a series of experiments on the alloys of iron and steel with various other metals, the object in view was twofold:

---

[*] Quarterly Journal of Science, x. 354.   [†] Ibid. ix. 319.

—first, to ascertain whether any alloy could be artificially formed, better, for the purpose of making cutting-instruments, than steel in its purest state; and secondly, whether any such alloys would, under similar circumstances, prove less susceptible of oxidation;—new metallic combinations for reflecting mirrors were also a collateral object of research.

Such a series of experiments were not commenced without anticipating considerable difficulties, but the facilities afforded us in the laboratory of the Royal Institution, where they were made, have obviated many of them. The subject was new, and opened into a large and interesting field. Almost an infinity of different metallic combinations may be made, according to the nature and relative proportions of the metals capable of being alloyed. It never has been shown by experiment, whether pure iron, when combined with a minute portion of carbon, constitutes the very best material for making edged tools; or whether any additional ingredient, such as the earths, or their bases, or any other metallic matter, may not be advantageously combined with the steel; and, if so, what the materials are, and what the proportion required to form the best alloy for this much desired and most important purpose. This is confessedly a subject of difficulty, requiring both time and patient investigation, and it will perhaps be admitted as some apology for the very limited progress as yet made.

By referring to the analysis of wootz, or Indian steel*, it will be observed that only a minute portion of the earths alumina and silex could be detected, these earths (or their bases) giving to the wootz its peculiar character. Being satisfied as to the constituent parts of this excellent steel, it was proposed to attempt making such a combination, and with this view various experiments were made. Many of them were fruitless: the successful method was the following. Pure steel in small pieces, and in some instances good iron mixed with charcoal powder, were heated intensely for a long time; in this way they formed carburets, which possessed a very dark metallic grey colour, something in appearance like the black ore of tellurium, and highly crystalline. When broken, the facets of small buttons, not weighing more than 500 grains, were frequently above

* Quarterly Journal of Science, vii. 288.

the eighth of an inch in width. The results of several experiments on its composition, which appeared very uniform, gave 94·36 iron + 5·64 carbon. This being broken and rubbed to powder in a mortar, was mixed with pure alumina, and the whole intensely heated in a close crucible for a considerable time. On being removed from the furnace and opened, an alloy was obtained of a white colour, a close granular texture, and very brittle: this, when analysed, gave 6·4 per cent. alumina, and a portion of carbon not accurately estimated. 700 grains of good steel, with 40 of the alumine alloy, were fused together, and formed a very good button, perfectly malleable; this, on being forged into a little bar and the surface polished, gave, on the application of dilute sulphuric acid, the beautiful damask which will presently be noticed as belonging peculiarly to wootz. A second experiment was made with 500 grains of the same steel and 67 of the alumine alloy, and this also proved good; it forged well, and gave the damask. This specimen has all the appreciable characters of the best Bombay wootz.

We have ascertained, by direct experiment, that the wootz, although repeatedly fused, retains the peculiar property of presenting a damask surface, when forged, polished, and acted upon by dilute acid. This appearance is apparently produced by a dissection of the crystals by the acid; for though by the hammering the crystals have been bent about, yet their forms may be readily traced through the curves which the twisting and hammering have produced. From this uniform appearance on the surface of wootz, it is highly probable that the much-admired sabres of Damascus are made from this steel; and if this be admitted, there can be little reason to doubt that the damask itself is merely an exhibition of crystallization. That on wootz it cannot be the effect of the mechanical mixture of two substances, as iron and steel, unequally acted upon by acid, is shown by the circumstance of its admitting re-fusion without losing this property. It is certainly true that a damasked surface may be produced by welding together wires of iron and steel; but if these welded specimens are fused, the damask does not again appear. Supposing that the damasked surface is dependent on the development of a crystalline structure, then the superiority of wootz in showing the effect, may fairly be considered as dependent on its power of crystallizing, when solidifying, in a more marked manner, and in more decided forms than the common steel. This can only be accounted for by some difference in the composition of the two bodies, and as it has been stated that only the earths in small quantities can be detected, it is reasonable to infer that the bases of these earths being combined with the iron and carbon render the mass more crystallizable, and that the structure drawn out by the hammer, and confused (though not destroyed), does actually occasion the damask. It is highly probable that the wootz is steel accidentally combined with the metal of the earths, and the irregularity observed in different cakes, and even in the same cake, is in accordance with this opinion. The earths may be in the ore, or they may be derived from the crucible in which the fusion is made.

In making the alumine alloy for the imitation of wootz, we had occasion to observe the artificial formation of plumbago. Some of the carburet of iron before mentioned having been pounded and mixed with fresh charcoal, and then fused, was found to have been converted into perfect plumbago. This had not taken place throughout the whole mass; the metal had soon melted and run to the bottom; but having been continued in the furnace for a considerable time, the surface of the button had received an additional portion of charcoal, and had become plumbago. It was soft, sectile, bright, stained paper, and had every other character of that body; it was indeed in no way distinguishable from it. The internal part of these plumbago buttons was a crystalline carburet: a portion of it having been powdered and fused several times with charcoal, at last refused to melt, and on the uncombined charcoal being burnt away by a low heat, it was found that the whole of the steel had been converted into plumbago: this powder we attempted to fuse, but were not successful.

It will appear by the following experiment, that we had formed artificial wootz, at a time when this certainly was not the object of research. In an attempt to reduce titanium and combine it with steel, a portion of menachanite was heated with charcoal, and a fused button obtained. A part of this button was next fused with some good steel; the proportions were 96 steel, 4 menachanite button. An alloy was formed, which worked well under the hammer; and the little bar obtained was evidently different from, and certainly superior to, steel. This was attributed to the presence of titanium, but none could be found in it; nor indeed was any found even in the menachanite button itself. The product was iron and carbon, combined with the earths or their bases, and was in fact excellent wootz. A beautiful damask was produced on this specimen by the action of dilute acid. Since this, many attempts have been made to reduce the oxide of titanium; it has been heated intensely with charcoal, oil, &c., but hitherto all have failed; the oxide has been changed into a black powder, but not fused. When some of the oxide was mixed with steel filings and a little charcoal added, on being intensely heated the steel fused, and ran into a fine globule which was covered by a dark-coloured transparent glass, adhering to the sides of the crucible. The steel contained no titanium; the glass proved to be oxide of titanium, with a little oxide of iron. These experiments have led us to doubt whether titanium has ever been reduced to the metallic state. From the effects of the heat upon the crucibles, which became soft and almost fluid, sometimes in fifteen minutes, we had in fact no reason to suppose the degree of heat inferior to any before obtained by a furnace:—that used in these last experiments was a blast furnace, supplied by a constant and powerful stream of air; the fuel good Staffordshire coke, with a little charcoal; both Hessian and Cornish crucibles were used, one being carefully luted into another,—and even three have been united, but they could not be made to stand the intense heat.

Meteoric iron is, by analysis, always found to contain nickel. The proportions are various, in the specimens that have been chemically examined. The iron from the Arctic regions was found to contain 3 per cent. only of nickel[*], while that from Siberia gave nearly 10 per cent. With the analysis of this last we are favoured by J. G. Children, Esq., and, having permission from that gentleman, we most willingly insert the account of this very accurate process :—

37 grains of Siberian meteoric iron gave 48·27 grains of peroxide of iron and 4·52 grains of oxide of nickel. Supposing

[*] Quarterly Journal of Science, vi. 369.

the equivalent number for nickel to be 28, these quantities are
equal to—

Iron	. . . .	33·69
Nickel	. . . .	3·56
		37·25

Supposing the quantities to be correctly

Iron	. . . .	33·5
Nickel	. . . .	3·5
		37·

the proportions per cent. are,—

Iron	. . . .	90·54
Nickel	. . . .	9·46
		100·00

A second experiment on 47 grains gave 61 grains of peroxide of iron = 42·57 iron. The ammoniacal solution of nickel
was lost by an accident; reckoning from the iron, the quantities per cent. are,—

Iron	. . . .	90·57
Nickel	. . . .	9·42
		99·99

A third experiment on 56 grains gave 73·06 grains peroxide
of iron = 50·99 iron, and 5·4 of oxide of nickel = 4·51 nickel;
or per cent.,—

Iron	. . . .	91·00
Nickel	. . . .	8·01
Loss	. . . .	0·99
		100·00

The mean of the three gives 8·96 per cent. of *nickel*.

The meteoric iron was dissolved in aqua regia, and the iron,
thrown down by pure ammonia, well washed, and heated red.

In the first experiment the ammoniacal solution was evaporated to dryness, the ammonia driven off by heat, and the oxide
of nickel re-dissolved in nitric acid and precipitated by pure
potassa, the mixture being boiled a few seconds.

In the third experiment the nickel was thrown down from
the ammoniacal solution at once by pure potassa. The first
method is best, for a minute portion of oxide of nickel escaped
precipitation in the last experiment, to which the loss is probably to be attributed.

All the precipitates were heated to redness.   J. G. C.

We attempted to make imitations of the meteoric irons with perfect success. To some good iron (horseshoe nails) were added 3 per cent. of pure nickel; these were enclosed in a crucible and exposed to a high temperature in the air-furnace for some hours. The metals were fused, and on examining the button, the nickel was found in combination with the iron. The alloy was taken to the forge, and proved under the hammer to be quite as malleable and pleasant to work as pure iron; the colour, when polished, rather whiter. This specimen, together with a small bar of meteoric iron, have been exposed to a moist atmosphere; they are both a little rusted. In this case it was omitted to expose a piece of pure iron with them; it is probable that, under these circumstances, the pure iron would have been more acted upon.

The same success attended in making the alloy to imitate the Siberian meteoric iron agreeably to Mr. Children's analysis. We fused some of the same good iron with 10 per cent. nickel; the metals were found perfectly combined, but less malleable, being disposed to crack under the hammer. The colour when polished had a yellow tinge. A piece of this alloy has been exposed to moist air for a considerable time, together with a piece of pure iron; they are both a little rusted, not, however, to the same extent; that with the nickel being but slightly acted upon, comparatively to the action on the pure iron: it thus appears that nickel, when combined with iron, has some effect in preventing oxidation, though certainly not to the extent that has at times been attributed to it. It is a curious fact, that the same quantity of the nickel alloyed with steel, instead of preventing its rusting, appeared to accelerate it very rapidly.

Platinum and rhodium have, in the course of these experiments, been alloyed with iron, but these compounds do not appear to possess any very interesting properties. With gold we have not made the experiment. The alloys of other metals with iron, as far as our experience goes, do not promise much usefulness. The results are very different when steel is used; it is only, however, of a few of its compounds that we are prepared to give any account.

Together with some others of the metals, the following have been alloyed with both English and Indian steel, and in various proportions: platinum, rhodium, gold, silver, nickel, copper and tin.

All the above-named metals appear to have an affinity for steel sufficiently strong to make them combine: alloys of platinum, rhodium, gold and nickel may be obtained when the heat is sufficiently high. This is so remarkable with platinum, that it will fuse when in contact with steel, at a heat at which the steel itself is not affected.

With respect to the alloy of silver, there are some very curious circumstances attending it. If steel and silver be kept in fusion together for a length of time, an alloy is obtained, which appears to be very perfect while the metals are in the fluid state, but on solidifying and cooling, globules of pure silver are expressed from the mass, and appear on the surface of the button. If an alloy of this kind be forged into a bar, and then dissected by the action of dilute sulphuric acid, the silver appears, not in combination with the steel, but in threads throughout the mass; so that the whole has the appearance of a bundle of fibres of silver and steel, as if they had been united by welding. The appearance of these silver fibres is very beautiful; they are sometimes $\frac{1}{8}$th of an inch in length, and suggest the idea of giving mechanical toughness to steel, where a very perfect edge may not be required.

At other times, when silver and steel have been very long in a state of perfect fusion, the sides of the crucible, and frequently the top also, are covered with a fine and beautiful dew of minute globules of silver; this effect can be produced at pleasure. At first we were not successful in detecting silver by chemical tests in these buttons; and finding the steel uniformly improved, were disposed to attribute its excellence to an effect of the silver, or to a quantity too small to be tested. By subsequent experiments we were, however, able to detect the silver, even to less than 1 in 500.

In making the silver alloys, the proportion first tried was 1 silver to 160 steel; the resulting buttons were uniformly steel and silver in fibres, the silver being likewise given out in globules during solidifying, and adhering to the surface of the fused button; some of these when forged gave out more globules of silver. In this state of mechanical mixture, the little bars, when exposed to a moist atmosphere, evidently produced voltaic action, and to this we are disposed to attribute the rapid destruction of the metal by oxidation, no such destructive action taking place when the two metals are chemically combined. These results indicated the necessity of diminishing the quantity of silver, and 1 silver to 200 steel was tried. Here, again, were fibres and globules in abundance; with 1 to 300, the fibres diminished, but still were present; they were detected even when the proportion of 1 to 400 was used. The successful experiment remains to be named. When 1 of silver to 500 steel were properly fused, a very perfect button was produced; no silver appeared on its surface; when forged and dissected by an acid, no fibres were seen, although examined by a high magnifying power. The specimen forged remarkably well, although very hard; it had in every respect the most favourable appearance. By a delicate test every part of the bar gave silver. This alloy is decidedly superior to the very best steel, and this excellence is unquestionably owing to combination with a minute portion of silver. It has been repeatedly made, and always with equal success. Various cutting tools have been made from it of the best quality. This alloy is perhaps only inferior to that of steel with rhodium, and it may be procured at a small expense; the value of silver, where the proportion is so small, is not worth naming; it will probably be applied to many important purposes in the arts. An attempt was made to procure the alloy of steel with silver by cementation: a small piece of steel wrapped in silver-leaf, being 1 to 160, was put into a crucible, which being filled up with pounded green glass, was submitted to a heat sufficient to fuse the silver; it was kept at a white heat for three hours. On examining it, the silver was found fused, and adhering to the steel; no part had combined. The steel had suffered by being kept so long at a high temperature. Although this experiment failed in effecting the alloy of steel with silver, there is reason to believe that with some other metals alloys may be obtained by this process; the following circumstance favours this suggestion. Wires of platinum and steel, of about equal diameter, were packed together, and, by an expert workman, were perfectly united by welding. This was effected with the same facility as could have been done with steel and iron. On being forged, the surface polished, and the steel slightly acted on by an acid, a very novel

F

and beautiful surface appeared, the steel and platinum forming dark and white clouds: if this can be effected with very fine wires, a damasked surface will be obtained of exquisite beauty. This experiment, made to ascertain the welding property of platinum, is only named here in consequence of observing that some of the largest of the steel clouds had much the appearance of being alloyed with a portion of the platinum. A more correct survey of the surface, by a high magnifying power, went far to confirm this curious fact: some more direct experiments are proposed to be made on this apparent alloy by cementation.

The alloys of steel with platinum, when both are in a state of fusion, are very perfect, in every proportion that has been tried. Equal parts by weight form a beautiful alloy, which takes a fine polish, and does not tarnish; the colour is the finest imaginable for a mirror. The specific gravity of this beautiful compound is 9·862.

90 of platinum with 20 of steel gave also a perfect alloy, which has no disposition to tarnish; the specific gravity 15·88: both these buttons are malleable, but have not yet been applied to any specific purpose.

10 of platinum to 80 of steel formed an excellent alloy. This was ground and very highly polished, to be tried as a mirror; a fine damask, however, renders it quite unfit for that purpose.

The proportions of platinum that appear to improve steel for edge instruments, are from 1 to 3 per cent. Experience does not yet enable us to state the exact proportion that forms the best possible alloy of these metals; 1·5 per cent. will probably be very nearly right. At the time of combining 10 of platinum with 80 steel, with a view to a mirror, the same proportions were tried with nickel and steel; this too had the damask, and consequently was unfit for its intention. It is curious to observe the difference between these two alloys, as to susceptibility for oxygen. The platinum and steel, after laying many months, had not a spot on its surface, while that with nickel was covered with rust; they were in every respect left under similar circumstances. This is given as an instance, showing that nickel with steel is much more subject to oxidation than when combined with iron.

The alloys of steel with rhodium are likely to prove highly valuable. The scarcity of that metal must, however, operate against its introduction to any great extent. It is to Dr. Wollaston we are indebted, not only for suggesting the trial of rhodium, but also for a liberal supply of the metal, as well as much valuable information relative to fuel, crucibles, &c.; this liberality enables us to continue our experiments on this alloy; these, with whatever else may be worth communicating, will be given in a future Number of this Journal. The proportions we have used are from 1 to 2 per cent. The valuable properties of the rhodium alloys are hardness, with sufficient tenacity to prevent cracking either in forging or in hardening. This superior hardness is so remarkable, that in tempering a few cutting articles made from the alloy, they required to be heated full 30° F. higher than the best wootz, wootz itself requiring to be heated full 40° above the best English cast steel. Thermometrical degrees are named, that being the only accurate method of tempering steel.

Gold forms a good alloy with steel. Experience does not yet enable us to speak of its properties. It certainly does not promise to be of the same value as the alloys of silver, platinum, and rhodium.

Steel with 2 per cent. of copper forms an alloy. Steel also alloys with tin. Of the value of these we have doubts. If, on further trial, they, together with other combinations, requiring more time than we have been able to bestow on them, should prove at all likely to be interesting and useful, the results will be frankly communicated.

Our experiments have hitherto been confined to small quantities of the metals, seldom exceeding 2000 grains in weight, and we are aware that the operations of the laboratory are not always successful when practised on a large scale. There does not, however, appear to be any good reason why equal success may not attend the working on larger masses of the metals, provided the same diligence and means are employed.

From the facility of obtaining silver, it is probable that its alloy with steel is the most valuable of those we have made. To enumerate its applications would be to name almost every edge-tool. It is also probable that it will prove valuable for making dies, especially when combined with the best Indian steel. Trial will soon be made with the silver in the large way, and the result, whatever it may be, will be candidly stated.

Table of Specific Gravities of Alloys, &c. mentioned in the preceding Paper.

Iron, unhammered . . . . 7·847
Wootz, unhammered (Bombay) . . . . . . 7·665
Wootz, tilted (Bombay) . . . . 7·6707
Wootz, in cake (Bengal) . . . . 7·730
Wootz, fused and hammered (Bengal) . . . 7·787
Meteoric iron, hammered . . . . 7·965
Iron, and 3 per cent. nickel . . . . 7·804
Iron, and 10 per cent. nickel . . . . . . 7·849
Steel, and 10 per cent. platinum (mirror) . . 8·100
Steel, and 10 per cent. nickel (mirror) . . . 7·684
Steel, and 1 per cent. gold, hammered . . . 7·870
Steel, and 2 per cent. silver, hammered . . . 7·808
Steel, and 1·5 per cent. platinum, hammered . 7·732
Steel, and 1·5 per cent. rhodium, hammered . 7·795
Steel, and 3 per cent. nickel, hammered . . . 7·750
Platinum 50, and steel 50, unhammered* . . 9·862
Platinum 90, and steel 20, unhammered† . . 15·880
Platinum, hammered and rolled . . . . . . 21·250

On the Alloys of Steel.   By STODART *and* FARADAY‡.
[Read March 21, 1822.]

THE alloys of steel made on a small scale in the laboratory of the Royal Institution proving to be good, and the experiments having excited a very considerable degree of interest both at home and abroad, gave encouragement to attempt the work on a more extended scale; and we have now the pleasure of stating, that alloys similar to those made in the Royal Institution have been made for the purpose of manufacture; and that they prove to be, in point of excellence, in every respect equal, if not superior to the smaller productions of the laboratory.   Previous, however, to extending the work, the former experiments were carefully repeated, and to the results were added some new combinations, namely, steel with palladium,

* The calculated mean specific gravity of this alloy is 11·2723, assuming the specific gravity of platinum and steel as expressed in this Table.
† The calculated mean specific gravity of this alloy is 16·0766.
‡ Philosophical Transactions, 1822, p. 253; also Phil. Mag. vol. lx. p. 363.

steel with iridium and osmium, and latterly, steel with chromium. In this last series of experiments we were particularly fortunate, having by practice acquired considerable address in the management of the furnaces, and succeeded in procuring the best fuel for the purpose. Notwithstanding the many advantages met with in the laboratory of the Royal Institution, the experiments were frequently rendered tedious from causes often unexpected, and sometimes difficult to overcome; among these, the failure of crucibles was perhaps the most perplexing. We have never yet found a crucible capable of bearing the high degree of temperature required to produce the perfect reduction of titanium; indeed we are rather disposed to question whether this metal has ever been so reduced : our furnaces are equal* (if any are) to produce this effect, but hitherto we have failed in procuring a crucible.

The metals that form the most valuable alloys with steel are silver, platinum, rhodium, iridium and osmium, and palladium; all of these have now been made in the large way, except indeed the last-named. Palladium has, for very obvious reasons, been used but sparingly; four pounds of steel with $\frac{1}{100}$th part of palladium has, however, been fused at once, and the compound is truly valuable, more especially for making instruments that require perfect smoothness of edge.

We are happy to acknowledge the obligations due from us to Dr. Wollaston, whose assistance we experienced in every stage of our progress, and by whom we were furnished with all the scarce and valuable metals; and that with a liberality which enabled us to transfer our operations from the laboratory of the chemist to the furnace of the maker of cast steel.

In making the alloys on a large scale, we were under the necessity of removing our operations from London to a steel furnace at Sheffield; and being prevented by other avocations from giving personal attendance, the superintendence of the work was consequently entrusted to an intelligent and confidential agent. To him the steel, together with the alloying metals in the exact proportion, and in the most favourable state for the purpose, was forwarded, with instructions to see the whole of the metals, and nothing else, packed into the cru-

* We have succeeded in fusing in these furnaces rhodium, and also, though imperfectly, platinum in crucibles.

cible, and placed in the furnace, to attend to it while there, and to suffer it to remain for some considerable time in a state of thin fusion, previous to its being poured out into the mould. The cast ingot was next, under the same superintendence, taken to the tilting mill, where it was forged into bars of a convenient size, at a temperature not higher than just to render the metal sufficiently malleable under the tilt hammer. When returned to us, it was subjected to examination both mechanical and chemical, as well as compared with the similar products of the laboratory. From the external appearance, as well as from the texture of the part when broken by the blow of the hammer, we were able to form a tolerably correct judgement as to its general merits; the hardness, toughness, and other properties were further proved by severe trials, after being fashioned into some instrument or tool, and properly hardened and tempered.

It would prove tedious to enter into a detail of experiments made in the Royal Institution; a brief notice of them will at present be sufficient. After making imitations of various specimens of meteoric iron, by fusing together pure iron and nickel, in proportions of 3 to 10 per cent., we attempted making an alloy of steel with silver, but failed, owing to a superabundance of the latter metal; it was found, after very many trials, that only the $\frac{1}{500}$th part of silver would combine with steel, and when more was used a part of the silver was found in the form of metallic dew, lining the top and sides of the crucible: the fused button itself was a mere mechanical mixture of the two metals, globules of silver being pressed out of the mass by contraction in cooling, and more of these globules being forced out by the hammer in forging; and further, when the forged piece was examined, by dissecting it with diluted sulphuric acid, threads or fibres of silver were seen mixed with the steel, having something of the appearance of steel and platinum when united by welding: but when the proportion of silver was only $\frac{1}{500}$th part, neither dew, globules, nor fibres appeared, the metals being in a state of perfect chemical combination, and the silver could only be detected by a delicate chemical test.

With platinum and rhodium, steel combines in every proportion; and this appears also to be the case with iridium and osmium: from 1 to 80 per cent. of platinum was perfectly combined with steel, in buttons of from 500 to 2000 grains. With rhodium, from 1 to 50 per cent. was successfully used. Equal parts by weight of steel and rhodium gave a button, which, when polished, exhibited a surface of the most exquisite beauty: the colour of this specimen is the finest imaginable for a metallic mirror, nor does it tarnish by long exposure to the atmosphere: the specific gravity of this beautiful compound is 9·176. The same proportion of steel and platinum gave a good button, but a surface highly crystalline renders it altogether unfit for a mirror. In the laboratory we ascertained that, with the exception of silver, the best proportion of the alloying metal, when the object in view was the improvement of edge-tools, was about $\frac{1}{100}$th part, and in this proportion they have been used in the large way. It may be right to notice, that in fusing the metals in the laboratory no flux whatever was used, nor did the use of any ever appear to be required.

Silver being comparatively of little value with some of the alloying metals, we were disposed to make trial with it as the first experiment in the large way. 8 lbs. of very good Indian steel was sent to our agent, and with it $\frac{1}{500}$th part of pure silver: a part of this was lost, owing to a defect in the mould; a sufficient quantity was, however, saved to satisfy us as to the success of the experiment. This, when returned, had the most favourable appearance both as to surface and fracture; it was harder than the best cast steel, or even than the Indian wootz, with no disposition whatever to crack, either under the hammer, or in hardening. Some articles, for various uses, have been made from this alloy; they prove to be of a very superior quality; its application will probably be ex-tended not only to the manufacture of cutlery, but also to various descriptions of tools; the trifling addition of price cannot operate against its very general introduction. The silver alloy may be advantageously used for almost every pur-pose for which good steel is required.

Our next experiment made in the large way was with steel and platinum. 10 lbs. of the same steel, with $\frac{1}{100}$th part of pla-tinum, the latter in the state produced by heating the ammonia muriate in a crucible to redness, was forwarded to our agent, with instructions to treat this in the same way as the last-named metals. The whole of this was returned in bars remarkable for smoothness of surface and beauty of fracture. Our own observation, as well as that of the workmen employed to make from it various articles of cutlery, was, that this alloy, though not so hard as the former, had considerably more toughness: this property will render it valuable for every purpose where tenacity, as well as hardness, is required; neither will the expense of platinum exclude it from a pretty general application in the arts; its excellence will much more than repay the extra cost.

The alloys of steel with rhodium have also been made in the large way, and are perhaps the most valuable of all; but these, however desirable, can never, owing to the scarcity of the metal, be brought into very general use. The compound of steel, iridium, and osmium, made in the large way, is also of great value; but the same cause, namely the scarcity and difficulty of procuring the metals, will operate against its very general introduction. A sufficient quantity of these metals may, perhaps, be obtained to combine with steel for the purpose of making some delicate instruments, and also as an article of luxury, when manufactured into razors. In the meantime, we have been enabled, repeatedly, to make all these alloys (that with palladium excepted) in masses of from 8 to 20 lbs. each; with such liberality were we furnished with the metals from the source already named.

A point of great importance in experiments of this kind was, to ascertain whether the products obtained were exactly such as we wished to produce. For this purpose, a part of each product was analysed, and in some cases the quantity ascertained; but it was not considered necessary in every case to verify the quantity by analysis, because, in all the experiments made in the laboratory, the button produced after fusion was weighed, and if it fell short of the weight of both metals put into the crucible, it was rejected as imperfect, and put aside. When the button gave the weight, and on analysis gave proofs of containing the metal put in to form the alloy, and also, on being forged into a bar and acted on by acids, presented a uniform surface, we considered the evidence of its composition as sufficiently satisfactory. The processes of analysis, though simple, we shall briefly state: the information may be desirable to others who may be engaged on similar experiments; and further, may enable every one to detect any attempt at imposition. It would be very desirable at present to possess a test as simple, by which we could distinguish the wootz, or steel of India, from that of Europe; but this, unfortunately, requires a much more difficult process of analysis.

To ascertain if platinum is in combination with steel, a small portion of the metal, or some filings taken from the bar, is to be put into dilute sulphuric acid; there will be rapid action; the iron will be dissolved, and a black sediment left, which will contain carbon, hydrogen, iron, and platinum; the carbon and hydrogen are to be burnt off, the small portion of iron separated by muriatic acid, and the residuum dissolved in a drop or two of nitromuriatic acid. If a piece of glass be moistened with this solution, and then heated by a spirit-lamp and the blowpipe, the platinum is reduced, and forms a metallic coating on the glass.

In analysing the alloy of steel and silver, it is to be acted on by dilute sulphuric acid, and the powder boiled in the acid; the silver will remain in such a minute state of division that it will require some time to deposit. The powder is then to be boiled in a small portion of strong muriatic acid *; this will dissolve the iron and silver, and the latter will fall down as a chloride of silver on dilution with water; or the powder may be dissolved in pure nitric acid, and tested by muriatic acid and ammonia.

The alloy of steel and palladium, acted on by dilute sulphuric acid, and boiled in that acid, left a powder, which, when the charcoal was burnt from it, and the iron partly separated by cold muriatic acid, gave on solution in hot muriatic acid, or in nitromuriatic acid, a muriate of palladium; the solution, when precipitated by prussiate of mercury, gave prussiate of palladium; and a glass plate moistened with it and heated to redness, became coated with metallic palladium.

The residuum of the rhodium alloy obtained by boiling in diluted sulphuric acid, had the combustible matter burnt off, and the powder digested in hot muriatic acid : this removed the iron; and by long digestion in nitromuriatic acid, a muriate

---

* Although it is a generally received opinion that muriatic acid does not act on silver, yet that is not the case; pure muriatic acid dissolves a small portion of silver very readily.

of rhodium was formed, distinguishable by its colour, and by the triple salt it formed with muriate of soda.

To analyse the compound of steel with iridium and osmium, the alloy should be acted on by dilute sulphuric acid, and the residuum boiled in the acid; the powder left is to be collected and heated with caustic soda in a silver crucible to dull redness for a quarter of an hour, the whole to be mixed with water, and having had excess of sulphuric acid added, it is to be distilled, and that which passes over condensed in a flask: it will be a solution of oxide of osmium, will have the peculiar smell belonging to that substance, and will give a blue precipitate with tincture of galls. The portion in the retort being then poured out, the insoluble part is to be washed in repeated portions of water, and then being first slightly acted on by muriatic acid to remove the iron, is to be treated with nitro-muriatic acid, which will give a muriate of iridium.

In these analyses, an experienced eye will frequently perceive, on the first action of the acid, the presence of the alloying metal. When this is platinum, gold, or silver, a film of the metal is quickly formed on the surface of the acid.

Of alloys of platinum, palladium, rhodium, and iridium and osmium, a ready test is offered when the point is not to ascertain what the metal is, but merely whether it be present or not. For this purpose we have only to compare the action of the same acid on the alloy and on a piece of steel; the increased action on the alloy immediately indicates the presence of the metal; and by the difference of action, which on experience is found to be produced with the different metals, a judgement may be formed even of the particular one present.

The order in which the different alloys stand with regard to this action is as follows: steel, chromium alloy, silver alloy, gold alloy, nickel alloy, rhodium alloy, iridium and osmium alloy, palladium alloy, platinum alloy. With similar acid the action on the pure steel was scarcely perceptible; the silver alloy gave very little gas, nor was the gold much acted on. All the others gave gas copiously, but the platinum alloy in most abundance.

In connexion with the analysis of these alloys, there are some very interesting facts to be observed during the action of acids on them, and perhaps none of these are more striking than those last referred to. When the alloys are immersed in diluted acid, the peculiar properties which some of them exhibit, not only mark and distinguish them from common steel, and from each other, but also give rise to some considerations on the state of particles of matter of different kinds when in intimate mixture or in combination, which may lead to clearer and more perfect ideas on this subject.

If two pieces, one of steel, and one of steel alloyed with platinum, be immersed in weak sulphuric acid, the alloy will be immediately acted on with great rapidity and the evolution of much gas, and will shortly be dissolved, whilst the steel will be scarcely at all affected. In this case it is hardly possible to compare the strength of the two actions. If the gas be collected from the alloy and from the steel for equal intervals of time, the first portions will surpass the second some hundreds of times.

A very small quantity of platinum alloyed with steel confers this property on it: $\frac{1}{400}$ increased the action considerably; with $\frac{1}{200}$ and $\frac{1}{100}$ it was powerful; with 10 per cent. of platinum it acted, but not with much power; with 50 per cent. the action was not more than with steel alone; and an alloy of 90 platinum with 20 steel was not affected by the acid.

The action of other acids on these alloys is similar to that of sulphuric acid, and is such as would be anticipated: dilute muriatic acid, phosphoric acid, and even oxalic acid, acted on the platinum alloy with the liberation of more gas than from zinc; and tartaric acid and acetic acid rapidly dissolved it. In this way chalybeate solutions, containing small portions of protoxide of iron, may be readily obtained.

The cause of the increased action of acids on this and similar alloys, is, as the President of this Society suggested to us, probably electrical. It may be considered as occasioned by the alloying metal existing in such a state in the mass, that its particles form voltaic combinations with the particles of steel, either directly or by producing a definite alloy, which is diffused through the rest of the steel; in which case the whole mass would be a series of such voltaic combinations: or it may be occasioned by the liberation, on the first action of the acid, of particles, which, if not pure platinum, contain, as has been shown, a very large proportion of that metal, and which, being in close contact with the rest of the mass, form voltaic combinations with it in a very active state: or, in the third place, it may result from the iron being mechanically divided by the platinum, so that its particles are more readily attacked by the acid, analogous to the case of protosulphuret of iron.

Although we have not been able to prove by such experiments, as may be considered strictly decisive, to which of these causes the action is owing, or how much is due to any of them, yet we do not hesitate to consider the second as almost entirely, if not quite, the one that is active. The reasons which induce us to suppose this to be the true cause of the action, rather than any peculiar and previous arrangement of the particles of steel and platinum, or than the state of division of the steel, are, that the two metals combine in every proportion we have tried, and do not in any case exhibit evidences of a separation between them, like those, for instance, which steel and silver exhibit; that when, instead of an acid, weaker agents are used, the alloy does not seem to act with them as if it were a series of infinitely minute voltaic combinations of steel and platinum, but exactly as steel alone would do; that the mass does not render platinum wire more negative than steel, as it probably in the third case would do; that it does not rust more rapidly in a damp atmosphere; and that when placed in saline solutions, as muriate of soda, &c., there is no action takes place between them. In such cases it acts just like steel; and no agent that we have as yet tried, has produced voltaic action that was not first able to set a portion of the platinum free by dissolving out the iron.

Other interesting phenomena exhibited by the action of acid on these steels, are the differences produced when they are hard and when soft. Mr. Daniel, in his interesting paper on the mechanical structure of iron, published in the Journal of Science, has remarked, that pieces of hard and soft steel being placed in muriatic acid, the first required fivefold the time of the latter to saturate the acid; and that when its surface was examined, it was covered with small cavities like worm-eaten wood, and was compact and not at all striated, and that the latter presented a fibrous and wavy texture.

The properties of the platinum alloy have enabled us to

observe other differences between hard and soft steel equally striking. When two portions of the platinum alloy, one hard and one soft, are put into the same diluted sulphuric acid and suffered to remain for a few hours, then taken out and examined, the hard piece presents a covering of a metallic black carbonaceous powder, and the surface is generally slightly fibrous; but the soft piece, on examination, is found to be covered with a thick coat of grey metallic plumbaginous matter, soft to the touch, and which may be cut with a knife, and its quantity seven or eight times that of the powder on the hard piece: it does not appear as if it contained any free charcoal, but considerably resembles the plumbaginous powder Mr. Daniel describes as obtained by the action of acid on cast iron.

The same difference is observed if pure steel be used, but it is not so striking; because, being much less rapidly attacked by the acid, it has to remain longer in it, and the powder produced is still further acted on.

The powder procured from the soft steel or alloy in these experiments, when it has not remained long in the acid, exactly resembles finely divided plumbago, and appears to be a carburet of iron, and probably of the alloying metal also. It is not acted on by water, but in the air the iron oxidates and discolours the substance. When it remains long in the acid, or is boiled in it, it is reduced to the same state as the powder from the hard steel or alloy.

When any of these residua are boiled in diluted sulphuric or muriatic acid, protoxide of iron is dissolved, and a black powder remains unalterable by the further action of the acid; it is apparently in greater quantity from the alloys than from pure steel, and when washed, dried, and heated to 300° or 400° in the air, burns like pyrophorus, with much fume: or if lighted, burns like bitumen, and with a bright flame; the residuum is protoxide of iron, and the alloying metal. Hence, during the action of the acid on the steel, a portion of hydrogen enters into combination with part of the metal and the charcoal, and forms an inflammable compound not acted upon by the acid.

Some striking effects are produced by the action of nitric acid on these powders. If that from pure steel be taken, it is entirely dissolved; and such is also the case if the powder be taken from an alloy the metal of which is soluble in nitric acid;

but if the powder is from an alloy the metal of which is not soluble in nitric acid, then a black residuum is left not touched by the acid; and which, when washed and carefully dried, is found, when heated, to be deflagrating; and with some of the metals, when carefully prepared, strongly explosive.

The fulminating preparation obtained from the platinum alloy, when dissolved in nitromuriatic acid, gave a solution containing much platinum and very little iron. When a little of it was wrapped in foil and heated, it exploded with much force, tearing open the foil, and evolving a faint light. When dropped on the surface of heated mercury, it exploded readily at 400° of Fahrenheit, but with difficulty at 370°. When its temperature was raised slowly, it did not explode, but was decomposed quietly. When detonated in the bottom of a hot glass tube, much water and fume were given off, and the residuum collected was metallic platinum with a very little iron and charcoal. We are uncertain how far this preparation resembles the fulminating platinum of Mr. Edmund Davy.

In these alloys of steel the differences of specific gravity are not great, and may probably be in part referred to the denser state of the metals from more or less hammering: at the same time it may be observed, that they are nearly in the order of the specific gravities of the respective alloying metals.

The alloys of steel with gold, tin, copper, and chromium, we have not attempted in the large way. In the laboratory, steel and gold were combined in various proportions;—none of the results were so promising as the alloys already named, nor did either tin or copper, as far as we could judge, at all improve steel. With titanium we failed, owing to the imperfection of crucibles. In one instance, in which the fused button gave a fine damask surface, we were disposed to attribute the appearance to the presence of titanium; but in this we were mistaken;—the fact was, we had unintentionally made wootz. The button, by analysis, gave a little silex and alumina, but not an atom of titanium; menachanite, in a particular state of preparation, was used: this might possibly contain the earths or their basis, or they may have formed a part of the crucible.

M. Berthier, who first made the alloy of steel and chromium*,

* Annales de Chimie, xvii. 55.

speaks very favourably of it.   We have made only two experiments.   1600 grains of steel, with 16 of pure chrome, were packed into one of the best crucibles, and placed in an excellent blast furnace : the metals were fused, and kept in that state for some time.   The fused button proved good and forged well : although hard, it showed no disposition to crack.   The surface being brightened, and slightly acted on by dilute sulphuric acid, exhibited a crystalline appearance; the crystals, being elongated by forging, and the surface again polished, gave, by dilute acid, a very beautiful damask.   Again, 1600 grains of steel with 48 of pure chrome were fused : this gave a button considerably harder than the former.   This too was as malleable as pure iron, and also gave a very fine damask.   Here a rather curious phenomenon was observed : the damask was removed by polishing, and restored by heat without the use of any acid.   The damasked surface, now coloured by oxidation, had a very novel appearance : the beauty was heightened by heating the metal in a way to exhibit all the colours caused by oxidation, from pale straw to blue, or from about 430° to 600° of Fahrenheit.   The blade of a sabre, or some such instrument made from this alloy, and treated in this way, would assuredly be beautiful, whatever its other properties might be; for of the value of the chrome alloy for edge-tools we are not prepared to speak, not having made trial of its cutting powers. The sabre blade, thus coloured, would amount to a proof of its being well tempered; the blue back would indicate the temper of a watch-spring, while the straw colour towards the edge would announce the requisite degree of hardness.   It is confessed, that the operation of tempering any blade of considerable length in this way would be attended with some difficulty.

In the account now given of the different alloys, only one triple compound is noticed, namely, steel, iridium and osmium; but this part of the subject certainly merits further investigation, offering a wide and interesting field of research.   Some attempts to form other combinations of this description proved encouraging, but we were prevented, at the time, by various other avocations, from bestowing on them that attention and labour they seemed so well to deserve *.

* It is our intention to continue these experiments at every opportunity; but they are laborious, and require much time and patience.

It is a curious fact, that when pure iron is substituted for steel, the alloys so formed are much less subject to oxidation. 3 per cent. of iridium and osmium, fused with some pure iron, gave a button, which, when forged and polished, was exposed, with many other pieces of iron, steel, and alloys, to a moist atmosphere: it was the last of all showing any rust. The colour of this compound was distinctly blue; it had the property of becoming harder when heated to redness, and quenched in a cold fluid. On observing this steel-like character, we suspected the presence of carbon: none, however, was found, although carefully looked for. It is not improbable that there may be other bodies, besides charcoal, capable of giving to iron the properties of steel; and though we cannot agree with M. Boussingault*, when he would replace carbon in steel by silica or its base, we think his experiments very interesting on this point, which is worthy further examination.

We are not informed as to what extent these alloys, or any of them, have been made at home, or to what uses they have been applied; their more general introduction in the manufacture of cutlery would assuredly add to the value, and consequently to the extension of that branch of trade. There are various other important uses to which the alloys of steel may advantageously be applied. If our information be correct, the alloy of silver, as well as that of platinum, has been to some considerable extent in use at His Majesty's Mint. We do know, that several of the alloys have been diligently and successfully made on the Continent, very good specimens of some of them having been handed to us; and we are proud of these testimonies of the utility of our endeavours.

To succeed in making and extending the application of these new compounds, a considerable degree of faithful and diligent attention will be required on the part of the operators. The purity of the metals intended to form the compound is essential; the perfect and complete fusion of both must in every case be ascertained: it is further requisite that the metals be kept for some considerable time in the state of thin fusion: after casting, the forging is to be attended to with equal care; the metal must on no account be overheated; and this is more particularly to be attended to when the alloying metal is fusible

* Annales de Chimie, xvi. 10.

at a low temperature, as silver. The same care is to be observed in hardening: the article is to be brought to a cherry-red colour, and then instantly quenched in the cold fluid.

In tempering, which is best performed in a metallic bath properly constructed, the bath will require to be heated for the respective alloys, from about 70° to 100° of Fahrenheit above the point of temperature required for the best cast steel. We would further recommend, that this act of tempering be performed twice; that is, at the usual time before grinding, and again just before the last polish is given to the blade. This second tempering may perhaps appear superfluous, but upon trial its utility will be readily admitted. We were led to adopt the practice by analogy, when considering the process of making and tempering watch-springs.

On Hydriodide of Carbon *.

In the 'Philosophical Transactions' for 1821, I have described a compound of chlorine and olefiant gas †, but had not at that time the means of ascertaining its composition. Since then I have obtained it in greater quantity, and analysed it. Four grains were passed in vapour over heated copper, in a green glass tube; iodide of copper was formed, and pure olefiant gas evolved, which amounted to 1·37 cubic inch. As 100 cubic inches of olefiant gas weigh about 30·15 grs., so 1·37 cubic inch will weigh 0·413 gr. Now 4 grains minus 0·413 leaves 3·587 iodine, and $3·587 : 0·413 :: 117·75 : 13·55$ nearly. Now 13·55 is so nearly the number of two proportions of olefiant gas, that the substance may be considered as composed of

$$\text{1 proportion of Iodine} \quad . \quad . \quad . \quad . \quad 117·75$$
$$\text{2 proportions of Olefiant gas} \quad . \quad . \quad 13·4$$

and is therefore analogous in its constitution to the compound of chlorine and olefiant gas, sometimes called chloric ether.

On Hydrate of Chlorine ‡.

It was generally considered before the year 1810, that chlorine gas was condensible by cold into a solid state; and we

---

* . Quarterly Journal of Science, xiii. 429.   † See page 51.

‡ Quarterly Journal of Science, xv. 71.

were first instructed by Sir Humphry Davy, in his admirable researches into the nature of that substance, published in the 'Philosophical Transactions' for 1810–11, that the solid body, obtained by cooling chlorine gas, was a compound with water; and that the dry gas could not be condensed at a temperature equal even to −40° Fahr., whilst on the contrary, moist gas, or a solution of chlorine in water, crystallized at the temperature of 40° Fahr.

M. Thénard, in his ' Traité de Chimie,' has described the deposition of the hydrate of chlorine by cold from an aqueous solution of the gas. It forms crystals of a bright yellow colour, which liquefy when their temperature is slightly raised, and in so doing give off abundance of gas.

This substance may be obtained well crystallized, by introducing into a clean bottle of the gas, a little water, but not sufficient to convert the whole into hydrate, and then placing the bottle in a situation the temperature of which is about or below freezing, for a few days: and I have constantly found the crystals better formed in the dark than in the light. The hydrate is produced in a crust or in dendritical crystals; but being left to itself, will in a few days sublime from one part of the bottle to another in the manner of camphor, and form brilliant and comparatively large crystals. These are of a bright yellow colour, and sometimes, though rarely, are delicate prismatic needles extending from half an inch to two inches into the atmosphere of the bottle: generally they are of shorter forms, and when most perfect and simple, have appeared to me to be acute flattened octahedra, the three axes of the octahedron having different dimensions.

Though a solution of chlorine deposits the hydrate when cooled, yet a portion remains in solution, and the crystals also dissolve slowly in water. It is therefore soluble, though not so much so as chlorine gas. When a solution of chlorine is cooled gradually till the whole is frozen, there is a perfect separation of the hydrate of chlorine from the rest of the water, or rather from the ice; for crystals of ice, formed in a solution of chlorine, when washed in pure water, and then dissolved, do not trouble nitrate of silver.

I neglected to ascertain the specific gravity of the crystals whilst the weather was cold and they were readily obtainable;

but I have endeavoured since to do so by means of cooling mixtures. The hydrate in thin plates was put into solutions of muriate of lime of different densities, but of the temperature of 32° Fahr. It seemed to remain in any part of a solution of specific gravity 1·2, but there was constantly a slight liberation of gas; and as minute and imperceptible bubbles may have adhered to the hydrate, the result can only be considered as a loose approximation. The solid crystals would probably be heavier than 1·2.

The hydrate of chlorine acts upon substances, as might be expected, from the action of chlorine upon the same substances, and it may perhaps now and then offer a convenient form for its application in experiment. When put into alcohol, an elevation of temperature amounting to 8° or 10° took place. There was rapid action, much ether, and muriatic acid formed, and a small portion of a triple compound of chlorine, carbon and hydrogen.

When put into solutions of ammoniacal salts it liberated nitrogen gas, formed muriatic acid, and also chloride of nitrogen, which remained undissolved at the bottom of the solution.

In aqueous solution of ammonia similar effects were produced, but less chloride of nitrogen was formed.

In order to arrive at a knowledge of the composition of this substance, I adopted the following process:—The crystals were collected together by a small quantity of solution of chlorine, then filtered and pressed between successive portions of bibulous paper at a temperature of 32° (care being taken to expose them as little as possible to the air), until as dry as they could be rendered by this means. A glass flask with a narrow neck, and containing a portion of water at 32°, having been previously counterpoised, a portion of the crystals were immediately after the last pressing introduced into it; they sank to the bottom of the water, and the flask being again weighed, the quantity of crystals introduced was ascertained. A weak solution of pure ammonia was then poured into the water in the flask, care being taken to add considerable excess over that required by the chlorine beneath. The whole was left for twenty-four hours, in which time the chlorine had had sufficient opportunity to act on the ammonia, and any portion of chloride of nitrogen that might at first have been formed would be resolved into its elements, and its chlorine be converted into muriatic acid. It was then slightly heated, neutralized by pure nitric acid, precipitated by nitrate of silver, and the chloride of silver obtained and weighed.

The following is an experiment conducted in this way:—
65 grains of the pressed crystals were put into the flask, and the ammonia added; at one time there was a faint smell of chloride of nitrogen for an instant at the mouth of the flask, and a little more ammonia was added. The next day 73·2 grs. of chloride of silver were obtained from the solution, and if this be considered as equivalent to 18 grs. of chlorine, then the 65 grs. of hydrate must have contained 47 grs. of water, or per cent.—

> Chlorine . . . . 27·7
> Water . . . . 72·3

This nearly accords with 10 proportionals of water to 1 of chlorine, and I have chosen it because it gave the largest proportion of chlorine of any experiment I made. It is evident that any loss or error either in the drying the crystals, or in the conversion of the chlorine into muriatic acid by the ammonia, would tend to diminish the proportion of that element; and it is even possible that the above proportion of chlorine is under-rated, but I believe it to be near the truth. The mean of several other experiments gave—

> Chlorine . . . . 26·3
> Water . . . . 73·6

Note.—Since writing the above, Mr. Faraday has succeeded in condensing chlorine into a liquid; for this purpose a portion of the solid and dried hydrate of chlorine is put into a small bent tube and hermetically sealed; it is then heated to about 100°, and a yellow vapour is formed which condenses into a deep yellow liquid heavier than water (sp. gr. probably about 1·3). Upon relieving the pressure by breaking the tube, the condensed chlorine instantly assumes its usual state of gas or vapour.

When perfectly dry chlorine is condensed into a tube by means of a syringe, a portion of it assumes the liquid form under a pressure equal to that of 4 or 5 atmospheres.

By putting some muriate of ammonia and sulphuric acid into the opposite ends of a bent glass tube, sealing it hermetically, and then suffering the acid to run upon the salt, muriatic acid is generated under such pressure as causes it to assume the liquid form; it is of an orange colour, lighter than sulphuric acid, and instantly assumes the gaseous state when the pressure is removed. Sir H. Davy has given an account of this experiment to the Royal Society. It is probable that by a similar mode of treatment several other gases may be liquefied.

On Fluid Chlorine*.

[Read March 13, 1823.]

It is well known that before the year 1810 the solid substance obtained by exposing chlorine, as usually procured, to a low temperature, was considered as the gas itself reduced into that form; and that Sir Humphry Davy first showed it to be a hydrate, the pure dry gas not being condensible even at a temperature of −40° F.

I took advantage of the late cold weather to procure crystals of this substance for the purpose of analysis. The results are contained in a short paper in the Quarterly Journal of Science, vol. xv.† Its composition is very nearly 27·7 chlorine, 72·3 water, or 1 proportional of chlorine and 10 of water.

The President of the Royal Society having honoured me by looking at these conclusions, suggested, that an exposure of the substance to heat under pressure would probably lead to interesting results; the following experiments were commenced at his request. Some hydrate of chlorine was prepared, and, being dried as well as could be by pressure in bibulous paper, was introduced into a sealed glass tube, the upper end of which was then hermetically closed. Being placed in water at 60°, it underwent no change; but when put into water at 100°, the substance fused, the tube became filled with a bright yellow atmosphere, and on examination was found to contain two fluid substances: the one, about three-fourths of the whole, was of a faint yellow colour, having very much the appearance of water; the remaining fourth was a heavy bright yellow fluid, lying at the bottom of the former, without any apparent tendency to mix with it. As the tube cooled, the yellow atmosphere condensed into more of the yellow fluid, which floated in a film on the pale fluid, looking very like chloride of nitrogen; and at 70° the pale portion congealed, although even at 32° the yellow portion did not solidify. Heated up to 100° the yellow fluid appeared to boil, and again produced the bright coloured atmosphere.

By putting the hydrate into a bent tube, afterwards hermetically sealed, I found it easy, after decomposing it by a heat of 100°, to distil the yellow fluid to one end of the tube, and to

* Phil. Trans. 1823, p. 160; also Phil. Mag. lxii. p. 413.   † See page 81.

separate it from the remaining portion. In this way a more complete decomposition of the hydrate was effected; and, when the whole was allowed to cool, neither of the fluids solidified at temperatures above 34°, and the yellow portion not even at 0°. When the two were mixed together, they gradually combined at temperatures below 60°, and formed the same solid substances as that first introduced. If, when the fluids were separated, the tube was cut in the middle, the parts flew asunder as if with an explosion, the whole of the yellow portion disappeared, and there was a powerful atmosphere of chlorine produced; the pale portion on the contrary remained, and when examined, proved to be a weak solution of chlorine in water, with a little muriatic acid, probably from the impurity of the hydrate used. When that end of the tube in which the yellow fluid lay was broken under a jar of water, there was an immediate production of chlorine gas.

I at first thought that muriatic acid and euchlorine had been formed; then, that two new hydrates of chlorine had been produced; but at last I suspected that the chlorine had been entirely separated from the water by the heat, and condensed into a dry fluid by the mere pressure of its own abundant vapour. If that were true, it followed that chlorine gas, when compressed, should be condensed into the same fluid; and as the atmosphere in the tube in which the fluid lay was not very yellow at 50° or 60°, it seemed probable that the pressure required was not beyond what could readily be obtained by a condensing syringe. A long tube was therefore furnished with a cap and stopcock, then exhausted of air and filled with chlorine, and being held vertically with the syringe upwards, air was forced in, which thrust the chlorine to the bottom of the tube, and gave a pressure of about 4 atmospheres. Being now cooled, there was an immediate deposit in films, which appeared to be hydrate, formed by water contained in the gas and vessels, but some of the yellow fluid was also produced. As this, however, might also contain a portion of the water present, a perfectly dry tube and apparatus were taken, and the chlorine left for some time over a bath of sulphuric acid before it was introduced. Upon throwing in air and giving pressure, there was now no solid film formed, but the clear yellow fluid was deposited, and more abundantly still upon cooling. After remaining some time it disappeared, having gradually mixed with the atmosphere above it; but every repetition of the experiment produced the same results.

Presuming that I had now a right to consider the yellow fluid as pure chlorine in the liquid state, I proceeded to examine its properties, as well as I could when obtained by heat from the hydrate. However obtained, it always appears very limpid and fluid, and excessively volatile at common pressure. A portion was cooled in its tube to $0°$; it remained fluid. The tube was then opened, when a part immediately flew off, leaving the rest so cooled, by the evaporation, as to remain a fluid under the atmospheric pressure. The temperature could not have been higher than $-40°$ in this case; as Sir Humphry Davy has shown that dry chlorine does not condense at that temperature under common pressure. Another tube was opened at a temperature of $50°$; a part of the chlorine volatilized, and cooled the tube so much as to condense the atmospheric vapour on it into ice.

A tube having the water at one end and the chlorine at the other was weighed, and then cut in two; the chlorine immediately flew off, and the loss being ascertained was found to be $1·6$ grain: the water left was examined and found to contain some chlorine: its weight was ascertained to be $5·4$ grains. These proportions, however, must not be considered as indicative of the true composition of hydrate of chlorine; for, from the mildness of the weather during the time when these experiments were made, it was impossible to collect the crystals of hydrate, press and transfer them, without losing much chlorine; and it is also impossible to separate the chlorine and water in the tube perfectly, or keep them separate, as the atmosphere within will combine with the water, and gradually re-form the hydrate.

Before cutting the tube, another tube had been prepared exactly like it in form and size, and a portion of water introduced into it, as near as the eye could judge, of the same bulk as the fluid chlorine; this water was found to weigh $1·2$ grain; a result, which, if it may be trusted, would give the specific gravity of fluid chlorine as $1·33$; and from its appearance in and on water, this cannot be far wrong.

Note on the Condensation of Muriatic Acid Gas into the liquid Form. By Sir H. DAVY, *Bart., Pres. R.S.*

In desiring Mr. Faraday to expose the hydrate of chlorine to heat in a closed glass tube, it occurred to me that one of three things would happen: that it would become fluid as a hydrate; or that a decomposition of water would occur, and euchlorine and muriatic acid be formed; or that the chlorine would separate in a condensed state. This last result having been obtained, it evidently led to other researches of the same kind. I shall hope, on a future occasion, to detail some general views on the subject of these researches. I shall now merely mention, that by sealing muriate of ammonia and sulphuric acid in a strong glass tube, and causing them to act upon each other, I have procured liquid muriatic acid: and by substituting carbonate for muriate of ammonia, I have no doubt that carbonic acid may be obtained, though in the only trial I have made the tube burst. I have requested Mr. Faraday to pursue these experiments, and to extend them to all the gases which are of considerable density, or to any extent soluble in water; and I hope soon to be able to lay an account of his results, with some applications of them that I propose to make, before the Society.

I cannot conclude this note without observing, that the generation of elastic substances in close vessels, either with or without heat, offers much more powerful means of approximating their molecules than those dependent upon the application of cold, whether natural or artificial; for as gases diminish only about $\frac{1}{480}$ in volume for every — degree of Fahrenheit's scale, beginning at ordinary temperatures, a very slight condensation only can be produced by the most powerful freezing mixtures—not half as much as would result from the application of a strong flame to one part of a glass tube, the other part being of ordinary temperature: and when attempts are made to condense gases into fluids by sudden mechanical compression, the heat, instantly generated, presents a formidable obstacle to the success of the experiment; whereas in the compression resulting from their slow generation in close vessels, if the process be conducted with common precautions, there is no source of difficulty or danger; and it may be easily assisted by artificial cold in cases when gases approach near to that point of compression and temperature at which they become vapours.

_____

On the Condensation of several Gases into Liquids*.

[Read April 10, 1823.]

I HAD the honour, a few weeks since, of submitting to the Royal Society a paper on the reduction of chlorine to the liquid state. An important note was added to the paper by the President, on the general application of the means used in this case to the reduction of other gaseous bodies to the liquid state; and in illustration of the process, the production of liquid muriatic acid was described. Sir Humphry Davy did me the honour to request I would continue the experiments, which I have done under his general direction, and the following are some of the results already obtained :—

Sulphurous Acid.—Mercury and concentrated sulphuric acid were sealed up in a bent tube, and, being brought to one end, heat was carefully applied, whilst the other end was preserved cool by wet bibulous paper. Sulphurous acid gas was produced where the heat acted, and was condensed by the sulphuric acid above; but when the latter had become saturated, the sulphurous acid passed to the cold end of the tube, and was condensed into a liquid. When the whole tube was cold, if the sulphurous acid were returned on to the mixture of sulphuric acid and sulphate of mercury, a portion was reabsorbed, but the rest remained on it without mixing.

Liquid sulphurous acid is very limpid and colourless, and highly fluid. Its refractive power, obtained by comparing it in water and other media with water contained in a similar tube, appeared to be nearly equal to that of water. It does not solidify or become adhesive at a temperature of 0° F. When a tube containing it was opened, the contents did not rush out as with explosion, but a portion of the liquid evaporated rapidly, cooling another portion so much as to leave it in the fluid state at common barometric pressure. It was however rapidly dissipated, not producing visible fumes, but producing the odour of pure sulphurous acid, and leaving the tube quite dry. A

* Philosophical Transactions, 1823, p. 189; also Phil. Mag. lxii. p. 416.

portion of the vapour of the fluid received over a mercurial bath, and examined, proved to be sulphurous acid gas. A piece of ice dropped into the fluid instantly made it boil, from the heat communicated by it.

To prove in an unexceptional manner that the fluid was pure sulphurous acid, some sulphurous acid gas was carefully prepared over mercury, and a long tube perfectly dry, and closed at one end, being exhausted, was filled with it; more sulphurous acid was then thrown in by a condensing syringe, till there were three or four atmospheres; the tube remained perfectly clear and dry; but on cooling one end to 0°, the fluid sulphurous acid condensed, and in all its characters was like that prepared by the former process.

A small gauge was attached to a tube in which sulphurous acid was afterwards formed, and at a temperature of 45° F. the pressure within the tube was equal to three atmospheres, there being a portion of liquid sulphurous acid present: but as the common air had not been excluded when the tube was sealed, nearly one atmosphere must be due to its presence; so that sulphurous acid vapour exerts a pressure of about two atmospheres at 45° F. Its specific gravity was nearly 1·42*.

Sulphuretted Hydrogen.—A tube being bent, and sealed at the shorter end, strong muriatic acid was poured in through a small funnel, so as nearly to fill the short leg without soiling the long one. A piece of platinum foil was then crumpled up

---

* I am indebted to Mr. Davies Gilbert, who examined with much attention the results of these experiments, for the suggestion of the means adopted to obtain the specific gravity of some of these fluids. A number of small glass bulbs were blown and hermetically sealed; they were then thrown into alcohol, water, sulphuric acid, or mixtures of these, and when any one was found of the same specific gravity as the fluid in which it was immersed, the specific gravity of the fluid was taken: thus a number of hydrometrical bulbs were obtained; these were introduced into the tubes in which the substances were to be liberated; and ultimately, the dry liquids obtained, in contact with them. It was then observed whether they floated or not, and a second set of experiments were made with bulbs lighter or heavier as required, until a near approximation was obtained. Many of the tubes burst in the experiments, and in others difficulties occurred from the accidental fouling of the bulb by the contents of the tube. One source of error may be mentioned in addition to those which are obvious, namely, the alteration of the bulk of the bulb by its submission to the pressure required to keep the substance in the fluid state.

and pushed in, and upon that were put fragments of sulphuret of iron, until the tube was nearly full. In this way action was prevented until the tube was sealed. If it once commences, it is almost impossible to close the tube in a manner sufficiently strong, because of the pressing out of the gas. When closed, the muriatic acid was made to run on to the sulphuret of iron, and then left for a day or two. At the end of that time, much protomuriate of iron had formed, and on placing the clean end of the tube in a mixture of ice and salt, warming the other end if necessary by a little water, sulphuretted hydrogen in the liquid state distilled over.

The liquid sulphuretted hydrogen was colourless, limpid, and excessively fluid. Ether, when compared with it in similar tubes, appeared tenacious and oily. It did not mix with the rest of the fluid in the tube, which was no doubt saturated, but remained standing on it. When a tube containing it was opened, the liquid immediately rushed into vapour; and this being done under water, and the vapour collected and examined, it proved to be sulphuretted hydrogen gas. As the temperature of a tube containing some of it rose from $0°$ to $45°$, part of the fluid rose in vapour, and its bulk diminished; but there was no other change: it did not seem more adhesive at $0°$ than at $45°$. Its refractive power appeared to be rather greater than that of water; it decidedly surpassed that of sulphurous acid. A small gauge being introduced into a tube in which liquid sulphuretted hydrogen was afterwards produced, it was found that the pressure of its vapour was nearly equal to 17 atmospheres at the temperature of $50°$.

The gauges used were made by drawing out some tubes at the blowpipe table until they were capillary, and of a trumpet form; they were graduated by bringing a small portion of mercury successively into their different parts; they were then sealed at the fine end, and a portion of mercury placed in the broad end; and in this state they were placed in the tubes, so that none of the substances used or produced could get to the mercury, or pass by it to the inside of the gauge. In estimating the number of atmospheres, one has always been subtracted for the air left in the tube.

The specific gravity of sulphuretted hydrogen appeared to be 0·9.

Carbonic Acid.—The materials used in the production of carbonic acid, were carbonate of ammonia and concentrated sulphuric acid; the manipulation was like that described for sulphuretted hydrogen. Much stronger tubes are however required for carbonic acid than for any of the former substances, and there is none which has produced so many or more powerful explosions. Tubes which have held fluid carbonic acid well for two or three weeks together, have, upon some increase in the warmth of the weather, spontaneously exploded with great violence; and the precautions of glass masks, goggles, &c., which are at all times necessary in pursuing these experiments, are particularly so with carbonic acid.

Carbonic acid is a limpid colourless body, extremely fluid, and floating upon the other contents of the tube. It distils readily and rapidly at the difference of temperature between $32°$ and $0°$. Its refractive power is much less than that of water. No diminution of temperature to which I have been able to submit it, has altered its appearance. In endeavouring to open the tubes at one end, they have uniformly burst into fragments, with powerful explosions. By enclosing a gauge in a tube in which fluid carbonic acid was afterwards produced, it was found that its vapour exerted a pressure of 36 atmospheres at a temperature of $32°$.

It may be questioned, perhaps, whether this and other similar fluids obtained from materials containing water, do not contain a portion of that fluid; inasmuch as its absence has not been proved, as it may be with chlorine, sulphurous acid, cyanogen, and ammonia. But besides the analogy which exists between the latter and the former, it may also be observed in favour of their dryness, that any diminution of temperature causes the deposition of a fluid from the atmosphere, precisely like that previously obtained; and there is no reason for supposing that these various atmospheres, remaining as they do in contact with concentrated sulphuric acid, are not as dry as atmospheres of the same kind would be over sulphuric acid at common pressure.

Euchlorine.—Fluid euchlorine was obtained by enclosing chlorate of potash and sulphuric acid in a tube, and leaving them to act on each other for twenty-four hours. In that time there had been much action, the mixture was of a dark reddish brown, and the atmosphere of a bright yellow colour. The mixture was then heated up to 100°, and the unoccupied end of the tube cooled to 0°; by degrees the mixture lost its dark colour, and a very fluid ethereal-looking substance condensed. It was not miscible with a small portion of the sulphuric acid which lay beneath it; but when returned on to the mass of salt and acid, it was gradually absorbed, rendering the mixture of a much deeper colour even than itself.

Euchlorine thus obtained is a very fluid transparent substance, of a deep yellow colour. A tube containing a portion of it in the clean end, was opened at the opposite extremity; there was a rush of euchlorine vapour, but the salt plugged up the aperture: whilst clearing this away, the whole tube burst with a violent explosion, except the small end in a cloth in my hand, where the euchlorine previously lay, but the fluid had all disappeared.

Nitrous Oxide.—Some nitrate of ammonia, previously made as dry as could be by partial decomposition by heat in the air, was sealed up in a bent tube, and then heated in one end, the other being preserved cool. By repeating the distillation once or twice in this way, it was found, on after-examination, that very little of the salt remained undecomposed. The process requires care. I have had many explosions with very strong tubes, and at considerable risk.

When the tube is cooled, it is found to contain two fluids, and a very compressed atmosphere. The heavier fluid on examination proved to be water, with a little acid and nitrous oxide in solution; the other was nitrous oxide. It appears in a very liquid, limpid, colourless state; and so volatile, that the warmth of the hand generally makes it disappear in vapour. The application of ice and salt condenses abundance of it into the liquid state again. It boils readily by the difference of temperature between 50° and 0°. It does not appear to have any tendency to solidify at −10°. Its refractive power is very much less than that of water, and less than any fluid that has yet been obtained in these experiments, or than any known fluid. A tube being opened in the air, the nitrous oxide immediately burst into vapour. Another tube was opened under water; the vapour being collected and examined proved to be nitrous oxide gas. A gauge being introduced into a tube, in which liquid nitrous oxide was afterwards produced, gave the pressure of its vapour as equal to above 50 atmospheres at 45°.

Cyanogen.—Some pure cyanuret of mercury was heated until perfectly dry. A portion was then enclosed in a green glass tube, in the same manner as in former instances, and being collected to one end, was decomposed by heat, whilst the other end was cooled. The cyanogen soon appeared as a liquid: it was limpid, colourless, and very fluid; not altering its state at the temperature of 0°. Its refractive power is rather less, perhaps, than that of water. A tube containing it being opened in the air, the expansion within did not appear to be very great; and the liquid passed with comparative slowness into the state of vapour, producing great cold. The vapour being collected over mercury, proved to be pure cyanogen.

A tube was sealed up with cyanuret of mercury at one end, and a drop of water at the other; the fluid cyanogen was then produced in contact with the water. It did not mix, at least in any considerable quantity, with that fluid, but floated on it, being lighter, though apparently not so much so as ether would be. In the course of some days, action had taken place, the water had become black, and changes, probably such as are known to take place in an aqueous solution of cyanogen, occurred. The pressure of the vapour of cyanogen appeared by the gauge to be 3·6 or 3·7 atmospheres at 45° F. Its specific gravity was nearly 0·9.

Ammonia.—In searching after liquid ammonia, it became necessary, though difficult, to find some dry source of that substance; and I at last resorted to a compound of it which I had occasion to notice some years since with chloride of silver*. When dry chloride of silver is put into ammoniacal gas, as dry as it can be made, it absorbs a large quantity of it; 100 grains condensing above 130 cubical inches of the gas; but the compound thus formed is decomposed by a temperature of 100° F. or upwards. A portion of this compound was sealed up in a bent tube and heated in one limb, whilst the other was cooled by ice or water. The compound thus heated under pressure fused at a comparatively low temperature, and

---

* Quarterly Journal of Science, v. 74; see also page 18.

boiled, giving off ammoniacal gas, which condensed at the opposite end into a liquid.

Liquid ammonia thus obtained was colourless, transparent, and very fluid. Its refractive power surpassed that of any other of the fluids described, and that also of water itself. From the way in which it was obtained, it was evidently as free from water as ammonia in any state could be. When the chloride of silver is allowed to cool, the ammonia immediately returns to it, combining with it, and producing the original compound. During this action a curious combination of effects takes place: as the chloride absorbs the ammonia, heat is produced, the temperature rising up nearly to 100°; whilst a few inches off, at the opposite end of the tube, considerable cold is produced by the evaporation of the fluid. When the whole is retained at the temperature of 60°, the ammonia boils till it is dissipated and re-combined. The pressure of the vapour of ammonia is equal to about 6·5 atmospheres at 50°. Its specific gravity was 0·76.

Muriatic Acid.—When made from pure muriate of ammonia and sulphuric acid, liquid muriatic acid is obtained colourless, as Sir Humphry Davy had anticipated. Its refractive power is greater than that of nitrous oxide, but less than that of water; it is nearly equal to that of carbonic acid. The pressure of its vapour at the temperature of 50° is equal to about 40 atmospheres.

Chlorine.—The refractive power of fluid chlorine is rather less than that of water. The pressure of its vapour at 60° is nearly equal to 4 atmospheres.

Attempts have been made to obtain hydrogen, oxygen, fluoboracic, fluosilicic, and phosphuretted hydrogen gases in the liquid state; but though all of them have been subjected to great pressure, they have as yet resisted condensation. The difficulty with regard to fluoboric gas consists probably in its affinity for sulphuric acid, which, as Dr. Davy has shown, is so great as to raise the sulphuric acid with it in vapour. The experiments will, however, be continued on these and other gases, in the hope that some of them, at least, will ultimately condense.

On the Liquefaction and Solidification of Bodies generally existing as Gases*.

[Received Dec. 19, 1844,—Read Jan. 9, 1845.]

THE experiments formerly made on the liquefaction of gases †, and the results which from time to time have been added to this branch of knowledge, especially by M. Thilorier ‡, have left a constant desire on my mind to renew the investigation. This, with considerations arising out of the apparent simplicity and unity of the molecular constitution of all bodies when in the gaseous or vaporous state, which may be expected, according to the indications given by the experiments of M. Cagniard de la Tour, to pass by some simple law into their liquid state, and also the hope of seeing nitrogen, oxygen, and hydrogen, either as liquid or solid bodies, and the latter probably as a metal, have lately induced me to make many experiments on the subject; and though my success has not been equal to my desire, still I hope some of the results obtained, and the means of obtaining them, may have an interest for the Royal Society; more especially as the application of the latter may be carried much further than I as yet have had opportunity of applying them. My object, like that of some others, was to subject the gases to considerable pressure with considerable depression of temperature. To obtain the pressure, I used mechanical force, applied by two air-pumps fixed to a table. The first pump had a piston of an inch in diameter, and the second a piston of only half an inch in diameter; and these were so associated by a connecting pipe, that the first pump forced the gas into and through the valves of the second, and then the second could be employed to throw forward this gas, already condensed to ten, fifteen, or twenty atmospheres, into its final recipient at a much higher pressure.

The gases to be experimented with were either prepared and retained in gas-holders or gas-jars, or else, when the pumps were dispensed with, were evolved in strong glass vessels, and sent under pressure into the condensing tubes. When the gases were over water, or likely to contain water,

---

* Phil. Trans. 1845, p. 155.   † Ibid. 1823, pp. 160, 189.
‡ Annales de Chimie, 1835, lx. 427, 432.

they passed, in their way from the air-holder to the pump, through a coil of thin glass tube retained in a vessel filled with a good mixture of ice and salt, and therefore at the temperature of 0° Fahr.; the water that was condensed here was all deposited in the first two inches of the coil.

Fig. 1.

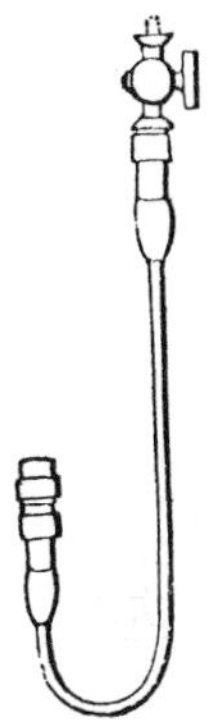

The condensing tubes were of green bottle glass, being from $\frac{1}{6}$th to $\frac{1}{4}$th of an inch external diameter, and from $\frac{1}{42}$nd to $\frac{1}{50}$th of an inch in thickness. They were chiefly of two kinds, about 11 and 9 inches in length; the one, when horizontal, having a curve downward near one end to dip into a cold bath, and the other, being in form like an inverted siphon, could have the bend cooled also in the same manner when necessary. Into the straight part of the horizontal tube, and the longest leg of the siphon tube, pressure-gauges were introduced when required.

Caps, stopcocks and connecting pieces were employed to attach the glass tubes to the pumps, and these, being of brass, were of the usual character of those employed for operations with gas, except that they were small and carefully made. The caps were of such size that the ends

Fig. 2.

of the glass tubes entered freely into them, and had rings or a female screw worm cut in the interior, against which the cement was to adhere. The ends of the glass tubes were roughened by a file, and when a cap was to be fastened on, both it and the end of the tube were made so warm, that the cement *, when applied, was thoroughly melted in contact with these parts, before the tube and cap were brought together and finally adjusted to each other. These junctions bore a pressure of thirty, forty, and fifty atmospheres, with only one failure in above one hundred instances; and that produced no complete separation of parts, but simply a small leak.

The caps, stopcocks, and connectors, screwed one into the

---

* Five parts of resin, one part of yellow bees'-wax, and one part of red ochre, by weight, melted together.

H

other, having one common screw thread, so as to be combined in any necessary manner. There were also screw plugs, some solid, with a male screw to close the openings or ends, of caps, &c., others with a female screw to cover and close the ends of stopcocks. All these screw joints were made tight by leaden washers; and by having these of different thickness, equal to from $\frac{4}{8}$ths to $\frac{12}{8}$ths of the distance between one turn of the screw thread and the next, it was easy at once to select the washer which should allow a sufficient compression in screwing up to make all air-tight, and also bring every part of the apparatus into its right position.

I have often put a pressure of fifty atmospheres into these tubes, and have had no accident or failure (except the one mentioned). With the assistance of Mr. Addams I have tried their strength by a hydrostatic press, and obtained the following results:—A tube having an external diameter of 0·24 of an inch and a thickness of 0·0175 of an inch, burst with a pressure of sixty-seven atmospheres, reckoning one atmosphere as 15 lbs. on the square inch. A tube which had been used, of the shape of fig. 1, its external diameter being 0·225 of an inch, and its thickness about 0·03 of an inch, sustained a pressure of 118 atmospheres without breaking, or any failure of the caps or cement, and was then removed for further use.

A tube such as I have employed for generating gases under pressure, having an external diameter of 0·6 of an inch, and a thickness of 0·035 of an inch, burst at twenty-five atmospheres.

Having these data, it was easy to select tubes abundantly sufficient in strength to sustain any force which was likely to be exerted within them in any given experiment.

The gauge used to estimate the degree of pressure to which the gas within the condensing tube was subjected was of the same kind as those formerly described *, being a small tube of glass closed at one end with a cylinder of mercury moving in it. So the expression of ten or twenty atmospheres, means a force which is able to compress a given portion of air into $\frac{1}{10}$th or $\frac{1}{20}$th of its bulk at the pressure of one atmosphere of 30 inches of mercury. These gauges had their graduation marked on them with a black varnish, and also with Indian ink:—there are several of the gases which, when condensed, cause the var-

* Philosophical Transactions, 1823, p. 192: see also p. 91.

nish to liquefy, but then the Indian ink stood. For further precaution, an exact copy of the gauge was taken on paper, to be
applied on the outside of the condensing tube. In most cases,
when the experiment was over, the pressure was removed from
the interior of the apparatus, to ascertain whether the mercury
in the gauge would return back to its first or starting-place.

For the application of cold to these tubes, a bath of Thilorier's mixture of solid carbonic acid and ether was used. An
earthenware dish of the capacity of 4 cubic inches or more
was fitted into a similar dish somewhat larger, with three or
four folds of dry flannel intervening, and then the bath mixture
was made in the inner dish. Such a bath will easily continue
for twenty or thirty minutes, retaining solid carbonic acid the
whole time; and the glass tubes used would sustain sudden
immersion in it without breaking.

But as my hopes of any success beyond that heretofore obtained depended more upon depression of temperature than on
the pressure which I could employ in these tubes, I endeavoured to obtain a still greater degree of cold. There are, in
fact, some results producible by cold which no pressure may
be able to effect. Thus, solidification has not as yet been conferred on a fluid by any degree of pressure. Again, that beautiful condition which Cagniard de la Tour has made known,
and which comes on with liquids at a certain heat, may have its
point of temperature for some of the bodies to be experimented
with, as oxygen, hydrogen, nitrogen, &c., below that belonging
to the bath of carbonic acid and ether; and, in that case, no
pressure which any apparatus could bear would be able to
bring them into the liquid or solid state.

To procure this lower degree of cold, the bath of carbonic
acid and ether was put into an air-pump, and the air and gaseous
carbonic acid rapidly removed. In this way the temperature fell so low, that the vapour of carbonic acid given off by
the bath, instead of having a pressure of one atmosphere, had
only a pressure of $\frac{1}{24}$th of an atmosphere, or 1·2 inch of mercury; for the air-pump barometer could be kept at 28·2 inches
when the ordinary barometer was at 29·4. At this low temperature the carbonic acid mixed with the ether was not more
volatile than water at the temperature of 86°, or alcohol at
ordinary temperatures.

In order to obtain some idea of this temperature, I had an alcohol thermometer made, of which the graduation was carried below 32° Fahr., by degrees equal in capacity to those between 32° and 212°. When this thermometer was put into the bath of carbonic acid and ether surrounded by the air, but covered over with paper, it gave the temperature of 106° below 0°. When it was introduced into the bath under the air-pump, it sank to the temperature of 166° below 0°; or 60° below the temperature of the same bath at the pressure of one atmosphere, *i. e.* in the air. In this state the ether was very fluid, and the bath could be kept in good order for a quarter of an hour at a time.

As the exhaustion proceeded I observed the temperature of the bath and the corresponding pressure at certain other points, of which the following may be recorded:—

The external barometer was 29·4 inches:

					inch.				Fahr.
when the mercury in the air-pump barometer was.. }					1	the bath temperature was.. }			−106°,
,,	,,	,,	,,	,,	10	,,	,,	,,	−112½,
,,	,,	,,	,,	,,	20	,,	,,	,,	−121,
,,	,,	,,	,,	,,	22	,,	,,	,,	−125,
,,	,,	,,	,,	,,	24	,,	,,	,,	−131,
,,	,,	,,	,,	,,	26	,,	,,	,,	−139,
,,	,,	,,	,,	,,	27	,,	,,	,,	−146,
,,	,,	,,	,,	,,	28	,,	,,	,,	−160,
,,	,,	,,	,,	,,	28·2	,,	,,	,,	−166;

but as the thermometer takes some time to acquire the temperature of the bath, and the latter was continually falling in degree; as also the alcohol thickens considerably at the lower temperature, there is no doubt that the degrees expressed are not so low as they ought to be, perhaps even by 5° or 6° in most cases.

With *dry* carbonic acid under the air-pump receiver I could raise the pump barometer to 29 inches when the external barometer was at 30 inches.

The arrangement by which this cooling power was combined in its effect on gases with the pressure of the pumps, was very simple in principle. An air-pump receiver open at the top was employed; the brass plate which closed the aperture had a small brass tube about 6 inches long, passing through it air-tight by means of a stuffing-box, so as to move easily up and down in a vertical direction. One of the glass condensing siphon tubes, already described, fig. 1, was screwed on to the lower end of the sliding tube, and the upper end of the latter was connected with a communicating tube in two lengths, reaching from it to the condensing pumps; this tube was small, of brass, and 9½ feet in length; it passed 6 inches horizontally from the condensing pumps, then rose vertically for 2 feet, afterwards proceeded horizontally for 7 feet, and finally turned down and was immediately connected with the sliding tube. By this means the latter could be raised and lowered vertically, without any strain upon the connexions, and the condensing tube lowered into the cold bath *in vacuo*, or raised to have its contents examined at pleasure. The capacity of the connecting tubes beyond the last condensing pump was only 2 cubic inches.

When experimenting with any particular gas, the apparatus was put together fast and tight, except the solid terminal screw-plug at the short end of the condensing tube, which, being the very extremity of the apparatus, was left a little loose. Then, by the condensing pumps, abundance of gas was passed through the apparatus to sweep out every portion of air, after which the terminal plug was screwed up, the cold bath arranged, and the combined effects of cold and pressure brought to unite upon the gas.

There are many gases which condense at less than the pressure of one atmosphere when submitted to the cold of a carbonic acid bath in air (which latter can upon occasions be brought considerably below −106° Fahr.). These it was easy, therefore, to reduce, by sending them through small conducting tubes into tubular receivers placed in the cold bath. When the receivers had previously been softened in a spirit-lamp flame, and narrow necks formed on them, it was not difficult, by a little further management, hermetically to seal up these substances in their condensed state. In this manner chlorine, cyanogen, ammonia, sulphuretted hydrogen, arseniuretted hydrogen, hydriodic acid, hydrobromic acid, and even carbonic acid, were obtained, sealed up in tubes in the liquid state; and euchlorine was also secured in a tube receiver with a cap and screw-plug. By using a carbonic acid bath, first cooled *in vacuo*, there is no doubt other condensed gases could be secured in the same way.

The fluid carbonic acid was supplied to me by Mr. Addams, in his perfect apparatus, in portions of about 220 cubic inches each. The solid carbonic acid, when produced from it, was preserved in a glass; itself retained in the middle of three concentric glass jars, separated from each other by dry jackets of woollen cloth. So effectual was this arrangement, that I have frequently worked for a whole day of twelve and fourteen hours, having solid carbonic acid in the reservoir, and enough for all the baths I required during the whole time, produced by one supply of 220 cubic inches*.

By the apparatus, and in the manner now described, all the gases before condensed were very easily reduced, and some new results were obtained. When a gas was liquefied, it was easy to close the stopcock, and then remove the condensing tube with the fluid from the rest of the apparatus. But in order to preserve the liquid from escaping as gas, a further precaution was necessary; namely, to cover over the exposed end of the stopcock by a blank female screw-cap and leaden washer, and also to tighten perfectly the screw of the stopcock plug. With these precautions I have kept carbonic acid, nitrous oxide, fluosilicon, &c. for several days.

Even with gases which could be condensed by the carbonic acid bath in air, this apparatus in the air-pump had, in one respect, the advantage; for when the condensing tube was lifted out of the bath into the air, it immediately became covered with hoar-frost, obscuring the view of that which was within; but *in vacuo* this was not the case, and the contents of the tube could be very well examined by the eye.

Olefiant Gas.—This gas condensed into a clear, colourless, transparent fluid, but did not become solid even in the carbonic acid bath *in vacuo*; whether this was because the temperature was not low enough, or for other reasons referred to in the account of euchlorine, is uncertain.

* On one occasion the solid carbonic acid was exceedingly electric, but I could not produce the effect again: it was probably connected with the presence of oil which was in the carbonic acid box; neither it nor the filaments of ice which formed on it in the air conducted, for when touched it preserved its electric state. Believing as yet that the account I have given of the cause of the electric state of an issuing jet of steam and water (Phil. Trans. 1843, p. 17) is the true one, I conclude that this also was a case of the production of electricity simply by friction, and unconnected with vaporization.

The pressure of the vapour of this substance at the temperature of the carbonic acid bath in air (−103° Fahr.) appeared singularly uncertain, being on different occasions, and with different specimens, 3·7, 8·7, 5 and 6 atmospheres. The Table below shows the tension of vapour for certain degrees below 0° Fahr., with two different specimens obtained at different times, and it will illustrate this point.

Fahr.	Atmospheres.	Atmospheres.
−100°	4·60	9·30
− 90	5·68	10·26
− 80	6·92	11·33
− 70	8·32	12·52
− 60	9·88	13·86
− 50	11·72	15·36
− 40	13·94	17·05
− 30	16·56	18·98
− 20	19·58	21·23
− 10		23·89
0		27·18
10		31·70
20		36·80
30		42·50

I have not yet resolved this irregularity, but believe there are two or more substances, physically, and perhaps occasionally chemically different, in olefiant gas; and varying in proportion with the circumstances of heat, proportions of ingredients, &c. attending the preparation.

The fluid affected the resin of the gauge graduation, and probably also the resin of the cap cement, though slowly.

Hydriodic Acid.—This substance was prepared from the iodide of phosphorus by heating it with a very little water. It is easily condensable by the temperature of a carbonic acid bath: it was redistilled, and thus obtained perfectly pure.

The acid may be obtained either in the solid or liquid, or (of course) in the gaseous state. As a solid it is perfectly clear, transparent and colourless; having fissures or cracks in it resembling those that run through ice. Its solidifying temperature is nearly −60° Fahr., and then its vapour has not the pressure of one atmosphere; at a point a little higher it becomes a clear liquid, and this point is close upon that which corresponds to a vaporous pressure of one atmosphere. The acid dissolves the cap cement and the bitumen of the gauge graduation; and appears also to dissolve and act on fat, for it leaked by the plug of the stopcock with remarkable facility. It acts on the brass of the apparatus, and also on the mercury in the gauge. Hence the following results as to pressures and temperatures are not to be considered more than approximations :—

At 0° Fahr. pressure was 2·90 atmospheres.
At 32° Fahr. pressure was 3·97 atmospheres.
At 60° Fahr. pressure was 5·86 atmospheres.

Hydrobromic Acid.—This acid was prepared by adding to perbromide of phosphorus * about one-third of its bulk of water in a proper distillatory apparatus formed of glass tube, and then applying heat to distil off the gaseous acid. This being sent into a very cold receiver, was condensed into a liquid, which being rectified by a second distillation, was then experimented with.

Hydrobromic acid condenses into a clear colourless liquid at 100° below 0°, or lower, and has not the pressure of one atmosphere at the temperature of the carbonic acid bath in air. It soon obstructs and renders the motion of the mercury in the air-gauge irregular, so that I did not obtain a measure of its elastic force; but it is less than that of muriatic acid. At and below the temperature of −124° Fahr. it is a solid, transparent, crystalline body. It does not freeze until reduced much lower than this temperature; but being frozen by the carbonic acid bath *in vacuo*, it remains a solid until the temperature in rising attains to −124°.

Fluosilicon.—I found that this substance in the gaseous state might be brought in contact with the oil and metal of the pumps, without causing injury to them, for a time sufficiently

---

* The bromides of phosphorus are easily made without risk of explosion. If a glass tube be bent so as to have two depressions, phosphorus placed in one and bromine in the other; then by inclining the tube, the vapour of bromine can be made to flow gradually on to, and combine with, the phosphorus. The fluid protobromide is first formed, and this is afterwards converted into solid perbromide. The excess of bromine may be dissipated by the careful application of heat.

long to apply the joint process of condensation already described. The substance liquefied under a pressure of about nine atmospheres at the lowest temperature, or at 160° below 0°; and was then clear, transparent, colourless, and very fluid, like hot ether. It did not solidify at any temperature to which I could submit it. I was able to preserve it in the tube until the next day. Some leakage had then taken place (for it ultimately acted on the lubricating fat of the stopcock), and there was no liquid in the tube at common temperatures; but when the bend of the tube was cooled to 32° by a little ice, fluid appeared: a bath of ice and salt caused a still more abundant condensation. The pressure appeared then to be above thirty atmospheres, but the motion of the mercury in the gauge had become obstructed through the action of the fluosilicon, and no confidence could be reposed in its indications.

Phosphuretted Hydrogen.—This gas was prepared by boiling phosphorus in a strong pure solution of caustic potassa, and the gas was preserved over water in a dark room for several days to cause the deposition of any mere vapour of phosphorus which it might contain. It was then subjected to high pressure in a tube cooled by a carbonic acid bath, which had itself been cooled under the receiver of the air-pump. The gas in its way to the pumps passed through a long spiral of thin narrow glass tube immersed in a mixture of ice and salt at 0°, to remove as much water from it as possible.

By these means the phosphuretted hydrogen was liquefied; for a pure, clear, colourless, transparent and very limpid fluid appeared, which could not be solidified by any temperature applied, and which, when the pressure was taken off, immediately rose again in the form of gas. Still the whole of the gas was not condensable into this fluid. By working the pumps the pressure would rise up to twenty-five atmospheres at this very low temperature, and yet at the pressure of two or three atmospheres and the same temperature, liquid would remain. There can be no doubt that phosphuretted hydrogen condensed; but neither can there be a doubt that some other gas, not so condensable, was also present, which perhaps may be either another phosphuretted hydrogen or hydrogen itself.

Fluoboron.—This substance was prepared from fluor spar, fused boracic acid and strong sulphuric acid, in a tube-generator such as that already described, and conducted into a condensing tube under the generating pressure. The ordinary carbonic acid bath did not condense it, but the application of one cooled under the air-pump caused its liquefaction, and fluoboron then appeared as a very limpid, colourless, clear fluid, showing no signs of solidification, but when at the lowest temperature mobile as hot ether. When the pressure was taken off, or the temperature raised, it returned into the state of gas.

The following are some results of pressure—all that I could obtain with the liquid in my possession; for, as the liquid is light and the gas heavy, the former rapidly disappears in producing the latter. They make no pretensions to accuracy, and are given only for general information.

Fahr.	Atmospheres.	Fahr.	Atmospheres.	Fahr.	Atmospheres.
$-100°$ . . 4·61		$-72°$ . . 9·23		$-62°$ . . 11·54	
-82 . . 7·5		-66 . . 10·00			

The preceding are, as far as I am aware, new results of the liquefaction and solidification of gases. I will now briefly add such other information respecting solidification, pressure, &c., as I have obtained with gaseous bodies previously condensed. As to pressure, considerable irregularity often occurred, which I cannot always refer to its true cause: sometimes a little of the compressed gas would creep by the mercury in the gauge, and increase the volume of enclosed air; and this varied with different substances, probably by some tendency which the glass had to favour the condensation of one (by something analogous to hygrometric action) more than another. But even when the mercury returned to its place in the gauge, there were anomalies which seemed to imply, that a substance, supposed to be one, might be a mixture of two or more. It is, of course, essential that the gauge be preserved at the same temperature throughout the observations.

Muriatic Acid.—This substance did not freeze at the lowest temperature to which I could attain. Liquid muriatic acid dissolves bitumen; the solution, liberated from pressure, boils, giving off muriatic acid vapour, and the bitumen is left in a solid frothy state, and probably altered, in some degree, chemically. The acid unites with and softens the resinous cap cement, but leaves it when the pressure is diminished. The following are certain pressures and temperatures, which, I believe, are not very far from truth; the marked numbers are from experiment.

Fahr.	Atmospheres.	Fahr.	Atmospheres.	Fahr.	Atmospheres.
—100	1·80	—53	5·83	— 5	13·88
— 92	2·28	—50	6·30	0	15·04
— 90	2·38	—42	7·40	10	17·74
— 83	2·90	—40	7·68	20	21·09
— 80	3·12	—33	8·53	25	23·08
— 77	3·37	—30	9·22	30	25·32
— 70	4·02	—22	10·66	32	26·20
— 67	4·26	—20	10·92	40	30·67
— 60	5·08	—10	12·82		

The result formerly obtained * was forty atmospheres at the temperature of 50° Fahr.

Sulphurous Acid.—When liquid, it dissolves bitumen. It becomes a crystalline, transparent, colourless, solid body at —105° Fahr.; when partly frozen the crystals are well formed. The solid sulphurous acid is heavier than the liquid, and sinks freely in it. The following is a Table of pressures in atmospheres of 30 inches mercury, of which the marked results are from many observations, the others are interpolated. They differ considerably from the results obtained by Bunsen †, but agree with my first and only result.

Fahr.	Atmospheres.	Fahr.	Atmospheres.	Fahr.	Atmospheres.
0	0·725	40	1·78	76·8	3·50
10	0·92	46·5	2·00	85	4·00
14	1·00	48	2·06	90	4·35
19	1·12	56	2·42	93	4·50
23	1·23	58	2·50	98	5·00
26	1·33	64	2·76	100	5·16
31·5	1·50	68	3·00	104	5·50
32	1·53	73·5	3·28	110	6·00
33	1·57				

Sulphuretted Hydrogen.—This substance solidifies at 122°

* Philosophical Transactions, 1823, p. 198; or p. 95.
† Bibliothèque Universelle, 1839, xxiii. p. 185.

Fahr. below 0°, and is then a white, crystalline, translucent substance, not remaining clear and transparent in the solid state like water, carbonic acid, nitrous oxide, &c., but forming a mass of confused crystals like common salt or nitrate of ammonia, solidified from the melted state. As it fuses at temperatures above −122°, the solid part sinks freely in the fluid, indicating that it is considerably heavier. At this temperature the pressure of its vapour is less than one atmosphere—not more, probably, than 0·8 of an atmosphere, so that the liquid allowed to evaporate in the air would not solidify as carbonic acid does.

The following is a Table of the tension of its vapour, the marked numbers being close to experimental results, and the rest interpolated. The curve resulting from these numbers, though coming out nearly identical in different series of experiments, is apparently so different in its character from that of water or carbonic acid, as to leave doubts on my mind respecting it, or else of the identity of every portion of the fluid obtained, yet the crystallization and other characters of the latter seemed to show that it was a pure substance.

Fahr.	Atmospheres.	Fahr.	Atmospheres.	Fahr.	Atmospheres.
−100	1·02	−50	2·35	0	6·10
− 94	1·09	−45	2·59	10	7·21
− 90	1·15	−40	2·86	20	8·44
− 83	1·27	−30	3·49	26	9·36
− 80	1·33	−24	3·95	30	9·94
− 74	1·50	−20	4·24	40	11·84
− 70	1·59	−16	4·60	48	13·70
− 68	1·67	−10	5·11	50	14·14
− 60	1·93	− 2	5·90	52	14·60
− 58	2·00				

Carbonic Acid.—The solidification of carbonic acid by M. Thilorier is one of the most beautiful experimental results of modern times. He obtained the substance, as is well known, in the form of a concrete white mass like fine snow, aggregated. When it is melted and resolidified by a bath of low temperature, it then appears as a clear, transparent, cystalline, colourless body, like ice; so clear indeed, that at times it was doubtful to the eye whether anything was in the tube, yet at the same time the part was filled with solid carbonic acid. It melts at the temperature of −70° or −72° Fahr., and the solid carbonic acid is heavier than the fluid bathing it. The solid or liquid carbonic acid at this temperature has a pressure of 5·33 atmospheres nearly. Hence it is easy to understand the readiness with which liquid carbonic acid, when allowed to escape into the air, exerting only a pressure of one atmosphere, freezes a part of itself by the evaporation of another part.

Thilorier gives −100° C. or −148° Fahr. as the temperature at which carbonic acid becomes solid. This however is rather the temperature to which solid carbonic acid can sink by further evaporation in the air, and is a temperature belonging to a pressure not only lower than that of 5·33 atmospheres, but even much below that of one atmosphere. This cooling effect to temperatures below the boiling-point often appears. A bath of carbonic acid and ether exposed to the air will cool a tube containing condensed solid carbonic acid, until the pressure within the tube is less than one atmosphere; yet, if the same bath be covered up so as to have the pressure of one atmosphere of carbonic acid vapour over it, then the temperature is such as to produce a pressure of 2·5 atmospheres by the vapour of the solid carbonic acid within the tube.

The estimates of the pressure of carbonic acid vapour are sadly at variance; thus, Thilorier[*] says it has a pressure of 26 atmospheres at −4° Fahr., whilst Addams[†] says that for that pressure it requires a temperature of 30°. Addams gives the pressure about 27½ atmospheres at 32°, but Thilorier and myself[‡] give it as 36 atmospheres at the same temperature. At 50° Brunel[§] estimates the pressure as 60 atmospheres, whilst Addams makes it only 34·67 atmospheres. At 86° Thilorier finds the pressure to be 73 atmospheres; at 4° more, or 90°, Brunel makes it 120 atmospheres; and at 10° more, or 100°, Addams makes it less than Thilorier at 86°, and only 62·32 atmospheres; even at 150° the pressure with him is not quite 100 atmospheres.

I am inclined to think that at about 90° Cagniard de la Tour's state comes on with carbonic acid. From Thilorier's data we may obtain the specific gravity of the liquid and the

---

[*] Ann. de Chim. 1835, lx. 427, 432.   [†] Report of Brit. Assoc. 1838, p. 70.
[‡] Phil. Trans. 1823, p. 193.   [§] Royal Institution Journal, xxi. 132.

vapour over it at the temperature of 86° Fahr., and the former is little more than twice that of the latter; hence a few degrees more of temperature would bring them together, and Brunel's result seems to imply that the state was then on, but in that case Addams's results could only be accounted for by supposing that there was a deficiency of carbonic acid. The following are the pressures which I have recently obtained :—

Fahr.	Atmospheres.	Fahr.	Atmospheres.	Fahr.	Atmospheres.
◡—111°	1·14	—60°	6·97	◡— 4°	21·48
—110	1·17	◡—56	7·70	0	22·84
◡—107	1·36	—50	8·88	◡ 5	24·75
—100	1·85	—40	11·07	◡ 10	26·82
◡— 95	2·28	◡—34	12·50	◡ 15	29·09
— 90	2·77	—30	13·54	20	30·65
◡— 83	3·60	◡—23	15·45	◡ 23	33·15
— 80	3·93	—20	16·30	30	37·19
◡— 75	4·60	◡—15	17·80	◡ 32	38·50
— 70	5·33	—10	19·38		

Carbonic acid is remarkable amongst bodies for the high tension of the vapour which it gives off whilst in the solid or glacial state. There is no other substance which at all comes near it in this respect, and it causes an inversion of what in all other cases is the natural order of events. Thus, if, as is the case with water, ether, mercury or any other fluid, that temperature at which carbonic acid gives off vapour equal in elastic force to one atmosphere, be called its boiling-point; or, if (to produce the actual effect of ebullition) the carbonic acid be plunged below the surface of alcohol or ether, then we shall perceive that the freezing and boiling points are inverted, *i. e.* that the freezing-point is the hotter, and the boiling-point the colder of the two, the latter being about 50° below the former.

Euchlorine.—This substance was easily converted from the gaseous state into a solid crystalline body, which, by a little increase of temperature, melted into an orange-red fluid, and by diminution of temperature again congealed; the solid euchlorine had the colour and general appearance of bichromate of potassa; it was moderately hard, brittle and translucent; and the crystals were perfectly clear. It melted at the temperature of 75° below 0°, and the solid portion was heavier than the liquid.

When in the solid state it gives off so little vapour, that the eye is not sensible of its presence by any degree of colour in the air over it when looking down a tube 4 inches in length, at the bottom of which is the substance. Hence the pressure of its vapour at that temperature must be very small.

Some hours after, wishing to solidify the same portion of euchlorine which was then in a liquid state, I placed the tube in a bath at −110°, but could not succeed, either by continuance of the tube in the bath, or shaking the fluid in the tube, or opening the tube to allow the full pressure of the atmosphere; but when the liquid euchlorine was touched by a platinum wire it instantly became solid, and exhibited all the properties before described. There are many similar instances amongst ordinary substances, but the effect in this case makes me hesitate in concluding that all the gases which as yet have refused to solidify at temperatures as low as 166° below 0°, cannot acquire the solid state at such a temperature.

Nitrous Oxide.—This substance was obtained *solid* by the temperature of the carbonic acid bath *in vacuo,* and appeared as a beautiful clear crystalline colourless body. The temperature required for this effect must have been very nearly the lowest, perhaps about 150° below 0°. The pressure of the vapour rising from the solid nitrous oxide was less than one atmosphere.

Hence it was concluded that liquid nitrous oxide could not freeze itself by evaporation at one atmosphere, as carbonic acid does; and this was found to be true, for when a tube containing much liquid was freely opened, so as to allow evaporation down to one atmosphere, the liquid boiled and cooled itself, but remained a liquid. The cold produced by the evaporation was very great, and this was shown by putting the part of the tube containing the liquid nitrous oxide, into a cold bath of carbonic acid, for the latter was like a hót bath to the former, and instantly made it boil rapidly.

I kept this substance for some weeks in a tube closed by stopcocks and cemented caps. In that time there was no action on the bitumen of the graduation, nor on the cement of the caps; these bodies remained perfectly unaltered.

Hence it is probable that this substance may be used in certain cases, instead of carbonic acid, to produce degrees of cold far below those which the latter body can supply. Down to a certain temperature, that of its solidification, it would not even require ether to give contact, and below that temperature it could easily be used mingled with ether; its vapour would do no harm to an air-pump, and there is no doubt that the substance placed *in vacuo* would acquire a temperature lower than any as yet known, perhaps as far below the carbonic acid bath *in vacuo* as that is below the same bath in air.

This substance, like olefiant gas, gave very uncertain results at different times as to the pressure of its vapour; results which can only be accounted for by supposing that there are two different bodies present, soluble in each other, but differing in the elasticity of their vapour. Four different portions gave at the same temperature, namely $-106°$ Fahr., the following great differences in pressure: 1·66; 4·4; 5·0; and 6·3 atmospheres; and this after the elastic atmosphere left in the tubes at the conclusion of the condensation had been allowed to escape, and be replaced by a portion of the respective liquids which then rose in vapour. The following Table gives certain results with a portion of liquid which exerted a pressure of six atmospheres at $-106°$ Fahr.

Fahr.	Atmospheres.	Atmospheres.
$-40°$ . . .	10·20	
-35 . . .	10·95	
-30 . . .	11·80	
-25 . . .	12·75	
-20 . . .	13·80	
-15 . . .	14·95	
-10 . . .	16·20	
-5 . . .	17·55	
0 . . .	19·05 . . .	24·40
5 . . .	20·70 . . .	26·08
10 . . .	22·50 . . .	27·84
15 . . .	24·45 . . .	29·68
20 . . .	26·55 . . .	31·62
25 . . .	28·85 . . .	33·66
30 . . .		. . . 35·82
35 . . .		. . . 38·10

The second column expresses the pressures given as the fluid was raised from low to higher temperatures. The third column shows the pressures given the next day with the same tube after it had attained to and continued at the atmospheric temperature for some hours. There is a difference of four or five atmospheres between the two, showing that in the first instance the previous low temperature had caused the solution of a more volatile part in the less volatile and liquid portion, and that the prolonged application of a higher temperature during the night had gradually raised it again in vapour. This result occurred again and again with the same specimen[*].

Cyanogen.—This substance becomes a solid transparent crystalline body, as Bunsen has already stated [†], which raised to the temperature of $-30°$ Fahr. then liquefies. The solid and liquid appear to be nearly of the same specific gravity, but the solid is perhaps the denser of the two.

The mixed solid and liquid substance yields a vapour of rather less pressure than one atmosphere. In accordance with this result, if the liquid be exposed to the air, it does not freeze itself as carbonic acid does.

The liquid tends to distil over and condense on the cap cement and bitumen of the gauge, but only slightly. When cyanogen is made from cyanide of mercury sealed up hermetically in a glass tube, the cyanogen distils back and condenses in the paracyanic residue of the distillation; but the pressure of the vapour at common temperatures is still as great, or very nearly so, as if the cyanogen were in a clean separate liquid state.

A measured portion of liquid cyanogen was allowed to escape and expand into gas. In this way one volume of liquid at the temperature of 63° Fahr. gave 393·9 volumes of gas at the same temperature and the barometric pressure of 30·2 inches. If 100 cubic inches of the gas be admitted to weigh

---

[*] This substance is one of those which I liquefied in 1823 (see Philosophical Transactions). Since writing the above, I perceive that M. Natterer has condensed it into the liquid state by the use of pumps only (see Comptes Rendus, 1844, Nov. 18, p. 1111), and obtained the liquid in considerable quantities. The non-solidification of it by exposure to the air perfectly accords with my own results.

[†] Bibliothèque Universelle, 1839, xxiii. p. 184.

I

55·5 grains, then a cubic inch of the liquid would weigh 218·6 grains. This gives its specific gravity as 0·866. When first condensed, I estimated it as nearly 0·9.

Cyanogen is a substance which yielded on different occasions results of vaporous tension differing much from each other, though the substance appeared always to be pure. The following are numbers in which I place some confidence, the pressures being in atmospheres of 30 inches of mercury, and the marked results experimental *.

Fahr.	Atmospheres.	Fahr.	Atmospheres.	Fahr.	Atmospheres.
0	1·25	◡38·5	2·72	77	5·00
8·5	1·50	◡44·5	3·00	◡ 79	5·16
◡10	1·53	◡48	3·17	83	5·50
15	1·72	◡50	3·28	88·3	6·00
◡20	1·89	◡52	3·36	◡ 93·5	6·50
22·8	2·00	54·3	3·50	◡ 95	6·64
◡27	2·20	◡63	4·00	98·4	7·00
◡32	2·37	◡70	4·50	◡103	7·50
34·5	2·50	◡74	4·79		

Ammonia.—This body may be obtained as a *solid*, white, translucent, crystalline substance, melting at the temperature of 103° below 0°; at which point the solid substance is heavier than the liquid. In that state the pressure of its vapour must be very small.

Liquid ammonia at 60° was allowed to expand into ammoniacal gas at the same temperature; one volume of the liquid gave 1009·8 volumes of the gas, the barometer being at the pressure of 30·2 inches. If 100 cubic inches of ammoniacal gas be allowed to weigh 18·28 grains, it will give 184·6 grains as the weight of a cubic inch of liquid ammonia at 60°. Hence its specific gravity at that temperature will be 0·731. In the old experiments I found by another kind of process that its specific gravity was 0·76 at 50°.

The following is a Table of the pressure of ammonia vapour, the marked results, as before, being those obtained by experiment :—

---

* See Bunsen's results, Bibliothèque Universelle, 1839, xxiii. p. 185.

Fahr.	Atmospheres.	Fahr.	Atmospheres.	Fahr.	Atmospheres.
0 . . 2·48	41 . . 5·10	61·3 . . 7·00
0·5 . . 2·50	44 . . 5·36	65·6 . . 7·50
9·3 . . 3·00	45 . . 5·45	67 . . 7·63
18 . . 3·50	45·8 . . 5·50	69·4 . . 8·00
21 . . 3·72	49 . . 5·83	73 . . 8·50
25·8 . . 4·00	51·4 . . 6·00	76·8 . . 9·00
26 . . 4·04	52 . . 6·10	80 . . 9·50
32 . . 4·44	55 . . 6·38	83 . . 10·00
33 . . 4·50	56·5 . . 6·50	85 . . 10·30
39·5 . . 5·00	60 . . 6·90	

Arseniuretted Hydrogen.—This body, liquefied by Dumas and Soubeiran, did not solidify at the lowest temperature to which I could submit it, *i. e.* not at 166° below 0° Fahr. In the following Table of the elasticity of its vapour the marked results are experimental, and the others interpolated :—

Fahr.	Atmospheres.	Fahr.	Atmospheres.	Fahr.	Atmospheres.
−75 . . 0·94	−30 . . 2·84	10 . . 6·24
−70 . . 1·08	−23 . . 3·32	20 . . 7·39
−64 . . 1·26	−20 . . 3·51	30 . . 8·66
−60 . . 1·40	−10 . . 4·30	32 . . 8·95
−52 . . 1·73	− 5 . . 4·74	40 . . 10·05
−50 . . 1·80	0 . . 5·21	50 . . 11·56
−40 . . 2·28	3 . . 5·56	60 . . 13·19
−36 . . 2·50		

The following bodies would not freeze at the very low temperature of the carbonic acid bath *in vacuo* (−166° Fahr.): —Chlorine, ether, alcohol, sulphuret of carbon, caoutchoucine, camphine or rectified oil of turpentine. The alcohol, caoutchoucine and camphine lost fluidity and thickened somewhat at −106°, and still more at the lower temperature of −166°. The alcohol then poured from side to side like an oil.

Dry yellow fluid nitrous acid when cooled below 0° loses the greater part of its colour, and then fuses into a white, crystalline, brittle and but slightly translucent substance, which fuses a little above 0° Fahr. The green and probably hydrated acid required a much lower temperature for its solidification, and then became a pale bluish solid. There were then evidently two bodies, the dry acid which froze out first, and then the hydrate, which requires at least −30° below 0° before it will solidify.

The following gases showed no signs of liquefaction when cooled by the carbonic acid bath *in vacuo*, even at the pressures expressed :—

	Atmospheres.
Hydrogen at	27
Oxygen	27
Nitrogen	50
Nitric oxide	50
Carbonic oxide	40
Coal-gas	32

The difference in the facility of leakage was one reason of the difference in the pressure applied. I found it impossible, from this cause, to raise the pressure of hydrogen higher than twenty-seven atmospheres by an apparatus that was quite tight enough to confine nitrogen up to double that pressure.

M. Cagniard de la Tour has shown that at a certain temperature, a liquid, under sufficient pressure, becomes clear transparent vapour or gas, having the same bulk as the liquid. At this temperature, or one a little higher, it is not likely that any increase of pressure, except perhaps one exceedingly great, would convert the gas into a liquid. Now the temperature of 166° below 0°, low as it is, is probably *above* this point of temperature for hydrogen, and perhaps for nitrogen and oxygen, and then no compression without the conjoint application of a degree of cold below that we have as yet obtained, can be expected to take from them their gaseous state. Further, as ether assumes this state before the pressure of its vapour has acquired thirty-eight atmospheres, it is more than probable that gases which can resist the pressure of from twenty-seven to fifty atmospheres at a temperature of 166° below 0° could never appear as liquids, or be made to lose their gaseous state at common temperatures. They may probably be brought into the state of very condensed gases, but not liquefied.

Some very interesting experiments on the compression of gases have been made by M. G. Aime *, in which oxygen,

---

* Annales de Chimie, 1843, viii. 275.

olefiant, nitric oxide, carbonic oxide, fluosilicon, hydrogen, and nitrogen gases were submitted to pressures, rising up to 220 atmospheres in the case of the last two; but this was in the depths of the sea where the results under pressure could not be examined. Several of them were diminished in bulk in a ratio far greater than the pressure put upon them; but both M. Cagniard de la Tour and M. Thilorier have shown that this is often the case whilst the substance retains the gaseous form. It is possible that olefiant gas and fluosilicon may have liquefied down below, but they have not yet been seen in the liquid state except in my own experiments, and in them not at temperatures above 40° Fahr. The results with oxygen are so unsteady and contradictory as to cause doubt in regard to those obtained with the other gases by the same process.

Thus, though as yet I have not condensed oxygen, hydrogen, or nitrogen, the original objects of my pursuit, I have added six substances, usually gaseous, to the list of those that could previously be shown in the liquid state, and have reduced seven, including ammonia, nitrous oxide, and sulphuretted hydrogen, into the solid form. And though the numbers expressing tension of vapour cannot (because of the difficulties respecting the use of thermometers and the apparatus generally) be considered as exact, I am in hopes they will assist in developing some general law governing the vaporization of all bodies, and also in illustrating the physical state of gaseous bodies as they are presented to us under ordinary temperature and pressure.

Royal Institution, Nov. 15, 1844.

Note.—Additional Remarks respecting the Condensation of Gases.

[Received February 20,—Read February 20, 1845.]

Nitrous Oxide.—Suspecting the presence on former occasions of nitrogen in the nitrous oxide, and mainly because of muriate in the nitrate of ammonia used, I prepared that salt in a pure state from nitric acid and carbonate of ammonia previously proved, by nitrate of silver, to be free from muriatic acid. After the nitrous oxide prepared from this salt had remained for some days in well-closed bottles in contact with a little water, I condensed it in the manner already described, and when condensed I allowed half the fluid to escape in vapour, that as much as possible of the less condensable portion might be carried off. In this way as much gas as would fill the capacity of the vessels twenty or thirty times or more was allowed to escape. Afterwards the following series of pressures was obtained :—

Fahr.°	Atmospheres.	Fahr.°	Atmospheres.	Fahr.°	Atmospheres.
−125	1·00	−70	4·11	−15	14·69
−120	1·10	−65	4·70	−10	16·15
−115	1·22	−60	5·36	− 5	17·70
−110	1·37	−55	6·09	0	19·34
−105	1·55	−50	6·89	5	21·07
−100	1·77	−45	7·76	10	22·89
− 95	2·03	−40	8·71	15	24·80
− 90	2·34	−35	9·74	20	26·80
− 85	2·70	−30	10·85	25	28·90
− 80	3·11	−25	12·04	30	31·10
− 75	3·58	−20	13·32	35	33·40

These numbers may all be taken as the results of experiments. Where the temperatures are not those actually observed, they are in almost all cases within a degree of it, and proportionate to the effects really observed. The departure of the real observations from the numbers given is very small. This Table I consider as far more worthy of confidence than the former, and yet it is manifest that the curve is not consistent with the idea of a pure single substance, for the pressures at the lowest temperature are too high. I believe that there are still two bodies present, and that the more volatile, as before said, is condensable in the liquid of the less volatile; but I think there is a far smaller proportion of the more volatile (nitrogen, or whatever it may be) than in the former case.

Olefiant Gas.—The olefiant gas condensed in the former experiment was prepared in the ordinary way, using excellent alcohol and sulphuric acid; then washed by agitation with about half its bulk of water, and finally left for three days over a thick mixture of lime and water with occasional agitation. In this way all the sulphurous and carbonic acids were removed, and I believe all the ether, except such minute portions as could not interfere with my results. In respect of the ether, I have since found that the process is satisfactory; for when I purposely added ether vapour to air, so as to increase its bulk by one-third, treatment like that above removed it, so as to leave the air of its original volume. There was yet a slight odour of ether left, but not so much as that conferred by adding one volume of the vapour of ether to 1200 or 1500 volumes of air. I find that when air is expanded $\frac{1}{4}$th or $\frac{1}{3}$rd more by the addition of the vapour of ether, washing first of all with about $\frac{1}{5}$th of its volume of water, then again with about as much water, and lastly with its volume of water, removes the ether to such a degree, that though a little smell may remain, the air is of its original volume.

As already stated, it is the presence of other and more volatile hydrocarbons than olefiant gas, which the tensions obtained seemed to indicate, both in the gas and the liquid resulting from its condensation. In a further search after these I discovered a property of olefiant gas which I am not aware is known (since I do not find it referred to in books), namely its ready solubility in strong alcohol, ether, oil of turpentine, and such like bodies*. Alcohol will take up two volumes of this gas; ether can absorb two volumes; oil of turpentine two volumes and a half; and olive oil one volume by agitation at common temperatures and pressure; consequently, when a vessel of olefiant gas is transferred to a bath of any of these liquids and agitated, absorption quickly takes place.

Examined in this way, I have found no specimen of olefiant gas that is entirely absorbed; a residue always remains, which, though I have not yet had time to examine it accurately, appears to be light carburetted hydrogen; and I have no doubt that this is the substance which has mainly interfered in my former results. This substance appears to be produced in every stage of the preparation of olefiant gas. On taking six different portions of gas at different equal intervals, from first to last, during one process of preparation, after removing the sulphurous and carbonic acid and the ether as before described, then the following was the proportion per cent. of insoluble

---

* Water, as Berzelius and others have pointed out, dissolves about $\frac{1}{8}$th its volume of olefiant gas, but I find that it also leaves an insoluble residue, which burns like light carburetted hydrogen.

gas in the remainder when agitated with oil of turpentine: 10·5; 10; 10·1; 13·1; 28·3; 61·8. Whether carbonic oxide was present in any of these undissolved portions I cannot at present say.

In reference to the part dissolved, I wish as yet to guard myself from being supposed to assume that it is one uniform substance; there is indeed little doubt that the contrary is true; for whilst a volume of oil of turpentine introduced into twenty times its volume of olefiant gas cleared from ether and the acids, absorbs $2\frac{1}{2}$ volumes of the gas, the same volume of fresh oil of turpentine brought into similar contact with abundance of the gas which remains when one-half has been removed by solution only dissolved 1·54 part, yet there was an abundant surplus of gas which would dissolve in fresh oil of turpentine at this latter rate. When two-thirds of a portion of fresh olefiant gas were removed by solution, the most soluble portion of that which remained required its bulk of fresh oil of turpentine to dissolve it. Hence at first one volume of camphine dissolved 2·50, but when the richer portion of the gas was removed, one volume dissolved 1·54 part; and when still more of the gas was taken away by solution, one volume of camphine dissolved only one volume of the gas. This can only be accounted for by the presence of various compounds in the soluble portion of the gas.

A portion of good olefiant gas was prepared, well agitated with its bulk of water in close vessels, left over lime and water for three days, and then condensed as before. When much liquid was condensed, a considerable proportion was allowed to escape to sweep out the uncondensed atmosphere and the more condensable vapours; and then the following pressures were observed:—

Fahr. °	Atmospheres.	Fahr. °	Atmospheres.	Fahr. °	Atmospheres.
−105	4·60	−65	8·30	−30	16·22
−100	4·82	−60	9·14	−25	17·75
− 95	5·10	−55	10·07	−20	19·38
− 90	5·44	−50	11·10	−15	21·11
− 85	5·84	−45	12·23	−10	22·94
− 80	6·32	−40	13·46	− 5	24·87
− 75	6·89	−35	14·79	0	26·90
− 70	7·55				

On examining the form of the curve given by these pressures, it is very evident that, as on former occasions, the pressures at low temperatures are too great to allow the condensed liquid to be considered as one uniform body, and the form of the curve at the higher pressures is quite enough to prove that no ether was present either in this or the former fluids.   On permitting the liquid in the tube to expand into gas, and treating 100 parts of that gas with oil of turpentine, eighty-nine parts were dissolved, and eleven parts remained insoluble. There can be no doubt that the presence of this latter substance, soluble as it is under pressure in the more condensable portions, is the cause of the irregularity of the curve, and the too high pressure at the lower temperatures.

The ethereal solution of olefiant gas being mixed with eight or nine times its volume of water, dissolved, and gradually minute bubbles of gas appeared, the separation of which was hastened by a little heat.   In this way about half the gas dissolved was re-obtained, and burnt like very rich olefiant gas. One volume of the alcoholic solution, with two volumes of water, gave very little appearance of separating gas.   Even the application of heat did not at first cause the separation, but gradually about half the dissolved olefiant gas was liberated.

The separation of the dissolved gas by water, heat, or change of pressure from its solutions, will evidently supply means of procuring olefiant gas in a greater state of purity than heretofore; the power of forming these solutions will also very much assist in the correct analysis of mixtures of hydrocarbons.   I find that light carburetted hydrogen is hardly sensibly soluble in alcohol or ether, and in oil of turpentine the proportion dissolved is not probably $\frac{1}{13}$th the volume of the fluid employed; but the further development of these points I must leave for the present.

Carbonic Acid.—This liquid may be retained in glass tubes furnished with cemented caps, and closed by plugs or stopcocks, as described; but it is important to remember the softening action on the cement, which, being continued, at last reduces its strength below the necessary point.   A tube of this kind was arranged on the 10th of January and left; on the 15th of February it exploded, not by any fracture of the tube, for that remained unbroken, but simply by throwing off the cap through a failure of the cement. Hence the cement joints should not be used for long experiments, but only for those enduring for a few days.

Oxygen.—Chlorate of potassa was melted and pulverized. Oxide of manganese was pulverized, heated red-hot for half an hour, mixed whilst hot with the chlorate, and the mixture put into a long strong glass generating tube with a cap cemented on, and this tube then attached to another with a gauge for condensation. The heat of a spirit-lamp carefully applied produced the evolution of oxygen without any appearance of water, and the tubes, both hot and cold, sustained the force generated. In this manner the pressure of oxygen within the apparatus was raised as high as 58·5 atmospheres, whilst the temperature at the condensing place was reduced as low as $-140°$ Fahr., but no condensation appeared. A little above this pressure the cement of two of the caps began to leak, and I could carry the observation no further with this apparatus.

From the former scanty and imperfect expressions of the elasticity of the vapour of the condensed gases, Dove was led to put forth a suggestion*, whether it might not ultimately appear that the same addition of heat (expressed in degrees of the thermometer) caused the same additional increase of expansive force for all gases or vapours in contact with their liquids, provided the observation began with the same pressure in all. Thus to obtain the difference between forty-four and fifty atmospheres of pressure, either with steam or nitrous oxide, nearly the same number of degrees of heat were required; to obtain the difference between twenty and twenty-five atmospheres, either with steam or muriatic acid, the same number were required. Such a law would of course make the rate of increasing expansive force the same for all bodies, and the curve laid down for steam would apply to every other vapour. This, however, does not appear to be the case. That the force of the vapour increases in a geometrical ratio for equal increments of heat is true for all bodies, but the ratio is not the same for all. As far as observations upon the follow-

* Poggendorff's Annalen, xxiii. 290; or Thomson on Heat and Electricity, p. 9.

ing substances, namely, water, sulphurous acid, cyanogen, ammonia, arseniuretted hydrogen, sulphuretted hydrogen, muriatic acid, carbonic acid, olefiant gas, &c., justify any conclusion respecting a general law, it would appear that the more volatile a body is, the more rapidly does the force of its vapour increase by further addition of heat, commencing at a given point of pressure for all; thus for an increase of pressure from two to six atmospheres, the following number of degrees require to be added for the different bodies named : water 69°, sulphurous acid 63°, cyanogen 64°·5, ammonia 60°, arseniuretted hydrogen 54°, sulphuretted hydrogen 56°·5, muriatic acid 43°, carbonic acid 32°·5, nitrous oxide 30°; and though some of these numbers are not in the exact order, and in other cases, as of olefiant gas and nitrous oxide, the curves sometimes even cross each other, these circumstances are easily accounted for by the facts already stated of irregular composition and the inevitable errors of first results.   There seems every reason therefore to expect that the increasing elasticity is directly as the volatility of the substance, and that by further and more correct observation of the forces, a general law may be deduced, by the aid of which, and only a single observation of the force of any vapour in contact with its fluid, its elasticity at any other temperature may be obtained.

Whether the same law may be expected to continue when the bodies approach near to the Cagniard de la Tour state is doubtful. That state comes on sooner in reference to the pressure required, according as the liquid is lighter and more expansible by heat and its vapour heavier, hence indeed the great reason for its facile assumption by ether. But though with ether, alcohol and water, that substance which is most volatile takes up this state with the lowest pressure, it does not follow that it should always be so; and in fact we know that ether takes up this state at a pressure between thirty-seven and thirty-eight atmospheres, whereas muriatic acid, nitrous oxide, carbonic acid and olefiant gas, which are far more volatile, sustain a higher pressure than this without assuming that peculiar state, and whilst their vapours and liquids are still considerably different from each other. Now whether the curve which expresses the elastic force of the vapour of a given fluid for increasing temperatures continues undisturbed after that fluid has passed the Cagniard de la Tour point or not is not known, and therefore it cannot well be anticipated whether the coming on of that state sooner or later with particular bodies will influence them in relation to the more general law referred to above.

The law already suggested gives great encouragement to the continuance of those efforts which are directed to the condensation of oxygen, hydrogen and nitrogen, by the attainment and application of lower temperatures than those yet applied. If to reduce carbonic acid from the pressure of two atmospheres to that of one, we require to abstract only about half the number of degrees that is necessary to produce the same effect with sulphurous acid, it is to be expected that a far less abstraction will suffice to produce the same effect with nitrogen or hydrogen, so that further diminution of temperature and improved apparatus for pressure may very well be expected to give us these bodies in the liquid or solid state.

Royal Institution, Feb. 19, 1845.

Historical Statement respecting the Liquefaction of Gases [*]

I was not aware at the time when I first observed the liquefaction of chlorine gas [†], nor until very lately, that any of the class of bodies called *gases*, had been reduced into the fluid form; but having during the last few weeks sought for instances where such results might have been afforded without the knowledge of the experimenter, I was surprised to find several recorded cases. I have thought it right therefore to bring these cases together, and only justice to endeavour to secure for them a more general attention than they appear as yet to have gained. I shall notice in chronological order, the fruitless, as well as the successful attempts, and those which probably occurred without being observed, as well as those which were remarked and described as such.

Carbonic Acid, &c.—The 'Philosophical Transactions' for 1797 contain, p. 222, an account of experiments made by Count Rumford, to determine the force of fired gunpowder.

[*] Quarterly Journal of Science, xvi. 229.
[†] Philosophical Transactions, 1823, pp. 160, 189—or page 85.

Dissatisfied both with the deductions drawn, and the means used previously, that philosopher proceeded to fire gunpowder in cylinders of a known diameter and capacity, and closed by a valve loaded with a weight that could be varied at pleasure. By making the vessel strong enough and the weight sufficiently heavy, he succeeded in confining the products within the space previously occupied by the powder. The Count's object induced him to vary the quantity of gunpowder in different experiments, and to estimate the force exerted only at the moment of ignition, when it was at its maximum. This force, which he found to be prodigious, he attributes to aqueous vapour intensely heated, and makes no reference to the force of the gaseous bodies evolved. Without considering the phenomena which it is the Count's object to investigate, it may be remarked, that in many of the experiments made by him, some of the gases, and especially carbonic acid gas, were probably reduced to the liquid state. The Count says,—

"When the force of the generated elastic vapour was sufficient to raise the weight, the explosion was attended by a very sharp and surprisingly loud report; but when the weight was not raised, as also when it was only a little moved, but not sufficiently to permit the leather stopper to be driven quite out of the bore, and the elastic fluid to make its escape, the report was scarcely audible at a distance of a few paces, and did not at all resemble the report which commonly attends the explosion of gunpowder. It was more like the noise which attends the breaking of a small glass tube, than anything else to which it could be compared. In many of the experiments, in which the elastic vapour was confined, this feeble report attending the explosion of the powder was immediately followed by another noise totally different from it, which appeared to be occasioned by the falling back of the weight upon the end of the barrel, after it had been a little raised, but not sufficiently to permit the leather stopper to be driven quite out of the bore. In some of these experiments a very small part only of the generated elastic fluid made its escape; in these cases the report was of a peculiar kind, and though perfectly audible at some considerable distance, yet not at all resembling the report of a musket. It was rather a very strong sudden hissing, than a clear, distinct and sharp report."

In another place it is said, " What was very remarkable in all these experiments, in which the generated elastic vapour was completely confined, was the small degree of expansive force which this vapour appeared to possess, after it had been suffered to remain a few minutes, or even only a few seconds, confined in the barrel; for upon raising the weight by means of its lever, and suffering this vapour to escape, instead of escaping with a loud report it rushed out with a hissing noise, hardly so loud or so sharp as the report of a common air-gun, and its effects against the leather stopper, by which it assisted in raising the weight, were so very feeble as not to be sensible." This the Count attributes to the formation of a hard mass, like a stone, within the cylinder, occasioned by the condensation of what was, at the moment of ignition, an elastic fluid.   Such a substance was always found in these cases; but when the explosion raised the weight and blew out the stopper, nothing of this kind remained.

The effects here described, both of elastic force and its cessation on cooling, may evidently be referred as much to carbonic acid and perhaps other gases as to water. The strong sudden hissing observed as occurring when only a little of the products escaped, may have been due to the passage of the gases into the air, with comparatively but little water, the circumstances being such as were not sufficient to confine the former, though they might the latter; for it cannot be doubted but that in similar circumstances the elastic force of carbonic acid would far surpass that of water.   Count Rumford says, that the gunpowder made use of, when well shaken together, occupied rather less space than an equal weight of water. The quantity of residuum before referred to, left by a given weight of gunpowder, is not mentioned, so that the actual space occupied by the vapour of water, carbonic acid, &c., at the moment of ignition, cannot be inferred; there can, however, be but little doubt that when perfectly confined they were in the state of the substances in M. Cagniard de la Tour's experiments[*].

When allowed to remain a few minutes, or even seconds, the expansive force at first observed diminished exceedingly, so as scarcely to surpass that of the air in a charged air-gun. Of course all that was due to the vaporization of water and some

* See Quarterly Journal of Science, vol. xv. p. 145.

of the other products would cease, as soon as the mass of metal had absorbed the heat, and they would concrete into the hard substance found in the cylinder : but it does not seem too much to suppose, that so much carbonic acid was generated in the combustion, as would, if confined, on the cooling of the apparatus, have been equal to many atmospheres, but that being condensable, a part became liquid, and thus assisted in reducing the force within to what it was found to be.

Ammonia.—I find the condensation of ammoniacal gas referred to in Thomson's ' System,' first edition, i. 405, and other editions ; Henry's ' Chemistry,' i. 237 ; Accum's ' Chemistry,' i. 310 ; Murray's ' Chemistry,' ii. 73 ; and Thenard's ' Traité de Chimie,' ii. 133.   Mr. Accum refers to the experiments of Fourcroy and Vauquelin, ' Ann. de Chimie,' xxix. 289, but has mistaken their object.   Those chemists used highly saturated solution of ammonia, see pp. 281, 286, and not the gas ; and their experiments on gases, namely, sulphurous acid gas, muriatic acid gas, and sulphuretted hydrogen gas, they state were fruitless, p. 287.   " All we can say is, that the condensation of most of these gases was above three-fourths of their volume."

Thomson, Henry, Murray, and, I suppose, Thenard, refer to the experiments of Guyton de Morveau, ' Ann. de Chimie,' xxix. 291, 297.   Thomson states the result of liquefaction at a temperature of $-45°$, without referring to the doubt, that Morveau himself raises, respecting the presence of water in the gas ; but Murray, Henry, and Thenard, in their statements notice its probable presence.   Morveau's experiment was made in the following manner :—A glass retort was charged with the usual mixture of muriate of ammonia and quick-lime, the former material being sublimed, and the latter carefully made from white marble, so as to exclude water as much as possible.   The beak of the retort was then adapted to an apparatus consisting of two balloons, and two flasks successively connected together, and luted by fat lute.   The balloons were empty ; the first flask contained mercury, the second water.   Heat was then applied to the retort, and the first globe cooled to $-21°·25$ C. ; aqueous vapours soon rose, which condensed as water in the neck of the retort, and as ice in the first balloon.   Continuing the heat, ammoniacal gas was disengaged, and it escaped by the last flask containing water, without anything being perceived in the second balloon. This balloon was then cooled to −43°·75 C., and then drops of a fluid lined its interior, and ultimately united at the bottom of the vessel. When the thermometer in the cooling mixture stood at −36°·25 C., the fluid already deposited preserved its state, but no further portions were added to it; reducing the temperature again to −41° C., and hastening the disengagement of ammoniacal gas, the liquid in the second balloon augmented in volume. Very little gas escaped from the last flask, and the pressure inwards was such as to force the oil of the lute into the balloon, where it congealed. Finally, the apparatus was left to regain the temperature of the atmosphere, and as it approached to it, the liquid of the second balloon became gaseous. The ice in the first balloon became liquid, as soon as the temperature had reached −21°·25 C.

M. Morveau remarks on this experiment, that it appears certain that ammoniacal gas made as dry as it can be, by passing into a vessel in which water would be frozen, and reduced to a temperature of −21° C., condenses into a liquid at the temperature of −48° C., and resumes its elastic form again as the temperature is raised; but he proposes to repeat the experiment and examine whether a portion of the gas so dried, when received over mercury, would not yield water to well-calcined potash, "for as it is seen that water charged with a little of the gas, remained liquid in the first balloon, at a temperature of −21°, it is possible that a much smaller quantity of water united to a much larger quantity of the gas, would become capable of resisting a temperature of −48° C."

Sir H. Davy, who refers to this experiment in his 'Elements of Chemical Philosophy,' p. 267, urges the uncertainty attending it, on the same grounds that Morveau himself had done; and now that the strength of the vapour of dry liquid ammonia is known, it cannot be doubted that M. Morveau had obtained in his second balloon only a very concentrated solution of ammonia in water. I find that the strength of the vapour of ammonia dried by potash, is equal to about that of 6·5 atmospheres at 50° F. *, and according to all analogy it would require a very intense degree of cold, and one at present beyond our means, to compensate this power and act as an equivalent to it.

* Philosophical Transactions, 1823, p. 197—or page 95.

Sulphurous Acid Gas.—It is said that sulphurous acid gas has been condensed into a fluid by Monge and Clouet, but I have not been able to find the description of their process. It is referred to by Thomson, in his 'System,' first edition, ii. 24, and in subsequent editions; by Henry, in his 'Elements,' i. 341; by Accum, in his 'Chemistry,' i. 319; by Aikin, 'Chemical Dictionary,' ii. 391; by Nicholson, 'Chemical Dictionary,' article Gas (sulphurous acid); and by Murray, in his 'System,' ii. 405. All these authors mention the simultaneous application of cold and pressure, but Thomson alone refers to any authority, and that is Fourcroy, ii. 74.

It is curious that Fourcroy does not, however, mention condensation as one of the means employed by Monge and Clouet, but merely says the gas is capable of liquefaction at 28° of cold. "This latter property," he adds, "discovered by citizens Monge and Clouet, and by which it is distinguished from all the other gases, appears to be owing to the water which it holds in solution, and to which it adheres so strongly as to prevent an accurate estimate of the proportions of its radical and acidifying principles."

Notwithstanding Fourcroy's objection, there can be but little reason to doubt that Monge and Clouet did actually condense the gas, for I have since found that from the small elastic force of its vapour at common temperatures (being equal to that of about two atmospheres only*), a comparatively moderate diminution of temperature is sufficient to retain it fluid at common pressure, or a moderate additional pressure to retain it so at common temperature; so that whether these philosophers applied cold only, as Fourcroy mentions, or cold and pressure, as stated by the other chemists, they would succeed in obtaining it in the liquid form.

Chlorine.—M. de Morveau, whilst engaged on the application of the means best adapted to destroy putrid effluvia and contagious miasmata, was led to the introduction of chlorine as the one most excellent for this purpose; and he proposed the use of phials, containing the requisite materials, as sources of the substance. One described in his 'Traité des Moyens de désinfecter l'air' (1801), was of the capacity of two cubical inches nearly; about 62 grains of black oxide of manganese in

* Philosophical Transactions, 1823, p. 191—or page 90.

K

coarse powder was introduced, and then the bottle two-thirds filled with nitromuriatic acid; it was shaken, and in a short time chlorine was abundantly disengaged. M. Morveau remarks upon the facility with which the chlorine is retained in these bottles; one, thus prepared, and forgotten, when opened at the end of eight years, gave an abundant odour of chlorine.

I had an impression on my mind that M. de Morveau had proposed the use of phials similarly charged, but made strong, well stoppered, and confined by a screw in a frame, so that no gas should escape, except when the screw and stopper were loosened; but I have searched for an account of such phials without being able to find any. If such have been made, it is very probable that in some circumstances, liquid chlorine has existed in them, for as its vapour at 60° F. has only a force of about four atmospheres*, a charge of materials might be expected frequently to yield much more chlorine than enough to fill the space, and saturate the fluid present; and the excess would of course take the liquid form. If such vessels have not been made, our present knowledge of the strength of the vapour of chlorine will enable us to construct them of a much more convenient and portable form than has yet been given to them.

Arseniuretted Hydrogen.—This is a gas which it is said has been condensed so long since as 1805. The experiment was made by Stromeyer, and was communicated, with many other results relating to the same gas, to the Göttingen Society, Oct. 12, 1805. See Nicholson's 'Journal,' xix. 382; also Thenard, 'Traité de Chimie,' i. 373; Brande's 'Manual,' ii. 212; and 'Annales de Chimie,' lxiv. 303. None of these contain the original experiment; but the following quotation is from Nicholson's 'Journal.' The gas was obtained over the pneumatic apparatus, by digesting an alloy of fifteen parts tin and one part arsenic, in strong muriatic acid. "Though the arsenicated hydrogen gas retains its aëriform state under every known degree of atmospheric temperature and pressure, Professor Stromeyer condensed it so far as to reduce it in part to a liquid, by immersing it in a mixture of snow and muriate of lime, in which several pounds of quicksilver had been frozen in the course of a few minutes." From the circumstance of its being reduced only in part to a liquid, we may be led to suspect that

* Philosophical Transactions, 1823, p. 198—or page 95.

it was rather the moisture of the gas that was condensed than the gas itself; a conjecture which is strengthened in my mind from finding that a pressure of three atmospheres was insufficient to liquefy the gas at a temperature of $0°$ F.

Chlorine.—The most remarkable and direct experiments I have yet met with in the course of my search after such as were connected with the condensation of gases into liquids, are a series made by Mr. Northmore, in the years 1805–6. It was expected by this gentleman " that the various affinities which take place among the gases under the common pressure of the atmosphere, would undergo considerable alteration by the influence of condensation;" and it was with this in view that the experiments were made and described. The results of liquefaction were therefore incidental, but at present it is only of them I wish to take notice. Mr. Northmore's papers may be found in Nicholson's ' Journal,' xii. 368, xiii. 232. In the first is described his apparatus, namely, a brass condensing pump; pear-shaped glass receivers, containing from three and a half to five cubic inches, and a quarter of an inch thick; and occasionally a siphon gauge. Sometimes as many as eighteen atmospheres were supposed to have been compressed into the vessel, but it is added, that the quantity cannot be depended on, as the tendency to escape even by the side of the piston, rendered its confinement very difficult.

Now that we know the pressure of the vapour of chlorine, there can be no doubt that the following passage describes a true liquefaction of that gas :—" Upon the compression of nearly two pints of oxygenated muriatic acid gas in a receiver, two and a quarter cubic inches capacity, it speedily became converted into a yellow *fluid*, of such extreme volatility, under the common pressure of the atmosphere, that it instantly evaporated upon opening the screw of the receiver; I need not add that this fluid, so highly concentrated, is of a most insupportable pungency. . . . There was a trifling residue of a yellowish substance left after the evaporation, which probably arose from a small portion of the oil and grease used in the machine," &c., xiii. 235.

Muriatic Acid.—Operating upon muriatic acid, Mr. Northmore obtained such results as induced him to state he could liquefy it in any quantity; but as the pressure of its vapour at

K 2

50° Fahr. is equal to about forty atmospheres *, he must have
been mistaken.   The following is his account:—"I now proceeded to the muriatic acid gas, and upon the condensation of
a small quantity of it, a beautiful green-coloured substance
adhered to the side of the receiver, which had all the qualities
of muriatic acid ; but upon a large quantity, four pints, being
condensed, the result was a yellowish-green glutinous substance, which does not evaporate, but is instantly absorbed by
a few drops of water; it is of a highly pungent quality, being
the essence of muriatic acid.   As this gas easily becomes fluid,
there is little or no elasticity, so that any quantity may be condensed without danger.   My method of collecting this and
other gases, which are absorbable by water, is by means of an
exhausted florence flask (and in some cases an empty bladder),
connected by a stopcock with the extremity of the retort."
xiii. 235.   It seems probable that the facility of condensation,
and even combination, possessed by muriatic acid gas in contact with oil of turpentine, may belong to it under a little
pressure, in contact with common oil, and thus have occasioned
the results Mr. Northmore describes.

Sulphurous Acid Gas.—With regard to this gas, Mr. Northmore says, "Having collected about a pint and a half of sulphurous acid gas, I proceeded to condense it in the three cubic
inch receiver; but after a very few pumps the forcing piston
became immoveable, being completely choked by the operation
of the gas.   A sufficient quantity had, however, been compressed to form vapour, and a thick slimy fluid, of a dark
yellow colour, began to trickle down the sides of the receiver,
which immediately evaporated with the most suffocating odour
upon the removal of the pressure." xiii. 236.   This experiment, Mr. Northmore remarks, corroborates the assertion of
Monge and Clouet, that by cold and pressure they had condensed this gas.   The fluid above described was evidently
contaminated with oil, but from its evaporation on removing
the pressure, and from the now ascertained low pressure of the
vapour of sulphurous acid, there can be no hesitation in admitting that it was sulphurous acid liquefied.

The results obtained by Mr. Northmore, with chlorine gas
and sulphurous acid gas, are referred to by Nicholson in his

* Philosophical Transactions, 1823, p. 198—or page 95.

'Chemical Dictionary,' 8vo, articles Gas (muriatic acid oxygenized) and Gas (sulphurous acid); and that of chlorine is referred to by Murray, in his 'System,' ii. 550; although at page 405 of the same volume, he says that only sulphurous acid "and ammonia of these gases that are at natural temperatures permanently elastic, have been found capable of this reduction."

Carbonic Acid.—Another experiment, in which it is very probable that liquid carbonic acid has been produced, is one made by Mr. Babbage about the year 1813. The object Mr. Babbage had in view, was to ascertain whether pressure would prevent decomposition; and it was expected that either that would be the case, or that decomposition would go on, and the rock be split by the expansive force of carbonic acid gas. The place was Chudley Rocks, Devonshire, where the limestone is dark and of a compact texture. A hole, about 30 inches deep and 2 inches in diameter, was made by the workmen in the usual way; it penetrated directly downwards into the rock; a quantity of strong muriatic acid, equal to perhaps a pint and a half, was then poured in, and immediately a conical wooden plug, that had previously been soaked in tallow, was driven hard into the mouth of the hole. The persons about then retired to a distance to watch the result, but nothing apparent happened, and, after waiting some time, they left the place. The plug was not loosened at the time, nor was any further examination of the state of things made; but it is very probable that if the rock were sufficiently compact in that part, the plug tight, and the muriatic acid in sufficient quantity, that a part of the carbonic acid had condensed into a liquid, and thus, though it permitted the decomposition, prevented that development of power which Mr. Babbage expected would have torn the rock asunder.

Oil-Gas Vapour.—An attempt has been made by Mr. Gordon, within the last few years, and is still continued, to introduce condensed gas into use in the construction of portable, elegant, and economical gas-lamps. Oil-gas has been made use of, and, I believe, as many as thirty atmospheres have been thrown into vessels, which, furnished with a stopcock and jet, have afterwards allowed of its gradual expansion and combustion. During the condensation of the gas in this manner, a liquid has been observed to deposit from it. It is not, however, a result of the liquefaction of the gas, but the deposition of a vapour (using the terms gas and vapour in their common acceptation) from it, and when taken out of the vessel it remains a liquid at common temperatures and pressures, may be purified by distillation in the ordinary way, and will even bear a temperature of 170° F. before it boils at ordinary pressure. It is the substance referred to by Dr. Henry in the 'Philosophical Transactions,' 1821, p. 159.

There is no reason for believing that oil-gas, or olefiant gas, has as yet been condensed into a liquid, or that it will take that form at common temperatures under a pressure of five, or ten, or even twenty atmospheres. If it were possible, a small, safe, and portable gas-lamp would immediately offer itself to us, which might be filled with liquid without being subject to any greater force than the strength of its vapour, and would afford an abundant supply of gas as long as any of the liquid remained. Immediately upon the condensation of cyanogen, which takes place at 50° Fahr. at a pressure under four atmospheres, I made such a lamp with it. It succeeded perfectly; but, of course, either the expense of the gas, the faint light of its flame, or its poisonous qualities, would preclude its application. But we may, perhaps, without being considered extravagant, be allowed to search in the products of oil, resins, coal, &c., distilled or otherwise treated, with this object in view, for a substance, which being a gas at common temperatures and pressure, shall condense into a liquid, by a pressure of from two to six or eight atmospheres, and which being combustible, shall afford a lamp of the kind described *.

Atmospheric Air.—As my object is to draw attention to the results obtained in the liquefaction of gases before the date of those described in the 'Philosophical Transactions' for 1823, I need not, perhaps, refer to the notice given in the 'Annals of Philosophy,' N. S. vi. 66, of the supposed liquefaction of atmospheric air, by Mr. Perkins, under a pressure of about 1100 atmospheres; but as such a result would be highly interesting, and is the only additional one on the subject I am acquainted with, I am desirous of doing so, as well also to point

* In reference to the probability of such results, see a paper " On Olefiant Gas," Annals of Philosophy, N. S. iii. 37.

out the remarkable difference between that result and those which are the subject of this and the other papers referred to. Mr. Perkins informed me that the air upon compression disappeared, and in its place was a small quantity of a fluid, which remained so when the pressure was removed, had little or no taste, and did not act on the skin. As far as I could by inquiry make out its nature, it resembled water; but if upon repetition it be found really to be the product of compressed common air, then its fixed nature shows it to be a result of a very different kind to those mentioned above, and necessarily attended by far more important consequences.

On the History of the Condensation of the Gases, in reply to Dr. Davy, *introduced by some Remarks on that of Electro-magnetic Rotation*.*

My dear Sir,   Royal Institution, May 10, 1836.

I have just concluded looking over Dr. Davy's Life of his bróther Sir Humphry Davy. In it, between pages 160 and 164 of the second volume, the author links together some account, with observations, of the discovery of electro-magnetic rotation, and that of the condensation of the gases, concluding at page 164 with these words: "I am surprised that Mr. Faraday has not come forward to do him [Sir Humphry Davy] justice. As I view the matter, it appears hardly less necessary to his own honest fame than his acknowledgement to Dr. Wollaston, on the subject of the first idea of the rotatory magnetic motion."

I regret that Dr. Davy by saying this has made that necessary which I did not before think so; but I feel that I cannot after his observation indulge my earnest desire to be silent on the matter without incurring the risk of being charged with something opposed to an *honest* character. This I dare not risk; but in answering for myself, I trust it will be understood that I have been driven unwillingly into utterance.

[The next three pages of this paper which refer to the electro-magnetic rotation have appeared in vol. ii. at p. 229, &c. of 'Experimental Researches in Electricity,' 8vo edition.

* Philosophical Magazine, 1836, vol. viii. p. 521.

At that time I did not reprint the part relating to the conden-sation of the gases. I find that this omission may be construed into a design to withdraw the statement from publication; I have no reason, however, to alter or weaken a word of it, and so am in a manner constrained to insert it here, where indeed it finds its proper place.]

With respect to the condensation of the gases, I have long ago done justice to those to whom it was really due [*], and now approach the subject again with considerable reluctance; for though I feel that there is some appearance of confusion, still I regret that Dr. Davy did not leave the matter as it stood. All my papers on the subject in the Transactions of the Royal Society had passed through the hands of Sir Humphry Davy, who had corrected them as he thought fit, and had presented them to that body. Again, all the facts that Dr. Paris has stated upon his own knowledge [†] are correct; he made that statement as his own voluntary act and without any previous communication with me, so that I think I might have been left in that silence which I so much desired.

The facts of the case, as far as I know them, are these:—In the spring of 1823, Mr. Brande was Professor of Chemistry, Sir Humphry Davy Honorary Professor of Chemistry, and I Chemical Assistant, in the Royal Institution. Having to give personal attendance on both the morning and afternoon che-mical lectures, my time was very fully occupied. Whenever any circumstance relieved me in part from the duties of my situation, I used to select a subject of research, and try my skill upon it. Chlorine was with me a favourite object, and having before succeeded in discovering new compounds of that element with carbon, I had considered that body more deeply, and resolved to resume its consideration at the first opportunity: accordingly, the absence of Sir Humphry Davy from town having relieved me from a part of the laboratory duty, I took advantage of the leisure and the cold weather and worked upon frozen chlorine, obtaining the results which are published in my paper in the 'Quarterly Journal of Science' for the 1st of April, 1823 [‡]. On Sir Humphry Davy's return to town, which I think must have been about the end of

[*] See page 124.   [†] Paris's Life of Davy, pp. 390, 391, 392.
[‡] Vol. xv. p. 71—or page 82.

February or the beginning of March, he inquired what I had been doing, and I communicated the results to him as far as I had proceeded, and said I intended to publish them in the ' Quarterly Journal of Science.' It was then that he suggested to me the heating of the crystals in a closed tube, and I proceeded to make the experiment which Dr. Paris witnessed, and has from his own knowledge described *. I did not at that time know what to anticipate, for Sir Humphry Davy *had not told me his expectations,* and I had not reasoned so deeply as he appears to have done. Perhaps he left me unacquainted with them to try my ability. How I should have proceeded with the chlorine crystals without the suggestion I cannot now say, but with the hint of heating the crystals in a close tube ended for the time Sir Humphry Davy's instructions to me, and I puzzled out for myself in the manner Dr. Paris describes, that the oil I had obtained was condensed chlorine. This is all very evident from the paper read to the Royal Society, though it may seem at first to stand opposed to the notes and papers that Sir Humphry Davy communicated in conjunction with and after mine. When my paper was written, it was, according to a custom consequent upon our relative positions, submitted to Sir Humphry Davy (as were all my papers for the ' Philosophical Transactions ' up to a much later period), and he altered it as he thought fit. This practice was one of great kindness to me, for various grammatical mistakes and awkward expressions were from time to time thus removed which might else have remained.

The passage at the commencement of the paper which I shall now quote was of Sir Humphry Davy's writing, and in fact contains everything that, and perhaps rather more than, he had said to me : " The President of the Royal Society having honoured me by looking at these conclusions, and suggested that an exposure of the substance to heat under pressure would probably lead to interesting results, the following experiments were commenced at his request †." I say " rather more,' because I believe *pressure* was not referred to in our previous verbal communication. However, I proceeded to make the

* Paris's Life, p. 391.

† Phil. Trans. 1823, p. 160, or Phil. Mag., First Series, vol. lxii. p. 413—or page 85.

experiment, and was making it when **Dr. Paris** came into the laboratory as he has described, and my thoughts at that moment are embodied and expressed in my paper in the following words: " I at first thought that muriatic acid and euchlorine had been formed; then that two new hydrates of chlorine had been produced; but at last I suspected that the chlorine had been entirely separated from the water by the heat, and condensed into a dry fluid by the mere pressure of its own abundant vapour *." I then describe an experiment entirely of my own, in which I proceed to verify this conjecture, and go on to say, " Presuming that I had now a right to consider the yellow fluid as pure chlorine in the liquid state, I proceeded to examine its properties, &c. &c. †"

To this paper Sir Humphry Davy added a note ‡, in which he says, " In desiring **Mr.** Faraday to expose the hydrate of chlorine to heat in a closed glass tube §, it *occurred to me* that one of three things would happen: that it would become fluid as a hydrate; or that a decomposition of water would occur, and euchlorine and muriatic acid be formed; or that the chlorine would separate in a condensed state." And then he makes the subject his own by condensing muriatic acid, and states that he had "requested" me (of course as Chemical Assistant) " to pursue these experiments, and to extend them to all the gases which are of considerable density, or to any extent soluble in water;" &c. This I did; and when he favoured me by requesting that I would write a paper on the results, I began it by stating " that Sir Humphry Davy did me the honour to request I would continue the experiments, which I have done under his general direction, and the following are some of the results already obtained‖:" and this paper being immediately followed by one on the application of these liquids as mechanical agents, by Sir Humphry Davy¶, he says in it, " One of the principal objects that I had in view in *causing experiments to be made* on the condensation of different gaseous bodies, by generating them under pressure, &c."

* Phil. Trans. 1823, p. 162.   † Ibid. p. 163.   ‡ Ibid. p. 164.
§ Observe, not "to heat under pressure." See my remarks in the preceding page.
‖ Phil. Trans. 1823, p. 189, or Phil. Mag., First Series, vol. lxii. p. 417—or page 89.   ¶ Phil. Trans. 1823, p. 199.

I certainly took up the subject of chlorine with the view of pursuing it as I could find spare time, and at the moments which remained to me after attending to the directions of my superiors. It however passed in the manner described into the hands of Sir Humphry Davy, and a comparison of the dates will readily show that I at least had no time of my own to pursue it. My original paper was published on the 1st of April, 1823, that being the first Number of the ' Quarterly Journal ' which *could* appear after the experiments had been made : but in the short time between the first experiment and the publication, much that I have referred to had occurred; for not only had I communicated my results to Sir Humphry Davy, and received from him the hint, but my paper on fluid chlorine had been read (13th of March), and his note also, of the same date, attached to it ; and the Editor of the ' Quarterly Journal,' Mr. Brande, had time, prior to the printing of my original paper, to attach a note to it stating the condensation of chlorine and muriatic acid, and expressing an expectation that several other gases would be liquefied by the same means *. On the 10th of April my paper on the condensation of several gases into liquids was read, on the 17th of April Sir Humphry Davy's on the application of condensed gases as mechanical agents, and on the 1st of May his Appendix to it on the changes of volume produced by heat.

I have never remarked upon or denied Sir Humphry Davy's right to his share of the condensation of chlorine or the other gases ; on the contrary, I think that I long ago did him full "justice" in the papers themselves. How could it be otherwise ? He saw and revised the manuscripts ; through his hands they went to the Royal Society, of which he was President at the time ; and he saw and revised the printer's proofs. Although he did not tell me of his expectations when he suggested the heating the crystals in a closed tube, yet I have no doubt that he had them† ; and though, perhaps, I regretted

---

* Quarterly Journal, xv. p. 74—or page 84.

† I perceive in a letter to Professor Edmund Davy, published by Dr. Davy in the ' Life,' vol. ii. p. 166, of the date of *September* 1, 1823, that Sir Humphry Davy said, " The experiments on the condensation of the gases were made under my direction, and *I had anticipated*, theoretically, all the results." It is evident that he considered the subject his own ; but I am glad that here, as

losing my subject, I was too much indebted to him for much previous kindness to think of saying that that was mine which he said was his. But *observe* (for my sake) that Sir Humphry Davy nowhere states that he told me what he expected, or contradicts the passages in the first paper of mine which describe my course of thought, and in which I claim the development of the actual results.

All this activity in the condensing of gases was simultaneous with the electro-magnetic affair already referred to, and I had learned to be cautious upon points of right and priority. When therefore I discovered in the course of the same year that *neither I nor Sir Humphry Davy* had the merit of first condensing the gases, and especially chlorine, I hastened to perform what I thought right, and had great pleasure in spontaneously doing justice and honour to those who deserved it*. I therefore published on the 1st of January of the following year (1824), a historical statement respecting the liquefaction of gases †, the beginning of which is as follows:—" *I was not aware* at the time when *I first observed* the liquefaction of chlorine gas, nor until very lately, that any of the class of bodies called gases had been reduced into the fluid form; but having during the last few weeks sought for instances where such results might have been afforded without the knowledge of the experimenter, I was surprised to find *several recorded cases*. I have thought it right, therefore, to bring these cases together, and *only justice* to endeavour to secure for them a more general attention than they appear as yet to have gained." Amongst other cases the liquefaction of *chlorine* is clearly described ‡. The value of this statement of mine has since been fully proved; for upon Mr. Northmore's complaint ten years after, with some degree of reason, that great injustice had been done to him in the affair of the condensation of gases, and his censure of " the conduct of Sir H. Davy, Mr. Faraday, and several other philosophers for withholding the name of the first discoverer,"

elsewhere, he never says that he had informed me of his expectations. In this, Sir Humphry Davy's negative, and Dr. Paris's positive testimony perfectly agree.

* Monge and Clouet had condensed sulphurous acid probably before the year 1800. Northmore condensed chlorine in the years 1805 and 1806.

† See page 124.   ‡ See page 131.

I was able by referring to the statement to convince him and his friend that if my papers had done him wrong, *I at least* had endeavoured also to do him right*.

Believing that I have now said enough to preserve my own "honest fame" from any injury it might have risked from the mistakes of Dr. Davy, I willingly bring this letter to a close, and trust that I shall never again have to address you on the subject.

I am, my dear Sir, yours, &c.,

M. FARADAY.

Richard Phillips, Esq., &c. &c.

Change of Musket Balls in Shrapnell Shells †.

MR. MARSH of Woolwich gave me some musket balls which had been taken out of Shrapnell shells. The shells had lain in the bottom of ships, and probably had sea-water amongst them. When the bullets are put in, the aperture is merely closed by a common cork. These bullets were variously acted upon: some were affected only superficially, others more deeply, and some were entirely changed. The substance produced is hard and brittle, it splits on the ball, and presents an appearance like some hard varieties of earthy hæmatite; its colour is brown, becoming, when heated, red; it fuses on platinum foil into a yellow flaky substance like litharge. Powdered and boiled in water, no muriatic acid or lead was found in solution. It dissolved in nitric acid without leaving any residuum, and the solution gave very faint indications only of muriatic acid. It is a protoxide of lead, perhaps formed in some way by the galvanic action of the iron shell and the leaden ball, assisted, probably, by the sea-water. It would be very interesting to know the state of the shells in which a change like this has taken place to any extent; it might have been expected, that as long as any iron remained, the lead would have been preserved in the metallic state.

* Phil. Mag. 1834, iv. p. 261.

† Quarterly Journal of Science, xvi. 163.

Action of Gunpowder on Lead*.

Mr. Marsh gave me also some balls from cartridges about fifteen years old, and which had probably been in a damp magazine. They were covered with white warty excrescences rising much above the surface of the bullet, and which, when removed, were found to have stood in small pits formed beneath them. These excrescences consist of carbonate of lead, and readily dissolve with effervescence in weak nitric acid, leaving the bullet in the corroded state which their formation has produced. It is evident there must have been a mutual action amongst the elements of the gunpowder itself, at the same time that it acted on the lead; and it would have been interesting, had the opportunity occurred, to have examined what changes the powder had suffered.

Purple Tint of Plate-glass affected by Light†.

It is well known that certain pieces of plate-glass acquire, by degrees, a purple tinge, and ultimately become of a comparatively deep colour. The change is known to be gradual, but yet so rapid as easily to be observed in the course of two or three years. Much of the plate-glass which was put a few years back into some of the houses in Bridge Street, Blackfriars, though at first colourless, has now acquired a violet or purple colour. Wishing to ascertain whether the sun's rays had any influence in producing this change, the following experiment was made :— Three pieces of glass were selected, which were judged capable of exhibiting this change; one of them was of a slight violet tint, the other two purple or pinkish, but the tint scarcely perceptible, except by looking at the edges. They were each broken into two pieces; three of the pieces were then wrapped up in paper and set aside in a dark place, and the corresponding pieces were exposed to air and sunshine. This was done in January last, and the middle of this month (September) they were examined. The pieces that were put away from light seemed to have undergone no change; those that were exposed to the sunbeams had increased in colour considerably;

* Quarterly Journal of Science, xvi. 163.   † Ibid. 164.

the two paler ones the most, and that to such a degree, that it would hardly have been supposed they had once formed part of the same pieces of glass as those which had been set aside. Thus it appears that the sun's rays can exert chemical powers even on such a compact body and permanent compound as glass.

On some Cases of the Formation of Ammonia, and on the Means of Testing the Presence of Minute Portions of Nitrogen in certain states*.

THE importance of the question relative to the simple or compound nature of any of the substances considered as elementary in the present state of chemical science, is such as to make any experimental information respecting it acceptable, however imperfect it may be. An opinion of this kind has induced me to draw up the following account of experiments relative to the formation of ammonia, by the action of substances apparently including no nitrogen. The experiments are not offered as satisfactory, even to myself, of the production of ammonia without nitrogen; indeed, I am inclined to believe the results all depend upon the difficulty of excluding that element perfectly, and the extreme delicacy of the test of its presence afforded by the formation of ammonia: yet as, on the contrary, notwithstanding my utmost exertions, I have failed to convince myself that ammonia could not be formed, except nitrogen were present, it has been supposed that the information obtained, though incomplete, might be interesting.

Having occasion, some time since, to examine an organic substance with reference to any nitrogen it might contain, I was struck with the difference in the results obtained, when heated alone in a tube, or when heated with hydrate of potassa: in the former case no ammonia was produced; in the latter, abundance. Supposing that the potash acted, by inducing the combination of the nitrogen in the substance with hydrogen, more readily than when no potash was present, and would therefore be useful as a delicate test of the presence of nitrogen in bodies, I was induced to examine its accuracy by heating it with substances containing no nitrogen, as lignine, sugar, &c.; and was

* Quarterly Journal of Science, xix. 16.

surprised to find that ammonia was still a result of the experiment. This led to trials with different vegetable substances, such as the proximate principles, acids, salts, &c., all of which yielded ammonia in greater or smaller quantity; and ultimately it was found, that even several metals when treated in the same way gave similar results; a circumstance which appeared considerably to simplify the experiment.

The experiment may be made in its simplest form in the following manner :—Put a small piece of clean zinc foil into a glass tube closed at one end, and about one-fourth of an inch in diameter; drop a piece of potash into the tube over the zinc; introduce a slip of turmeric paper slightly moistened at the extremity with pure water, retaining it in the tube in such a position that the wetted portion may be about 2 inches from the potash; then holding the tube in an inclined position, apply the flame of a spirit-lamp, so as to melt the potash that it may run down upon the zinc, and heat the two whilst in contact, taking care not to cause such ebullition as to drive up the potash; in a second or two the turmeric paper will be reddened at the moistened extremity, provided that part of the tube has not been heated. On removing the turmeric paper and laying the reddened portion upon the hot part of the tube, the original yellow tint will be restored: from which it may be concluded that ammonia has been formed; a result confirmed by other modes of examination to be hereafter mentioned.

The first source of nitrogen which suggested itself was the atmosphere: the experiment was therefore repeated, very carefully, in hydrogen gas, but the same results were obtained.

The next opinion entertained was, that the potash might have been touched accidentally by animal or other substances, which had adhered to it in sufficient quantity to produce the ammonia: the alkali was therefore heated red-hot, as a preparatory step, and afterwards allowed to touch nothing but clean glass or metals; but still the same effects were produced. The zinc used was selected from a compact piece of foil, was well rubbed with tow dipped in alkali, washed in alkaline solution, afterwards boiled repeatedly in distilled water, and dried, not by wiping, but in a hot atmosphere; and yet the same products were obtained.

All these precautions, with regard to impurity from fingering, were found to be essentially requisite, in consequence of the delicacy of the means afforded by heat and turmeric paper for testing the presence of ammonia, or rather, of matter containing its elements. As a proof of this, it may be mentioned, that some sea sand was heated red-hot for half an hour in a crucible, and then poured out on to a copper-plate, and left to cool; when cold, a portion of it (about 12 grains) was put into a clean glass tube; another equal portion was put into the palm of the hand, and looked at for a few moments, being moved about by a finger, and then introduced by platina foil into another tube, care being taken to transfer no animal substance but what had adhered to the grains of sand: the first tube when heated yielded no signs of ammonia to turmeric paper, the second a very decided portion.

As a precaution, with regard to adhering dirt, the tubes used in precise experiments were not cleaned with a cloth or tow, but were made from new tube, the tube being previously heated red-hot, and air then drawn through it; and no zinc or potash was used in these experiments, except such as had been previously tried by having portions heated in a tube to ascertain whether when alone they gave ammonia.

It was then thought probable that the alkali might contain a minute quantity of some nitrous compound, or of a cyanide, introduced during its preparation. A carbonate of potash was therefore prepared from pure tartar, rendered caustic by lime calcined immediately preceding its use, the caustic solution separated by decantation from the carbonate of lime, not allowed to touch a filter or anything else animal or vegetable, and boiled down in clean flasks; but the potash thus obtained, though it yielded no appearance of ammonia when heated alone, always gave it when heated with zinc.

The water used in these experiments was distilled, and in cases where it was thought necessary was distilled a second, and even a third time. The experiments of Sir Humphry Davy[*] show how tenaciously small portions of nitrogen are held by water, and that in certain circumstances the nitrogen may produce ammonia. I am not satisfied that I have been able to avoid this source of error.

At last, to avoid every possible source of impurity in the pot

* Phil. Trans. 1807, p. 11.

ash, a portion of that alkali was prepared from potassium; and as the experiment made with it includes all the precautions taken to exclude nitrogen, I will describe it rather minutely, as illustrative of the way in which the other numerous experiments were made. A piece of new glass tube, about half an inch in diameter, was first wiped clean, and then heated red-hot, a current of air passing at the same time through it; about six inches in length was drawn off at the blowpipe lamp, and sealed at one extremity. Some distilled water was put into a new glass retort, and heated by a lamp; when about one-half had distilled over, the beak of the retort was introduced into the tube before-mentioned, and a small portion of water (about fifty grains) condensed into it. A solid compact piece of potassium was then chosen, and having been wiped with a linen cloth, was laid on a clean glass plate, the exterior to a considerable depth removed by a sharp lancet, and portions taken from the interior by metallic forceps, and dropped successively into the tube containing the water before-mentioned. Of course the water was decomposed, and the tube filled with hydrogen; and when a sufficient quantity of solution of potash had been thus formed, the tube was heated in a lamp, and drawn out to a capillary opening, about two inches from the closed extremity (fig. 2, plate I). The tube now formed almost a close vessel; and being heated, as the water became vapour, it passed off at the minute aperture, and ultimately a portion of pure fused hydrate of potassa remained in the bottom of the tube. The aperture of the tube was now closed, and the whole set aside to cool.

A piece of new glass tube was selected about 0·3 of an inch in diameter; it was heated to dull redness, and air passed through it: about 10 inches of it was then cut off, and being softened near to one end by heat, it was drawn out at that part until of small diameter (*a*, fig. 3, plate I): that part was then fixed into a cap, by which it could afterwards be attached to a receiver containing hydrogen. The tube containing the potassium potash being now broken in an agate mortar, a piece or two of the potash were introduced by metallic forceps into the tube at the open end, so as to pass on to the contracted part; a roll of zinc-foil, about one grain in weight, cleaned with all the precautions already described, was afterwards introduced, and then more of the potash. The tube was then bent near the middle to a right angle; a slip of turmeric paper introduced, so as just to pass the bend, and thus prepared, it was ready to be filled with hydrogen.

The precautions taken with regard to the purity of the hydrogen, were as follows :—A quantity of water had been put into a close copper boiler, and boiled for some hours, after which it had been left all night in the boiler to cool. A pneumatic trough was filled with this water just before it was required for use. The hydrogen was prepared from clean zinc, which being put into a gas bottle, the latter was filled entirely with the boiled water, and then sulphuric acid being poured in through the water, the gas was collected, the excess of liquid being allowed to boil over. The hydrogen was received in the usual manner into jars filled with the water of the trough, the transferring jar, when filled, being entirely immersed in the water, so as to exclude the air from every part, even of the stopcock. The first jar of gas was thrown away, and only the latter portions used.

The gas being ready, the experimental tube was attached to the transferring jar by a connecting piece, so that the part of it containing the zinc and potash was horizontal, whilst the other portion descended directly downwards. A cup of clean mercury, the metal being about an inch in depth, was then held under the open end of the tube, and by lowering the jar containing the hydrogen in the water of the pneumatic trough, so as to give sufficient pressure, and opening the stopcock, the hydrogen in the jar was made to pass through the tube, and sweep all the common air before it. When from 100 to 150 cubic inches, or from 200 to 300 times the contents of the tube had passed through, the cup of mercury was raised as high as it could be, so as to prevent the passage of any more gas, the pressure from the jar in the water-trough was partly removed, and the stopcock closed; then, by lowering the cup of mercury a little, the surface of the metal in it was made lower than that within the tube, and in this state of things the flame of a spirit-lamp applied to the contracted part of the tube (*a*, fig. 3, plate I), sealed it hermetically, without the introduction of any air, and separated the apparatus from the jar on the water-trough.

In this way every precaution was taken that I could devise for the exclusion of nitrogen; yet, when a lamp was applied to the potash and zinc, the alkali no sooner melted down and mingled with the metal, than ammonia was developed, which rendered the turmeric paper brown, the original yellow re-appearing by the application of heat to the part.

Still anxious to obtain a potash which should be unexceptionably free from any source of nitrogen, I heated a portion of potash with zinc, endeavouring to exhaust anything it might contain which could give rise to the formation of ammonia: it was then dissolved in pure water, allowed to settle, the clear portion poured off and evaporated in a flask by boiling; but the potash thus prepared gave ammonia, when heated with zinc, in hydrogen gas.

With regard to the evidence of the nature of the substance produced, it was concluded to be ammonia in the experiments made in hydrogen, from its changing the colour of turmeric paper to reddish-brown; from the disappearance of the reddish-brown tint and reproduction of yellow colour by heat; from its solubility in water, as evinced by the greater depth of colour on moist turmeric paper than on dry; from its odour; and from its yielding white fumes with the vapour of muriatic acid. When formed in open tubes, its nature was still further tested by its neutralizing acids and restoring the blue colour of reddened litmus paper; by its rendering a minute drop of sulphate of copper on a slip of white paper deep blue; and also, at the suggestion of Dr. Paris, by introducing into it a slip of paper moistened in a mixed solution of nitrate of silver and arsenious acid, the yellow tint of arsenite of silver being immediately produced.

These experiments upon the production of ammonia from substances apparently containing no nitrogen, will call to mind that made by Mr. Woodhouse, of Philadelphia, on the action of water on a calcined mixture of charcoal and potash, during which much ammonia was produced [*]; and also the strict investigation of that experiment made by the President of the Royal Society during his inquiries into the nature of elementary bodies [†]. Sir Humphry Davy found that when one part of potash and four of charcoal were ignited in close vessels cooled out of contact of the atmosphere, pure water admitted to the mixture, and the whole distilled, small quantities of ammonia were produced. That when the operation was repeated

Nicholson's Journal, xxi. 290.   † Phil. Trans. 1809, p. 100; 1810, p. 43.

upon the same mixture ignited a second time, the proportion diminished; in a third operation it was sensible; in a fourth barely perceptible. The same mixture, however, by the addition of a new quantity of potash, again gained the power of producing ammonia in two or three successive operations; and when any mixture had ceased to give ammonia, the power was not restored by cooling it in contact with air.

Sir Humphry Davy refrains from drawing conclusions from these processes, observing with regard to the composition of nitrogen in these experiments, that till the weight of the substances concerned and produced in these operations are compared, no correct decision on the question can be made: I am anxious to be understood as imitating the caution of one whose judgment stands so high in chemical science; and therefore draw no positive conclusion from the experiment I have described, or from the results I have yet to mention. As, however, I think they may lead to elucidations of the question, I shall venture to give them, not with the minute detail of the preceding experiment, but in a more general manner.

Potash is not the only substance which produces this effect with the metals and vegetable substances. Soda produces it; so also does lime and baryta, the latter not being so effective as the former, or producing the phænomena so generally. The common metallic oxides, as those of manganese, copper, tin, lead, &c., do not act in this manner.

Water or its elements appear to be necessary to the experiment. Potash or soda in the state of hydrates generally contain the water necessary. Potash, dried as much as could be by heat, produced little or no ammonia with zinc; but re-dissolved in pure water and evaporated, more water being left in it than before, it was found to produce it as usual. Pure caustic lime, with very dry linen, produced scarcely a trace of ammonia, whilst the same portion of linen with hydrate of lime yielded it readily.

The metals when mixed with the potash appear to act by, or according to their power of absorbing oxygen. Potassium, iron, zinc, tin, lead, and arsenic evolve much ammonia, whilst spongy platina, silver, gold, &c., produce no effect of the kind. A small portion of fine clean iron wire dropped into potash melted at the bottom of a tube, caused the evolution of some ammonia, but it soon ceased, and the wire blackened upon its surface; the introduction of a second portion of clean wire caused a second evolution of ammonia. Clean copper wire, in fused potash, caused a very slight evolution of ammonia, and became tarnished.

The following, among other vegetable substances supposed to contain no nitrogen, have been tried with potash in tubes open to the air:—Lignine, prepared by boiling linen in weak solution of potash, then in water, afterwards in weak acid, and finally in water again; oxalate of potassa, oxalate of lime, tartrate of lead, acetate of lime, asphaltum, gave very striking quantities to turmeric and litmus paper: acetate of potash, acetate of lead, tartrate of potash, benzoate of potash, oxalate of lead, sugar, wax, olive oil, naphthaline, produced ammonia, but in smaller quantity: resin appeared to yield none, nor when potash was heated in the vapour of alcohol or ether, or in olefiant gas, could any ammonia be detected.

It may be remarked, that much appeared to depend upon the quantity of potash used; sugar, for instance, which with a little potash would with difficulty yield traces of ammonia, does so very readily when the quantity of potash is doubled or trebled; and linen, which with potash gives ammonia very readily, yields it the more readily, and in greater quantity, as the proportion of potash is increased.

The experiments with the substances which contain carbon, assimilate, in consequence of the presence of that body, with the one by Mr. Woodhouse. Whether the substances act exactly as charcoal does, probably cannot be decided until the correct nature of the action is ascertained; but there are apparently some very evident differences. The ammonia in the charcoal experiment does not exist until after the ignition, nor before the addition of water; but in several experiments of the nature of those described in this paper, the ammonia is evolved before the substances acting or acted upon are charred. Thus, if linen fibre, cut small, be mixed in the tube with hydrate of lime, and heated, ammonia is evolved before the heat has risen so high as to render the linen more than slightly brown; and oxalate of potash, in a tube with potash, when heated, gives much ammonia before any blackening is produced.

Mr. Woodhouse's experiment may be very readily repeated,

though not in an exact way, by heating a little tartrate of lead with potash in a tube in the flame of a spirit-lamp, driving off the water and first products, and raising the residue to dull redness. If a drop of water be allowed to flow down on to the residue when cold, and it be then heated, ammonia will be found to rise with the water.

I was induced in the course of these experiments to try again and again, whether the potash or lime would not yield ammonia when heated alone; but when well prepared, and the tubes experimented in perfectly clean, they gave no indications of it. By exposure to air for three days in a room, hydrate of lime appeared to have acquired the power of evolving a little ammonia when heated, and caustic lime so exposed gave still stronger traces of it. Potash also exhibited an effect of this kind, and potash which had been heated with zinc, and contained oxide of zinc, most decidedly. Some potash and zinc were heated together; a part was immediately put into a clean close bottle; another part was dissolved in pure water, decanted, the solution evaporated in a covered Wedgewood's basin, and then also set aside in a close vessel for twenty-four hours: at the end of that time the first portion, heated in a tube, gave no decided trace of ammonia, but the latter yielded very distinct evidence of its presence, having apparently absorbed the substance which was its source from the atmosphere during the operations it had been submitted to. White Cornish clay being heated red-hot, and then exposed to the air for a week, gave plenty of ammonia when heated in a tube. When the substances were preserved in well-stoppered phials, these effects were not produced.

Such are the general and some of the particular facts which I have observed relative to this anomalous production of ammonia. I have refrained from all reasoning upon the probability of the compound nature of nitrogen ; or upon what might be imagined to be its elements, not seeing sufficient to justify more than private opinion on that matter. I have endeavoured to make the principal experiments as unexceptionable as possible, by excluding every source of nitrogen, but I must confess I have not convinced myself I have succeeded. The results seem to me of such a nature as to deserve attention, and if it should hereafter be proved that nitrogen had entered in some unperceived way into the experiments, they will still show the extreme delicacy of heat, or heat and potash, as a test of its presence by the formation of ammonia.

With respect to the delicacy of the test, it may be observed that it offers many facilities to the detection of nitrogen when in certain states of combination, which chemists probably were not before aware of. A portion of asbestos, which had been heated red-hot, was introduced into a tube by metallic forceps and heated; it gave no ammonia; another similar portion, compressed together, and introduced by the fingers, gave ammonia when heated. A very minute particle of nitre was dropped into hydrate of potassa, and heated to dull redness; it gave no ammonia; a small piece of zinc-foil, dropped in and the heat applied, caused an abundant evolution of that substance.

The circumstance also of absorption by lime and other bodies, of something from inhabited atmospheres, which yields ammonia when thus tested, is very interesting; and Dr. Paris has suggested to me that this power may probably be applicable to the examination of the atmosphere of infected and inhabited places, and may perhaps furnish the means of investigating such atmospheres upon correct principles.

February 17, 1825.

On the Substitution of Tubes for Bottles, in the preservation of certain Fluids, such as Chloride of Sulphur, Protochlorides of Phosphorus and Carbon, &c.[*]

THERE are many fluids in the laboratory which are much more conveniently retained in tubes, such as that depicted, fig. 5, plate I, than in bottles, and from which they may be taken in a less wasteful manner when required for the purpose of experiment. A piece of glass tube, a quarter of an inch or more in diameter, being selected, is to be closed at one end by the blowpipe; and then, being softened near the other end, is to be drawn out obliquely, so as to form the long narrow neck represented in the figure, but to which, in the first case, the short piece of tube is to be left attached; this forms a funnel, into which the preparation to be preserved is to be put. Then, warming the body of the tube, the expanding air passes

[*] Quarterly Journal of Science, xix. 149.

out through the fluid; and afterwards, on cooling the vessel, the liquid descends into it. A small spirit-lamp flame being now applied at the upper part of the long neck, softens the glass, which is then to be drawn out to a fine point and sealed. In this state the substance may be preserved clean and pure for any length of time.

If a small portion be required for an experiment, the extreme point of the neck is to be opened by pinching it off, the tube is then to be inclined until the quantity required has entered the neck, where, by capillary attraction, it will form a small column, and the tube being warmed by the hand, the atmosphere within it will expand and expel the portion of fluid on to the place required. A very little practice will enable the experimenter to judge of the quantity he is forcing out, and in this way he may take a portion not larger than the 1-20th of a common drop, or he may take the whole contents of the tube. When the quantity required has been taken out, the tube is to be placed in an upright position, and the flame of a lamp or candle, or even a piece of paper, closes the aperture in a moment and as perfectly as before.

I have found these tubes very serviceable when working with substances either very small in quantity or obtained with great difficulty, in consequence of the entire prevention of waste resulting from their use. They are easily labeled by scratching the name of the substance with a diamond on them, and may conveniently be retained by putting several of them together into a tumbler, or other glass of that kind.

Composition of Crystals of Sulphate of Soda*.

It is known that when a hot strong solution of sulphate of soda is put into a vessel and closed up, it may be reduced to common temperatures without crystallizing, although, if the vessel be opened, abundance of crystals will immediately form. It has also frequently been observed, that in some circumstances crystals would form in the solution during cooling, even though the vessel had not been opened or agitated. These crystals, when observed in the solution, are very transparent and of a large size; they are quadrangular prisms, with dihedral summits.

* Quarterly Journal of Science, xix. 153.

Upon opening the vessel, the surrounding solution crystallizes rapidly, enveloping the first-formed set of crystals with others, which, however, are very readily distinguished from them in consequence of their immediately assuming a white opake appearance. Upon taking out the crystals, those first formed are found to be much harder than the usual crystals of sulphate of soda, and, when broken, it is found that the opacity is not merely superficial, but that it penetrates them to a considerable depth, and even at times throughout.

These harder and peculiar crystals are readily obtained by closing up a solution of sulphate of soda, saturated at 180°, in a Florence flask, boiling the solution in the flask so as to expel the air before closing it. Upon standing twenty-four hours, fine groups of crystals are formed. When the flask is opened, the solution deposits fresh crystals; but on breaking the flask, the latter may be scraped off by a knife in consequence of the superior hardness of the first set.

The hard crystals when separated are found to be efflorescent, like those of the usual kind, and they ultimately give off all their water, leaving only dry sulphate of soda. When a given weight was heated in a platina crucible, one half their weight passed off as water, the rest being dry salt; they consequently contain eight proportionals of water, or 72 sulphate of soda, and $8 \times 9 = 72$ water. The usual crystals of sulphate of soda contain 10 proportionals of water.

When crystallized sulphate of soda is heated in a flask, a part of it dissolves in the water present, whilst the rest is thrown down in an anhydrous state. The solution at 180° appears to contain one proportional of salt 72, and 18 proportionals of water 162; from which, if correct, it would result, that when the crystals are heated to 180° $\frac{5}{9}$ of the salt take all the water, whilst $\frac{4}{9}$ separate in the dry state.

On new Compounds of Carbon and Hydrogen, and on certain other Products obtained during the Decomposition of Oil by Heat*.
[Read June 16, 1825.]

THE object of the paper which I have the honour of submitting at this time to the attention of the Royal Society, is to describe

* Phil. Trans. 1825. p. 440; and Phil. Mag. lxvi. p. 180.

particularly two new compounds of carbon and hydrogen, and generally, other products obtained during the decomposition of oil by heat. My attention was first called to the substances formed in oil at moderate and at high temperatures, in the year 1820; and since then 1 have endeavoured to lay hold of every opportunity for obtaining information on the subject. A particularly favourable one has been afforded me lately through the kindness of Mr. Gordon, who has furnished me with considerable quantities of a fluid obtained during the compression of oil-gas, of which I had some years since possessed small portions, sufficient to excite great interest, but not to satisfy it.

It is now generally known, that in the operations of the Portable Gas Company, when the oil-gas used is compressed in the vessels, a fluid is deposited, which may be drawn off and preserved in the liquid state. The pressure applied amounts to 30 atmospheres; and in the operation, the gas previously contained in a gasometer over water, first passes into a large strong receiver, and from it, by pipes, into the portable vessels. It is in the receiver that the condensation principally takes place; and it is from that vessel that the liquid I have worked with has been taken. The fluid is drawn off at the bottom by opening a conical valve: at first a portion of water generally comes out, and then the liquid. It effervesces as it issues forth; and by the difference of refractive power it may be seen that a dense transparent vapour is descending through the air from the aperture. The effervescence immediately ceases; and the liquid may be readily retained in ordinary stoppered, or even corked bottles, a thin phial being sufficiently strong to confine it. I understand that 1000 cubical feet of good gas yield nearly one gallon of the fluid.

The substance appears as a thin light fluid; sometimes transparent and colourless, at others opalescent, being yellow or brown by transmitted, and green by reflected light. It has the odour of oil-gas. When the bottle containing it is opened, evaporation takes place from the surface of the liquid; and it may be seen by the striæ in the air that vapour is passing off from it. Sometimes in such circumstances it will boil, if the bottle and its contents have had their temperature raised a few degrees. After a short time this abundant evolution of vapour ceases, and the remaining portion is comparatively fixed.

The specific gravity of this substance is 0·821. It does not solidify at a temperature of 0° F. It is insoluble, or nearly so, in water; very soluble in alcohol, ether, and volatile and fixed oils. It is neutral to test colours. It is not more soluble in alkaline solutions than in water; and only a small portion is acted upon by them. Muriatic acid has no action upon it. Nitric acid gradually acts upon it, producing nitrous acid, nitric oxide gas, carbonic, and sometimes hydrocyanic acid, &c., but the action is not violent. Sulphuric acid acts upon it in a very remarkable and peculiar manner, which I shall have occasion to refer to more particularly presently.

This fluid is a mixture of various bodies; which, though they resemble each other in being highly combustible, and throwing off much smoke when burnt in large flame, may yet by their difference of volatility be separated in part from each other. Some of it drawn from the condenser, after the pressure had been repeatedly raised to 30 atmospheres, and at a time when it was at 28 atmospheres, then introduced rapidly into a stoppered bottle and closed up, was, when brought home, put into a flask and distilled, its temperature being raised by the hand. The vapour which came off, and which caused the appearance of boiling, was passed through a glass tube at 0°, and then conducted to the mercurial trough; but little uncondensed vapour came over, not more than thrice the bulk of the liquid; a portion of fluid collected in the cold tube, which boiled and evaporated when the temperature was allowed to rise; and the great bulk of the liquid which remained might now be raised to a comparatively high point, before it entered into ebullition.

A thermometer being introduced into another portion of the fluid, heat was applied, so as to keep the temperature just at the boiling-point. When the vessel containing it was opened, it began to boil at 60° F. As the more volatile portions were dissipated, the temperature rose: before a tenth part had been thrown off, the temperature was above 100°. The heat continued gradually to rise, and before the substance was all volatilized it had attained 250°.

With the hope of separating some distinct substances from this evident mixture, a quantity of it was distilled, and the vapours condensed at a temperature of 0° into separate portions, the receiver being changed with each rise of 10° in the retort, and the liquid retained in a state of incipient ebullition. In this way a succession of products were obtained, but they were by no means constant; for the portions, for instance, which came over when the fluid was boiling from 160° to 170°, when re-distilled began to boil at 130°, and a part remained which did not rise under 200°. By repeatedly rectifying all these portions, and adding similar products together, I was able to diminish these differences of temperature, and at last bring them more nearly to resemble a series of substances of different volatility. During these operations I had occasion to remark, that the boiling-point was more constant at or between 176° and 190°, than at any other temperature, large quantities of fluid distilling over without any change in the degree, whilst in other parts of the series it was constantly rising. This induced me to search in the products obtained between these points for some definite substance; and I ultimately succeeded in separating a new compound of carbon and hydrogen, which I may by anticipation distinguish as bicarburet of hydrogen.

Bicarburet of Hydrogen.—This substance was obtained in the first instance in the following manner:—Tubes containing portions of the above rectified products were introduced into a freezing mixture at 0°; many of them became turbid, probably from the presence of water; one received at 176° (by which is meant that that was the boiling-point of the contents of the retort when it came over) became partly solid, crystals forming round the side, and a fluid remaining in the centre; whilst two other portions, one received at 186° and the other at 190°, became quite hard. A cold glass rod being introduced into one of these tubes, the mass within was found to resist considerable pressure; but by breaking it down, a solid part was thrust to the bottom of the tube, whilst a fluid remained above: the fluid was poured off, and in this way the solid portion partly purified. The contents of the tube were then allowed to fuse, were introduced into a larger and stronger tube, furnished with another which entered loosely within it, both being closed of course at the lower end; then again lowering the temperature of the whole to 0°, bibulous paper was introduced, and pressed on to the surface of the solid substance in the large tube by the end of the smaller one. In this way much fluid was removed by successive portions of paper, and a solid substance remained, which did not become fluid until raised to 28° or 29°. To complete the separation of the permanently fluid part, the substance was allowed to melt, then cast into a cake in a tin-foil mould, and pressed between many folds of bibulous paper in a Bramah's press, care having been taken to cool the paper, tin-foil, flannel, boards, and other things used, as near to 0° as possible, to prevent solution of the solid substance in the fluid part to be removed. It was ultimately distilled from off caustic lime, to separate any water it might contain.

The general process, which appears to me to be the best for the preparation of this substance only, is to distil a portion of the fluid deposited during the condensation of oil-gas, to set aside the product obtained before the temperature rises to 170°, to collect that which comes over by 180°, again separately that which comes over by 190°, and also the portion up to 200° or 210°. That before 170° will upon re-distillation yield portions to be added to those of 180° and 190°; and the part obtained from 190° upwards will also, when re-distilled, yield quantities boiling over at 180°, 190°, &c. Having then these three portions obtained at 180°, 190°, and 200°, let them be rectified one after the other, and the products between 175° and 195° received in three or four parts at successive temperatures. Then proceed with these as before described.

It will sometimes happen, when the proportion of bicarburet of hydrogen is small in the liquid, that the rectifications must be many times repeated before the fluids at 185° and 190° will deposit crystals on cooling; that is to say, before sufficient of the permanently fluid part at low temperatures has been removed, to leave a solution so saturated as to crystallize at 0°.

Bicarburet of hydrogen appears in common circumstances as a colourless transparent liquid, having an odour resembling that of oil-gas, and partaking also of that of almonds. Its specific gravity is nearly 0·85 at 60°. When cooled to about 32° it crystallizes, becoming solid; and the portions which are on the sides of the glass exhibit dendritical forms. By leaving tubes containing thin solid films of it in ice-cold water, and allowing the temperature to rise slowly, its fusing-point was found to be very nearly 42° F.; but when liquid, it may, like water and some saline solutions, be cooled much below that point before any part becomes solid. It contracts very much on congealing, 9 parts in bulk becoming 8 very nearly; hence its specific gravity in that state is about 0·956. At 0° it appears as a white or transparent substance, brittle, pulverulent, and of the hardness nearly of loaf-sugar.

It evaporates entirely when exposed to the air. Its boiling-point in contact with glass is 186°. The specific gravity of its vapour, corrected to a temperature of 60°, is nearly 40, hydrogen being 1; for 2·3 grains became 3·52 cubic inches of vapour at 212°. Barometer 29·98. Other experiments gave a mean approaching very closely to this result.

It does not conduct electricity.

This substance is very slightly soluble in water; very soluble in fixed and volatile oils, in ether, alcohol, &c.; the alcoholic solution being precipitated by water. It burns with a bright flame and much smoke. When admitted to oxygen gas, so much vapour rises as to make a powerful detonating mixture. When passed through a red-hot tube, it gradually deposits carbon, yielding carburetted hydrogen gas.

Chlorine introduced to the substance in a retort exerted but little action until placed in sunlight, when dense fumes were formed, without the evolution of much heat; and ultimately much muriatic acid was produced, and two other substances, one a solid crystalline body, the other a dense thick fluid. It was found by further examination, that neither of these were soluble in water; that both were soluble in alcohol—the liquid readily, the solid with more difficulty. Both of them appeared to be triple compounds of chlorine, carbon, and hydrogen;— but I reserve the consideration of these, and of other similar compounds, to another opportunity.

Iodine appears to exert no action upon the substance in several days in sunlight; it dissolves in the liquid in small quantity, forming a crimson solution.

Potassium heated in the liquid did not lose its brilliancy, or exert any action upon it, at a temperature of 186°.

Solution of alkalies, or their carbonates, had no action upon it.

Nitric acid acted slowly upon the substance and became red, the fluid remaining colourless. When cooled to 32°, the substance became solid and of a fine red colour, which disappeared upon fusion. The odour of the substance with the acid was exceedingly like that of almonds, and it is probable that hydrocyanic acid was formed. When washed with water, it appeared to have undergone little or no change.

Sulphuric acid added to it over mercury exerted a moderate action upon it, little or no heat was evolved, no blackening took place, no sulphurous acid was formed; but the acid became of a light yellow colour, and a portion of a clear colourless fluid floated, which appeared to be a product of the action. When separated, it was found to be bright and clear, not affected by water or more sulphuric acid, solidifying at about 34°, and being then white, crystalline, and dendritical. The substance was lighter than water, soluble in alcohol, the solution being precipitated by a small quantity of water, but becoming clear by great excess*.

With regard to the composition of this substance, my experiments tend to prove it a binary compound of carbon and hydrogen, two proportionals of the former element being united to one of the latter. The absence of oxygen is proved by the inaction of potassium, and the results obtained when passed through a red-hot tube.

The following is a result obtained when it was passed in

---

* The action of sulphuric acid on this and the other compounds to be described is very remarkable. It is frequently accompanied with heat; and large quantities of those bodies which have elasticity enough to exist as vapours when alone at common pressures, are absorbed. No sulphurous acid is produced; nor when the acid is diluted does any separation of the gas, vapour, or substance take place, except of a small portion of a peculiar product resulting from the action of the acid on the substances, and dissolved by it. The acid combines directly with carbon and hydrogen; and I find when united with bases forms a peculiar class of salts, somewhat resembling the sulphovinates, but still different from them. I find also that sulphuric acid will condense and combine with olefiant gas, no carbon being separated, or sulphurous or carbonic acid being formed; and this absorption has in the course of eighteen days amounted to 84·7 volumes of olefiant gas to one volume of sulphuric acid. The acid produced combines with bases, &c., forming peculiar salts, which I have not yet had time, but which it is my intention to examine, as well as the products formed by the action of sulphuric acid on naphtha, essential oils, &c., and even upon starch and lignine, in the production of sugar, gum, &c., where no carbonization takes place, but where similar results seem to occur.

vapour over heated oxide of copper:—0·776 grain of the substance produced 5·6 cubic inches of carbonic acid gas, at a temperature of 60°, and pressure 29·98 inches; and 0·58 grain of water was collected. The 5·6 cubic inches of gas are equivalent to 0·711704 grain of carbon by calculation, and the 0·58 grain of water to 0·064444 of hydrogen.

Carbon . . . . . 0·711704 or 11·44
Hydrogen . . . . . 0·064444 or 1·

These quantities nearly equal in weight the weight of the substance used; and making the hydrogen 1, the carbon is not far removed from 12, or two proportionals.

Four other experiments gave results all approximating to the above. The mean result was 1 hydrogen, 11·576 carbon.

Now considering that the substance must, according to the manner in which it was prepared, still retain a portion of the body boiling at 186°, but remaining fluid at 0°, and which substance I find, as will be seen hereafter, to contain less carbon than the crystalline compound (only about 8·25 to 1 of hydrogen), it may be admitted, I think, that the constant though small deficit of carbon found in the experiments is due to the portion so retained; and that the crystalline compound would, if pure, yield 12 of carbon for each 1 of hydrogen, or two proportionals of the former element and one of the latter.

2 proportionals Carbon   12 ⎫
1   ,,   Hydrogen   1 ⎬13 bicarburet of hydrogen.
                             ⎭

This result is confirmed by such data as I have been able to obtain by detonating the vapour of the substance with oxygen. Thus in one experiment 8092 mercury grain measures of oxygen at 62° had such quantity of the substance introduced into it as would entirely rise in vapour; the volume increased to 8505: hence the vapour amounted to 413 parts, or $\frac{1}{20.6}$ of the mixture nearly. Seven volumes of this mixture were detonated in a eudiometer tube by an electric spark, and diminished in consequence nearly to 6·1: these, acted upon by potash, were further diminished to 4, which were pure oxygen. Hence 3 volumes of mixture had been detonated, of which nearly 0·34 was vapour of the substance, and 2·65 oxygen. The carbonic acid amounted to 2·1 volumes, and must have consumed an equal bulk of oxygen gas; so that 0·55 remain

M

as the quantity of oxygen which has combined with the hydrogen to form water, and which with the 0·34 of vapour nearly make the diminution of 0·9.

It will be seen at once that the oxygen required for the carbon is four times that for the hydrogen; and that the whole statement is but little different from the following theoretical one, deduced partly from the former experiments:—1 volume of vapour requires 7·5 volumes of oxygen for its combustion; 6 of the latter combine with carbon to form 6 of carbonic acid, and the 1·5 remaining combine with hydrogen to form water. The hydrogen present therefore in this compound is equivalent to 3 volumes, though condensed into one volume in union with the carbon; and of the latter elements there are present six proportionals, or 36 by weight. A volume therefore of the substance in vapour contains—

$$\begin{aligned} \text{Carbon} \quad & . \quad . \quad . \quad . \quad 6 \times 6 = 36 \\ \text{Hydrogen} \quad & . \quad . \quad . \quad 1 \times 3 = 3 \\ \hline & 39 \end{aligned}$$

and its weight or specific gravity will be 39, hydrogen being 1. Other experiments of the same kind gave results according with these.

Among the liquid products obtained from the original fluid was one which, procured as before mentioned, by submitting to 0° the portion distilling over at 180° or 190°, corresponded with the substance already described, as to boiling-points, but differed from it in remaining fluid at low temperatures; and I was desirous of comparing the two together. I had no means of separating this body from the bicarburet of hydrogen, of which it would of course be a saturated solution at 0°. Its boiling-point was very constantly 186°. In its general characters of solubility, combustibility, action of potassium, &c., it agreed with the substance already described. Its specific gravity was 0·86 at 60°. When raised in vapour, 1·11 grain of it gave 1·573 cubic inch of vapour at 212°, equal to 1·212 cubic inch at 60°. Hence 100 cubic inches would weigh about 91·6 grains, and its specific gravity would be 43·25 nearly. In another experiment, 1·72 grain gave 2·4 cubic inches at 212°, equal to 1·849 cubic inch at 60°; from which the weight of 100 cubic inches would be deduced as 93 grains; and its specific gravity to hydrogen as 44 to 1. Hence probably the reason why, experimentally, the specific gravity of bicarburet of hydrogen in vapour was found higher than by theory it would appear to be when pure.

Sulphuric acid acted much more powerfully upon this substance than upon the bicarburet; great heat was evolved, much discoloration occasioned, and a separation took place into a thick black acid, and a yellow lighter liquid, resisting any further action at common temperatures.

0·64 grain of this substance was passed over heated oxide of copper: 4·51 cubic inches of carbonic acid gas were obtained, and 0·6 grain of water. The carbonic acid and water are equivalent to—

Carbon	. . . . .	0·573176, or 8·764
Hydrogen	. . . .	0·066666, or 1·

but as the substance must have contained much bicarburet of hydrogen, it is evident that, if in a pure state, the carbon would fall far short of the above quantity, and the compound would approximate of course to a simple carburet of hydrogen containing single proportionals.

New Carburet of Hydrogen.—Of the various other products from the condensed liquor, the next most definite to the bicarburet of hydrogen appears to be that which is most volatile. If a portion of the original liquid be warmed by the hand or otherwise, and the vapour which passes off be passed through a tube at 0°, very little uncondensed vapour will go on to the mercurial trough; but there will be found after a time a portion of fluid in the tube, distinguished by the following properties. Though a liquid at 0°, it upon slight elevation of temperature begins to boil, and before it has attained 32° is all resolved into vapour or gas, which may be received and preserved over mercury.

This gas is very combustible, and burns with a brilliant flame. The specific gravity of the portion I obtained was between 27 and 28, hydrogen being 1; for 39 cubic inches introduced into an exhausted glass globe were found to increase its weight 22·4 grains at 60° F., bar. 29·94. Hence 100 cubic inches weigh nearly 57·44 grains.

M 2

When cooled to 0° it condensed again; and enclosed in this state in a tube of known capacity, and hermetically sealed up, the bulk of a given weight of the substance at common temperatures was ascertained. This compared with water gave the specific gravity of the liquid as 0·627 at 54°. It is therefore among solids or liquids the lightest body known.

This gas or vapour when agitated with water is absorbed in small quantities. Alcohol dissolves it in large quantity; and a solution is obtained, which, upon the addition of water, effervesces, and a considerable quantity of the gas is liberated. The alcoholic solution has a peculiar taste, and is neutral to test papers.

Olive oil dissolves about six volumes of the gas.

Solution of alkali does not affect it; nor does muriatic acid.

Sulphuric acid condenses the gas in very large quantity, one volume of the acid condensing above 100 volumes of the vapour. Sometimes the condensation is perfect; at other times a small quantity of residual gas is left, which burns with a pale blue flame, and seems to be a product of too rapid action. Great heat is produced during the action; no sulphurous acid is formed; the acid is much blackened, has a peculiar odour, and upon dilution generally becomes turbid, but no gas is evolved. A permanent compound of the acid with carbon and hydrogen is produced, and enters as before mentioned into combination with bases.

A mixture of two volumes of this vapour with fourteen volumes of pure oxygen was made, and a portion detonated in a eudiometer tube. 8·8 volumes of the mixture diminished by the spark to 5·7 volumes, and these by solution of potash to 1·4 volume, which were oxygen. Hence 7·4 volumes had been consumed, consisting of—

Vapour of substance	1·1
Oxygen	6·3
Carbonic acid formed	4·3
Oxygen in carbonic acid	4·3
Oxygen combining with hydrogen	2·0
Diminution by spark	3·1

This is nearly as if one volume of the vapour or gas had

required six volumes of oxygen, had consumed four of them in producing four of carbonic acid gas, and had occupied the other two by four of hydrogen to form water. Upon which view, four volumes or proportionals of hydrogen $=4$, are combined with four proportionals of carbon $=24$, to form one volume of the vapour, the specific gravity of which would therefore be 28. Now this is but little removed from the actual specific gravity obtained by the preceding experiments; and knowing that this vapour must contain small portions of other substances in solution, there appears no reason to doubt that, if obtained pure, it would be found thus constituted.

As the proportions of the elements in this vapour appear to be the same as in olefiant gas, it became interesting to ascertain whether chlorine had the same action upon it as on the latter body. Chlorine and the vapour were therefore mixed in an exhausted retort: rapid combination took place, much heat was evolved, and a liquor produced resembling hydrochloride of carbon, or the substance obtained by the same process from olefiant gas. It was transparent, colourless, and heavier than water. It had the same sweet taste, but accompanied by an after aromatic bitterness, very persistent. Further, it was composed of nearly equal volumes of the vapour and chlorine: it could not therefore be the same as the hydrochloride of carbon from olefiant gas, since it contained twice as much carbon and hydrogen. It was therefore treated with excess of chlorine in sunlight: action slowly took place, more chlorine combined with the substance, muriatic acid was formed, and ultimately a fluid tenacious triple compound of chlorine, carbon, and hydrogen was obtained; but no chloride of carbon. This is a remarkable circumstance, and assists in showing that though the elements are the same, and in the same proportions as in olefiant gas, they are in a very different state of combination.

The tension of the most volatile part of the condensed oil-gas liquid, and indeed of the substance next beneath olefiant gas in elasticity existing in the mixture constituting oil-gas, appears to be equal to about four atmospheres at the temperature of 60°. To ascertain this a tube was prepared, like the

one * delineated in the sketch, fig. 1, containing a mercurial
gauge at *a*, *c*, and the extremities being open. It was then
cooled to 0° from *a* to *b*, and in that state made the receiver
into which the first product from a portion of the original fluid
was distilled. The part at *b* was then closed by a spirit-lamp;

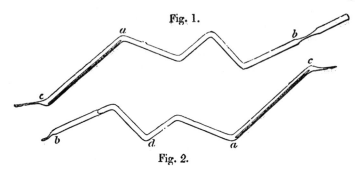

Fig. 1.

Fig. 2.

and having raised enough vapour to make it issue at *c*, that
was also closed. The apparatus now placed as at fig. 2, had
a and *d* cooled to 0°, whilst the fluid collected in *b* was warmed
by the hand or the air; and when a portion had collected in *d*
sufficient for the purpose, the whole instrument was immersed
in water at 60°; and before the vapour had returned and been
all dissolved by the liquid at *b*, the pressure upon the gauge
within was noted. Sometimes the fluid at *d* was rectified by
warming that part of the tube and cooling *a* only, the re-
absorption at *b* being prevented or rather retarded in consequence of the superior levity of the fluid at *d*; so that the
first portions which returned to *b* lay upon it in a stratum, and
prevented sudden solution in the mass below. This difference
in specific gravity was easily seen upon agitation, in consequence of the striæ produced during the mixture.

Proceeding in this way, it was found, as before stated, that
the highest elastic power that could be obtained from the substances in the tube was about four atmospheres at 60°; and as
there seems no reason to doubt but that portions of the most
volatile substances in oil-gas beneath olefiant gas were con-

* The particular inclination of the parts of the tube one to another was
given, that the fluid, when required, might be returned from *a* to *d* without
passing on to *b*.

tained in the fluid, inasmuch as even olefiant gas itself is dissolved by it in small proportions, it may be presumed that there is no substance in oil-gas much more volatile than the one requiring a pressure of four atmospheres at 60°, except the well-known compounds; or, in other words, that there is not a series of substances passing upwards from this body to olefiant gas, and possessing every intermediate degree of elasticity, as there seems to be from this body downwards, to compounds requiring 250° or 300° for their ebullition.

In reference to these more volatile products, I may state that I have frequently observed a substance come over in small quantity, rising with the vapour which boils off at 50° or 60°, and crystallizing in spiculæ in the receiver at 0°. A temperature of 8° or 10° causes its fusion and disappearance. It is doubtless a peculiar and definite body, but the quantity is extremely small, or else it is very soluble in the accompanying fluids. I have not yet been able to separate it, or examine it minutely.

I ventured some time since upon the condensation of various gases *, to suggest the possibility of forming a vapour lamp, which containing a brilliantly combustible substance (liquid at a pressure of two, three, or four atmospheres at common temperatures, but a vapour at less pressure), should furnish a constant light for a length of time, without requiring high or involving inconstant pressure. Such a lamp I have now formed, feeding it with the substance just described; and though at present it is only a matter of curiosity, and perhaps may continue so, yet there is a possibility that processes may be devised, by which the substance may be formed in larger quantity, and render an application of this kind practically useful.

On the remaining portions of the condensed Oil-gas Liquor.—It has been before mentioned, that by repeated distillations various products were obtained, boiling within limits of temperature which did not vary much, and which when distilled were not resolved into other portions, differing far from each other in volatility, as always happened in the earlier distillations. Though conscious that these were mixtures, perhaps of unknown bodies, and certainly in unknown proportions, yet experiments were made on their composition by passing

* Quarterly Journal of Science, xvi. 240, and page 134.

them over oxide of copper, in hopes of results which might assist in suggesting correct views of their nature. They all appeared to be binary compounds of carbon and hydrogen, and the following Table exhibits the proportions obtained; the first column expressing the boiling temperature at which the products were distilled, as before mentioned; the second the hydrogen, made a constant quantity; and the third the carbon.

140°	1	7·58
150°	1	8·38
160°	1	7·90
176°	1	8·25
190°	1	8·76
200°	1	9·17
210°	1	8·91
220°	1	8·46

These substances generally possess the properties before described, as belonging to the bicarburet of hydrogen. They all resist the action of alkali, even that which requires a temperature above 250° for its ebullition; and in that point are strongly distinguished from the oils from which they are produced. Sulphuric acid acts upon them instantly with phænomena already briefly referred to.

———————

Dr. Henry, whilst detailing the results of his numerous and exact experiments in papers laid before the Royal Society, mentions in that read February 22, 1821 *, the discovery made by Mr. Dalton, of a vapour in oil-gas of greater specific gravity than olefiant gas, requiring much more oxygen for its combustion, but yet condensable by chlorine. Mr. Dalton appears to consider all that was condensable by chlorine as a new and constant compound of carbon and hydrogen; but Dr. Henry, who had observed that the proportion of oxygen required for its combustion varied from 4·5 to 5 volumes, and the quantity of carbonic acid produced, from 2·5 to 3 volumes, was inclined to consider it as a mixture of the vapour of a highly volatile oil with the olefiant and other combustible gases: and he further mentions, that naphtha in contact with

* Phil. Trans. for 1821, p. 136.

hydrogen gas will send up such a vapour; and that he has been informed, that when oil-gas was condensed in Gordon's lamp, it deposited a portion of highly volatile oil.

A writer in the 'Annals of Philosophy,' N. S. iii. 37, has deduced from Dr. Henry's experiments, that the substance, the existence of which was pointed out by Mr. Dalton, was not a new gas *sui generis,* " but a modification of olefiant gas, constituted of the same elements as that fluid, and in the same proportions; with this only difference, that the compound atoms are triple instead of double:" and Dr Thomson has adopted this opinion in his ' Principles of Chemistry.' This, I believe, is the first time that two gaseous compounds have been supposed to exist, differing from each other in nothing but density; and though the proportion of 3 to 2 is not confirmed, yet the more important part of the statement is, by the existence of the compound described at page 163; which, though composed of carbon and hydrogen in the same proportion as in olefiant gas, is of double the density*.

It is evident that the vapour observed by Mr. Dalton and

* In reference to the existence of bodies composed of the same elements and in the same proportions, but differing in their qualities, it may be observed, that now we are taught to look for them, they will probably multiply upon us. I had occasion formerly to describe a compound of olefiant gas and iodine (Phil. Trans. cxi. 72), which upon analysis yielded one proportional of iodine, two proportionals of carbon, and two of hydrogen (Quarterly Journal, xiii. 429). M. Serrulas, by the action of potassium upon an alcoholic solution of iodine, obtained a compound decidedly different from the preceding in its properties; yet when analysed, it yielded the same elements in the same proportions (Ann. de Chimie, xx. 245 ; xxii. 172).

Again: MM. Liebig and Gay-Lussac, after an elaborate and beautiful investigation of the nature of fulminating compounds of silver, mercury, &c., were led to the conclusion that they were salts, containing a new acid, and owed their explosive powers to the facility with which the elements of this acid separated from each other (Annales de Chimie, xxiv. 294 ; xxv. 285). The acid itself, being composed of one proportional of oxygen, one of nitrogen, and two of carbon, is equivalent to a proportional of oxygen + a proportional of cyanogen, and is therefore considered as a true cyanic acid. But M. Wöhler, by deflagrating together a mixture of ferroprussiate of potash and nitre, has formed a salt, which, according to his analysis, is a true cyanate of potash. The acid consists of one proportional of oxygen, one of nitrogen, and two of carbon. It may be transferred to various other bases; as the earths, the oxides of lead, silver, &c.; but the salts formed have nothing in common with

Dr. Henry must have contained not only this compound, and a portion of the bicarburet of hydrogen, but also portions of the other (as yet apparently indefinite) substances; and there can be no doubt that the quantity of these vapours will vary from the point of full saturation of the gas, when standing over water and oil, to unknown, but much smaller proportions. It is therefore an object in the analysis of oil and coal-gas, to possess means by which their presence and quantity may be ascertained; and this I find may be done with considerable exactness by the use of sulphuric acid, oil, &c., in consequence of their solvent power over them.

Sulphuric acid is in this respect a very excellent agent. It acts upon all these substances instantly, evolving no sulphurous acid; and though when the quantity of substance is considerable as compared with the acid, a body is left undecomposed by or uncombined with the acid, and volatile, so as constantly to afford a certain portion of vapour, yet when the original substance is in small quantity, as where it exists in vapour in a given volume of gas, this does not interfere, in consequence of the solubility of the vapour of the new compound produced by the action of the acid in the acid itself in small quantities: and I found that when 1 volume of the vapour of any of the products of the oil-gas liquor was acted upon, either alone, or mixed with 1, 2, 3, 4, up to 12 volumes of air, oxygen, or hydrogen, by from half a volume to a volume of sulphuric acid, it was entirely absorbed and removed.

When olefiant gas is present, additional care is required in analytical experiments, in consequence of the gradual combination of the olefiant gas with the sulphuric acid. I found that 1 volume of sulphuric acid in abundance of olefiant gas, absorbed about 7 volumes in twenty-four hours in the dull light of a room; sunshine seemed to increase the action a little. When the olefiant gas was diluted with air or hydrogen, the quantity absorbed in a given time was much diminished; and in those cases it was hardly appreciable in two hours,

the similar salts of MM. Liebig and Gay-Lussac, except their composition (Gilbert's Annalen, lxxviii. 157; Ann. de Chimie, xxvii. 190). M. Gay-Lussac observes, that if the analysis be correct, the difference can only be accounted for by admitting a different mode of combination.

a length of time which appears to be quite sufficient for the removal of any of the peculiar vapours from oil or coal-gas.

My mode of operating was generally in glass tubes over clean mercury*, introducing the gas, vapour or mixture, and then throwing up the sulphuric acid by means of a bent tube with a bulb blown in it, passing the acid through the mercury by the force of the mouth. The following results are given as illustrations of the process :—

Oil-gas from a Gasometer.

	Sul. Acid.	in 8'.	in 1 hour.	2 hours.	Dimin. per cent.
188 vol. +	9·5 vol. diminished to	155·0	148·5	146·4	22·12
107 +	13·0	88·5	84·5	82·0	23·33
138 +	5·2	113·7	108·0	106·5	22·82

Oil-gas from Gordon's Lamp.

	Sul. Acid.	15'	30'	3 hours.	Dimin. per. cent.
214 vol. +	6·8 vol. diminished to	183·3	180·8	176·0	17·75
159 +	5·9	137·5	136·0	130·4	17·98
113 +	12·2	98·0	96·0	92·0	18·58

Coal-gas of poor quality.

548·6	+ 27·6	533·3	529·2	529·0	3·57
273·6	+ 27·8	267·9	266·0	266·0	2·78
190·6	+ 13·1	186·0	184·2	184·1	3·41

Oil may also be used in a similar manner for the separation of these vapours. It condenses about 6 volumes of the most elastic vapour at common temperatures, and it dissolves with greater facility the vapour of those liquids requiring higher temperatures for their ebullition. I found that in mixtures made with air or oxygen for detonation, I could readily separate the vapour by means of olive oil; and when olefiant and other gases were present, its solvent power over them was prevented by first agitating the oil with olefiant gas or with a portion of the gas to saturate it, and then using it for the removal of the vapours.

---

* If the mercury contain oxidizable metals, the sulphuric acid acts upon it, and evolves sulphurous acid gas. It may be cleaned sufficiently by being left in contact with sulphuric acid for twenty-four hours, agitating it frequently at intervals.

In the same way some of the more fixed essential oils may be used, as *dry* oil of turpentine; and even a portion of the condensed liquor itself, as that part which requires a temperature of 220° or 230° for its ebullition; care being taken to estimate the expansion of the gas by the vapour of the liquid, which may readily be done by a known portion of common *air preserved over* the liquid as a standard.

With reference to the proportions of the different substances in the liquid as obtained by condensation of oil-gas, it is extremely difficult to obtain anything like precise results, in consequence of the immense number of rectifications required to separate the more volatile from the less volatile portions; but the following Table will furnish an approximation. It contains the loss of 100 parts by weight of the original fluid by evaporation in a flask, for every 10° in elevation of temperature, the substance being retained in a state of ebullition.

100 parts at 58°	parts.	differences.
had lost at 70 . . .	1·1	
		1·9
80 . . .	3·0	
		2·2
90 . . .	5·2	
		2·5
100 . . .	7·7	
		2·4
110 . . .	10·1	
		3·1
120 . . .	13·2	
		2·9
130 . . .	16·1	
		3·2
140 . . .	19·3	
		3·1
150 . . .	22·4	
		3·2
160 . . ·	25·6	
		3·4
170 . . .	29·0	
		15·7
180 . . .	44·7	
		23·4
190 . . .	68·1	
		16·1
200 . . .	84·2	
		7·4
210 . . .	91·6	
		3·7
220 . . .	95·3	
		1·3
230 . . .	96·6	

The residue, 3·4 parts, was dissipated before 250° with slight decomposition. The third column expresses the quantity volatilized between each 10°, and indicates the existence of what has been described as bicarburet of hydrogen in considerable quantity.

The importance of these vapours in oil-gas, as contributing to its very high illuminating powers, will be appreciated, when it is considered that with many of them, and those of the denser kind, it is quite saturated. On distilling a portion of liquid, which had condensed in the pipes leading to an oil-gas gasometer, and given to me by Mr. Hennel, of the Apothecaries' Hall, I found it to contain portions of the bicarburet of hydrogen. It was detected by submitting the small quantity of liquid which distilled over before 190° to a cold of 0°, when the substance crystallized from the solution. It is evident, therefore, that the gas from which it was deposited must have been saturated with it. On distilling a portion of recent coal-gas tar,—as was expected, none could be detected in it; but the action of sulphuric acid is sufficient to show the existence of some of these bodies in the coal-gas itself.

With respect to the probable uses of the fluid from compressed oil-gas, it is evident in the first place, that being thus volatile, it will, if introduced into gas, which burns with a pale flame, give such quantity of vapour as to make it brightly illuminating; and even the vapour of those portions which require temperatures of 170°, 180°, or higher, for their ebullition, is so dense as to be fully sufficient for this purpose in small quantities. A taper was burnt out in a jar of common air over water; a portion of fluid boiling at 190° was thrown up into it, and agitated: the mixture then burnt from a large aperture with the bright flame and appearance of oil-gas, though of course many times the quantity that would have been required of oil-gas for the same light was consumed; at the same time there was no mixture of blueness with the flame, whether it were large or small. Mr. Gordon has, I understand, proposed using it in this manner.

The fluid is also an excellent solvent of caoutchouc, surpassing every other substance in this quality. It has already been applied to this purpose.

It will answer all the purposes to which the essential oils are applied as solvents,—as in varnishes, &c.; and in some cases where volatility is required, when rectified it will far surpass them.

It is possible that, at some future time, when we better understand the minute changes which take place during the decomposition of oil, fat, and other substances by heat, and have more command of the process, that this substance, among others, may furnish the fuel for a lamp, which remaining a fluid at the pressure of two or three atmospheres, but becoming a vapour at less pressure, shall possess all the advantages of a gas lamp, without involving the necessity of high pressure.

Royal Institution, June 7, 1825.

On Pure Caoutchouc, and the Substances by which it is accompanied in the State of Sap or Juice*.

I have had an opportunity lately, through the kindness of Mr. Thomas Hancock, of examining the chemical properties of caoutchouc in its pure form, as well as of ascertaining the nature and proportions of the other substances with which it is mixed, when it exudes as sap or juice from the tree. At present much importance attaches to this substance, in consequence of its many peculiar and excellent qualities, and its increasing applications to useful purposes. I have thought, therefore, that a correct account of its chemical nature would possess some interest.

The extensive uses, both domestic and scientific, to which Mr. Hancock has applied common caoutchouc, in consequence of his peculiar mode of liquefying it, are well known. Hence he was fully alive to the importance of its applications, when in its original state of division. When he gave me the substance, he communicated many of his observations upon it, which, with others of my own, form the present paper.

The fluid, I understood, had been obtained from the southern part of Mexico, and was very nearly in the state in which it came from the tree; it had been altered simply by the formation of a slight film of solid caoutchouc on the surface of the cork which closed the bottle. The caoutchouc thus removed was not a 500th part of the whole. The fluid was a pale yellow, thick, creamy-looking substance, of uniform consistency. It had a disagreeable acescent odour, something resembling that

* Quarterly Journal of Science, xxi. 19.

of putrescent milk; its specific gravity was 1011·74. When exposed to the air in thin films, it soon dried, losing weight, and leaving caoutchouc of the usual appearance and colour, and very tough and elastic: 202·4 grains of the liquid dried in a Wedgewood basin, at 100° Fahr., became in a few days 94·4 grains, and the solid piece formed being then removed from the capsule, and exposed on all sides to the air until quite dry, became 91 grains: hence 100 parts of sap left nearly 45 of solid matter.

Heat caused immediate coagulation of the sap, the caoutchouc separating in the solid form, and leaving an aqueous solution of the other substances existing with it in its first state.

Alcohol poured into the sap in sufficient quantity caused a coagulum and a precipitate, both of which were caoutchouc of considerable purity. The alcohol retained in solution the extraneous matters, which, possessing peculiar properties, will be hereafter described.

Solution of alkali added to the sap evolved a very fetid odour, but did not appear to exert any particular action on the caoutchouc.

The sap, left to itself for several days, gradually separated into two parts; the opake portion contracted upwards, leaving beneath a deep brown, but transparent solution, evidently containing substances very different in their nature from caoutchouc itself, and which, considering the specific gravity of the sap and of pure caoutchouc (the latter being lighter than water), were probably present in considerable quantity.

It was found that, by mixing the sap with water, no other change took place than mere dilution. The mixture was uniform, and had all the properties of a weak or thin sap. Heat, evaporation, acids, and alkali, produced the same effects, generally, as before.

When the diluted sap was suffered to remain at rest, a separation soon took place, similar to that which occurred with the native juice, but to a greater extent; a creamy portion rose to the top, whilst a clear aqueous solution remained beneath. Hence it was found easy to wash the caoutchouc, and remove from it other principles which had been generally involved in it to a greater or smaller extent during its coagulation. For this purpose a portion of the sap was mixed with about four volumes of water, and the mixture put into a funnel, stopped below by a cork; in the course of eighteen or twenty-four hours, when the caoutchouc had risen to the top and occupied about its original volume, the aperture at the bottom of the funnel was opened and the solution drawn off; more water was then added to, and mixed with, the caoutchouc, and the operation repeated, and this was done four or five times, until the water came away nearly pure. During the latter washings, the caoutchouc required a longer time to rise to the surface, in consequence of the decreasing specific gravity of the solution in which it was suspended. This was obviated at times, according to the experiments for which the caoutchouc was required, by performing the first washings with solutions of common salt, muriatic acid, &c., and ultimately finishing with pure water.

In this way the caoutchouc was purified without any alteration of its original state. It now appeared in its state of mixture with water perfectly white: portions of it left for a twelvemonth over water underwent no change in that time, except coagulation and a slight film upon the surface; the rest was as miscible with the water as at first, and when coagulated, equally elastic. The sap or the washed caoutchouc is much more easily preserved in the diluted than in the concentrated state.

It produced no particular appearance with the solutions of iron or other metals.

When evaporated, either on paper, in a capsule, or otherwise, the caoutchouc was left in its elastic state, and perfectly unaltered, except with respect to purity. When put on to absorbent surfaces, as bibulous paper, chalk, or plaster of Paris, the water was rapidly abstracted, and the caoutchouc almost immediately united into a mass, retaining the form of the thing on which it was cast. In this way Mr. Hancock has made beautiful medallions with the sap. Poured on to a filter, the water passes through, and the caoutchouc coagulates.

When aggregated in any of these ways, the caoutchouc appears at first as a soft white solid, almost like curd, which by pressure exudes much water, contracts, becomes more compact, has acquired elasticity, but is still soft, white and opake. It also attains this state without pressure, if time be permitted for the water to evaporate. The opacity belonging to it is not an essential property of the body, but due to water enclosed within its mass; further exposure to air allows of the gradual dissipation of this water, and then the caoutchouc appears in its pure and dry state, as a perfectly transparent, colourless and elastic body, except it be in thick masses, when a trace of colour is perceived. The change from first to last is best seen by pouring enough of the pure mixture into a Wedgewood or glass basin, to form ultimately a plate of $\frac{1}{10}$th or $\frac{1}{12}$th of an inch in thickness, and leaving it exposed to air at common temperatures undisturbed.

No appearance of texture can be observed in the pure transparent caoutchouc; it resembles exactly a piece of clear strong jelly. All the phenomena dependent upon its elasticity, which are known to belong to common caoutchouc, are well exhibited by it. When very much extended, it assumes a beautiful pearly or fibrous appearance, probably belonging to the effects which Dr. Brewster has observed elastic bodies to produce, when in a state of tension, upon light. When it has been extended and doubled several times, until further extension in the same direction is difficult, it is found to possess very great strength. Its specific gravity is 0·925, and no reduplication and pressure of it in a Bramah's press was found permanently to alter it. It is evidently pervious to water in a slight degree, or otherwise the interior of a piece of caoutchouc coagulated from the sap would always remain opake. It is equally evident that water passes but very slowly through it, from the time it takes to evaporate that which lies in the middle of a thin cake. It is a non-conductor of electricity.

The pure caoutchouc has a very adhesive surface, which it retains after many months' exposure to air. Its fresh cut surfaces pressed together also adhere with a force equal to that of any other part of the piece.

A strip of it boiled in solution of potash, so strong as to be solid when cold, was not at all affected by it, except that its surface assumed a pearly or tendinous appearance; no swelling or softening, above what would have been produced by water, occurred.

The combustibility of caoutchouc is very well known. When the pure substance is heated in a tube, it is resolved into substances more or less volatile, with the deposition of only a small trace of charcoal; at a higher temperature it is resolved into

N

charcoal and compounds of carbon and hydrogen; it yields no ammonia by destructive distillation, nor any compounds of oxygen, and my experiments agree with those of Dr. Ure, in indicating carbon and hydrogen as its only elements. I have not, however, been able to verify his proportions, which are 90 carbon, 9·11 hydrogen, or by theory nearly 3 proportionals of carbon to 2 of hydrogen, and have never obtained quite so much as 7 carbon to 1 hydrogen by weight. The mean of my experiments gives,—

$$\left.\begin{array}{ll}\text{Carbon} & . \quad . \quad 6\text{·}812 \\ \text{Hydrogen} & . \quad 1\text{·}000\end{array}\right\} \text{ or } \left\{\begin{array}{l}8 \text{ proportionals nearly.} \\ 7.\end{array}\right.$$

No means which have yet been discovered seem competent, when the caoutchouc has once been aggregated, to restore it to its pristine state. Previous to its aggregation it may be either scented or coloured. A solution of camphor in alcohol was added to water, so as to precipitate the camphor in a flocculent state; a little of this was added to some of the pure caoutchouc in water well agitated, and then coagulation caused by heat or absorption; the caoutchouc obtained was highly odorous.

In the trials made to give it colour, the body colours were found to answer best—indigo, cinnabar, chrome-yellow, carmine, lake, &c., were rubbed very fine with water; then mixed well with the pure caoutchouc, in a somewhat diluted state, and coagulation induced either upon an absorbent surface or otherwise. Perfectly-coloured specimens were thus obtained.

The aqueous liquid obtained either by letting the sap stand for some time, or by the first and second washing, was of a brown colour, bitter, acid to litmus in consequence of the presence of acetic acid, due apparently to spontaneous changes in the substances present. It was difficult to filter. Being boiled, acid vapours rose, a precipitate fell to the bottom, and now the solution (*a*) became clear, either by standing or filtration, and could be separated from the solid matter.

The precipitate or substance thus obtained was dark brown, glossy and brittle, much heavier than water, not soluble in alcohol, ether, water, essential or fixed oils. Weak solution of alkali dissolved it, forming a deep brown solution, precipitable by dilute muriatic acid. It burnt upon platina-foil, like animal matter, with flame, leaving a bulky charcoal. When heated in a tube, it became charred, yielding much ammonia. It resembles albumen more than any other substance, and is the source of the nitrogen or ammonia obtained by the distillation of common caoutchouc.

The brown aqueous solution (*a*) became frothy on agitation; alkali rendered it of a deep yellow colour, and produced a putrescent odour, similar to that evolved by alkali or quick-lime from white of egg or blood. It was remarkably distinguished by the deep-green colour it produced with persalts of iron, especially when a little alkali was present, and the dense yellow precipitates it formed with muriate of zinc and nitrate of lead; indeed, precipitates were produced in solutions of most of the metals by it. The colour produced with iron does not seem to be a precipitate.

With the hope of obtaining something peculiar from this solution, a quantity of it was precipitated by nitrite of lead; a colourless solution and a yellowish-green precipitate were obtained. The latter, being well washed, was next diffused through water, and sulphuretted hydrogen passed through it; by filtration a deep brown solid was obtained and a yellowish solution. The precipitate when washed and dried was brittle and hard; on platina-foil it at first burnt with flame, swelling much, and giving out odour of ammonia like animal matter; after that sulphurous acid burnt off, and ultimately lead and oxide of lead remained; hence it was a combination of sulphuret of lead, and a highly azoted substance. Heated in a tube, it gave out much ammonia; digested in alcohol, scarcely a trace of matter was removed.

The sulphuretted hydrogen solution, being boiled and evaporated, left a yellow varnish-like substance, not deliquescent, soluble in water, acid to taste and to litmus, the acid not being sulphuric: it rendered persulphate of iron green, precipitated nitrate of lead, and gave no ammonia by heat.

The concentrated solution (*a*) acted upon by alcohol, had an insoluble matter thrown down, which being separated and well washed with alcohol, was afterwards treated with water, a deep brown aqueous solution (*b*) was obtained, and a small insoluble portion left; this was almost black when dried, tasteless, brittle, burning with difficulty, and when heated in a tube giving much ammonia.

The solution (*b*) was almost tasteless, and when dried left a green, shining, brittle substance, resoluble in water, and of course precipitable by alcohol. It colours solution of persulphate of iron green; but if its strong aqueous solution be treated with muriatic acid, a reddish-brown precipitate is formed, which, when separated, dissolves in water, does not colour persalts of iron, and when evaporated yields a pulverulent substance, burning, but not with facility, and producing a little ammonia when heated in a tube.

The alcoholic solution from which these matters had been separated contained the particular principle which colours persalts of iron green. When evaporated it left a brown, brittle, transparent substance, becoming soft by exposure to moist air. It is very bitter, soluble in water, &c., slightly acid. When heated on platina-foil, it does not burn easily, but runs out into a bulky charcoal, much like animal matter; at the same time it does not yield ammonia when heated in a tube *per se*, though the smell is very like that of animal substance.

Ether warmed with it dissolved a small portion of matter, and the solution, upon evaporation, left globules, which, in all their characters, corresponded with wax; its quantity was but small.

Nine hundred and eighty-one grains of the original sap were washed in water several times. The washed caoutchouc, being coagulated by heat and perfectly dried, weighed 311 grains. The aqueous solutions, upon being boiled, yielded sufficient of the heavy precipitate to equal, when dried, 18·6 grains. The clear solution was now evaporated to dryness and digested in alcohol, 28·5 grains of insoluble matter were left, and the solution, upon evaporation, afforded 70 grains of dry matter. Hence the following are the contents nearly of 1000 parts of the original sap:—

Caoutchouc	317·0
Albuminous precipitate	19·0
Peculiar bitter colouring matter, a highly azotated substance . . ⎱ Wax . . . . ⎰	71·3
Substance soluble in water, not in alcohol	29·0
Water, acid, &c.	563·7
	1000·0

Thinking it probable that whilst in its natural state of division the caoutchouc would combine more intimately or readily with fixed and volatile oils than when aggregated, as it generally is in commerce, an experiment or two were made in consequence. A portion of well-washed milky caoutchouc being added to olive oil, and the two beaten well together, a singularly adhesive stringy substance was produced, which holding the water diffused through it, assumed a very pearly aspect, stiffened, and was almost solid; upon being heated so as to drive off the water, it became oily, fluid and clear, and was then a solution of caoutchouc in the fixed oil. On adding water and stirring considerably, it again became adhesive as before. Thus introduced, caoutchouc would probably be a useful element in varnishes.

Oil of turpentine being added to a mixture of one volume of sap and one volume of water, and well agitated with it, was found to be only imperfectly miscible; after standing twenty-four hours, three portions were formed: the lower, the usual aqueous solution; the upper, oil of turpentine, holding a little caoutchouc in solution; the intervening part a clot or tenacious mass, soft and adhesive, like bird-lime, consisting of caoutchouc, with some oil of turpentine. It was very difficult to dry, and always remained adhesive at the surface; but experiments of this kind were not pursued, for want, at that time, of further quantities of the original sap.

Such is a general view of the nature of the sap from which the substance is obtained, and of the substance itself. I have not endeavoured to give an accurate account of the properties or quantities of the other substances present, because there is reason to believe that both vary in different specimens, probably according to the age of the tree, the time of the year, or the manner in which the sap is drawn; nor have I dwelt upon the inaccuracies of former accounts, inasmuch as they are evidently referable to the impurity of the substance examined.

Those who wish to look to former accounts of the chemical or physical qualities of this remarkable substance, will perhaps find the following references useful :—

1751. De la Condamine on an Elastic Resin, newly discovered at Cayenne, by M. Fresneau; and on the Use of various Milky Saps from Trees of

Guiane or France Equinoctiale.—*Mém. de l'Acad. Royale*, 1751. pp. 17, 319.

1763. MM. Herissant and Macquer on Solution of Caoutchouc.—*Mém. de l'Académie*, 1763, p. 49.

1768. Macquer, Memoir on the Means of dissolving the Resin Caoutchouc, known by the name Elastic Resin of Cayenne, and making it appear with all its properties.—*Mém. de l'Académie*, 1768, pp. 58, 208.

1781. Berniard, Memoir on Caoutchouc, known by the name of Elastic Gum.—*Journ. de Physique*, xvii. 265.

1790. Fourcroy on the Sap furnishing Elastic Gum.—*Ann. de Chim.* xi. 225; again, *Connaissances Chimiques*, viii. 36.

1791. Grossart on the Means of making Instruments of Gum Elastic, with the Bottles obtained from Brazil.—*Ann. de Chim.* xi. 143.

1791. Fabbroni on Solution of Caoutchouc in repeatedly-rectified Petroleum. —*Ann. de Chim.* xi. 195; xii. 156.

Pelletier on Solution of Elastic Gum in Sulphuric Ether.—*Mém. de l'Institut*, i. 56.

1801. Howison on the Elastic Gum Vine of Prince of Wales's Island, and of Experiments made on the Milky Juice which it produces, with Hints respecting the useful Purposes to which it may be applied.—*Asiatic Researches*, v. 157.

1801. Roxburg, Dr., Botanical Description of *Urceola Elastica*, or Caoutchouc Vine of Sumatra and Pulo-Penang, with an Account of the Properties of its inspissated Juice, compared with those of the American Caoutchouc.—*Asiatic Researches*, v. 167.

1803. Gough, Description of a Property of Caoutchouc or Indian Rubber, with Reflections on the Cause of the Elasticity of this Substance.— *Manchester Memoirs*, N. S. i. 288.

1805. Simple Method of making Tubes of Elastic Gum Caoutchouc, avoiding the expense of Ether.—*Phil. Mag.* xxii. 340.

1807. Murray's Chemistry, iv. 177, contains a compendium of what was then known respecting this substance.

Royal Institution, January 1826.

On the Mutual Action of Sulphuric Acid and Naphthaline*.

[Read February 16, 1826.]

IN a paper "On new Compounds of Carbon and Hydrogen," lately honoured by the Royal Society with a place in the Philosophical Transactions, I had occasion briefly to notice the peculiar action exerted on certain of those compounds by sulphuric acid†. During my attempts to ascertain more minutely

---

* Philosophical Transactions, 1826, p. 140.   † pp. 160, 163, 164, 170.

the general nature of this action, I was led to suspect the occasional combination of the hydrocarbonaceous matter with the acid, and even its entrance into the constitution of the salts which the acid afterwards formed with bases. Although this opinion proved incorrect, relative to the peculiar hydrocarbons forming the subject of that paper, yet it led to experiments upon analogous bodies, and amongst others, upon naphthaline, which terminated in the production of the new acid body and salts now to be described.

Some of the results obtained by the use of the oil-gas products are very peculiar. If, when completed, I find them sufficiently interesting, I shall think it my duty to place them before the Royal Society, as explicatory of that action of sulphuric acid which was briefly noticed in my last paper.

Most authors who have had occasion to describe naphthaline, have noticed its habitudes with sulphuric acid. Mr. Brande several years since * stated that naphthaline dissolved in heated sulphuric acid " in considerable abundance, forming a deep violet-coloured solution, which bears diluting with water without decomposition. The alkalies produce in this solution a white flaky precipitate, and if diluted the mixture becomes curiously opalescent, in consequence of the separation of numerous small flakes." The precipitate by alkali was probably one of the salts to be hereafter described.

Dr. Kidd observes †, that " it blackens sulphuric acid when boiled with it; the addition of water to the mixture having no other effect than to dilute the colour, neither does any precipitation take place upon saturating the acid with ammonia."

Mr. Chamberlain states ‡, that sulphuric acid probably decomposes naphthaline, for that it holds but a very small quantity in solution. The true interpretation of these facts and statements will be readily deduced from the following experimental details.

1. *Production and properties of the new acid formed from Sulphuric Acid and Naphthaline.*—Naphthaline, which had been almost entirely freed from naphtha by repeated sublimation and pressure, was pulverized; about one part with three or

---

* Quart. Journ. of Science, 1819, viii. 289.   † Phil. Trans. 1821, p. 216.
‡ Annals of Philosophy, 1823, N.S. vi. p. 136.

four parts by weight of cold sulphuric acid were put into a bottle, well shaken, and left for thirty-six hours. The mixture then contained a tenacious deep red fluid, and a crystalline solid; it had no odour of sulphurous acid. Water being added, all the liquid and part of the solid was dissolved; a few fragments of naphthaline were left, but the greater part was retained in solution. The diluted fluid being filtered was of a light brown tint, transparent, and of an acid and bitter taste.

For the purpose of combining as much naphthaline as possible with the sulphuric acid, 700 grains with 520 grains of oil of vitriol were warmed in a Florence flask until entirely fluid, and were well shaken for about 30 minutes. The mixture was red; and the flask being covered up and left to cool, was found after some hours to contain, at the bottom, a little brownish fluid, strongly acid, the rest of the contents having solidified into a highly crystalline mass. The cake was removed, and its lower surface having been cleaned, it was put into another Florence flask with 300 grains more of naphthaline, the whole melted and well shaken together, by which a uniform mixture was obtained, but opake and dingy in colour. It was now poured into glass tubes, in which it could be retained and examined without contact of air. In these the substance was observed to divide into two portions, which could easily be distinguished from each other, whilst both were retained in the fluid state. The heavier portion was in the largest quantity; it was of a deep red colour, opake in tubes half an inch in diameter, but in small tubes could be seen through by a candle, or sunlight, and appeared perfectly clear. The upper portion was also of a deep red colour, but clear, and far more transparent than the lower: the line of separation very defined. On cooling the tubes, the lighter substance first solidified, and after some time the heavier substance also became solid. In this state, whilst in the tube, they could with great difficulty be distinguished from each other.

These two substances were separated, and being put into tubes, were further purified by being left in a state of repose at temperatures above their fusing-points, so as to allow of separation; and when cold, the lower part of the lighter substance and the upper, as well as the lower part of the heavier substance, were set aside for further purification.

The *heavier substance* was a red crystalline solid, soft to the nail like a mixture of wax and oil. Its specific gravity was from 1·3 to 1·4, varying in different specimens; its taste sour, bitter, and somewhat metallic. When heated in a tube, it fused, forming, as before, a clear but deep red fluid. Further heat decomposed it, naphthaline, sulphurous acid, charcoal, &c. being produced. When heated in the air it burnt with much flame. Exposed to air it attracted moisture rapidly, became brown and damp upon the surface, and developed a coat of naphthaline. It dissolved entirely in alcohol, forming a brown solution. When rubbed in water a portion of naphthaline separated, amounting to 27 per cent., and a brown acid solution was obtained. This was found by experiments to contain a peculiar acid mixed with a little free sulphuric acid, and it may conveniently be called *the impure acid.*

The *lighter substance* was much harder than the former, and more distinctly crystalline. It was of a dull red colour, easily broken down in a mortar, the powder being nearly white, and adhesive like naphthaline. It was highly sapid, being acid, bitter, and astringent. When heated in a tube it melted, forming a clear red fluid, from which by a continued heat much colourless naphthaline sublimed, and a black acid substance was left, which at a high temperature gave sulphurous acid and charcoal. When heated in the air it took fire and burnt like naphthaline. Being rubbed in a mortar with water, a very large portion of it proved to be insoluble; this was naphthaline; and on filtration the solution contained the peculiar acid found to exist in the *heavier substance*, contaminated with very little sulphuric acid. More minute examination proved that this *lighter substance* in its fluid state was a solution of a small quantity of the dry peculiar acid in naphthaline; and that the *heavier substance* was a union of the peculiar acid in large quantity with water, free sulphuric acid, and naphthaline.

It was easy by diminishing the proportion of naphthaline to make the whole of it soluble, so that when water was added to the first result of the experiment, nothing separated; and the solution was found to contain sulphuric acid with the peculiar acid. But reversing the proportions, no excess of naphthaline was competent, at least in several hours, to cause the entire disappearance of the sulphuric acid. When the experiment was carefully made with pure naphthaline, and either at common or slightly elevated temperatures, no sulphurous acid appeared to be formed, and the action seemed to consist in a simple union of the concentrated acid and the hydrocarbon.

Hence it appears, that when concentrated sulphuric acid and naphthaline are brought into contact at common or moderately elevated temperatures, a peculiar compound of sulphuric acid with the elements of the naphthaline is produced, which possesses acid properties; and as this exists in large quantity in the heavier of the bodies above described, that product may conveniently be called the *impure solid acid*. The experiments made with it, and the mode of obtaining the pure acid from it, are now to be described.

Upon applying heat and agitation to a mixture of one volume of water and five volumes of impure solid acid, the water was taken up to the exclusion of nearly the whole of the free naphthaline present; the latter separating in a colourless state from the red hydrated acid beneath it. As the temperature of the acid diminished, crystallization in tufts commenced here and there, and ultimately the whole became a brownish yellow solid. A sufficient addition of water dissolved nearly the whole of this hydrated acid, a few flakes only of naphthaline separating.

A portion of the impure acid in solution was evaporated at a moderate temperature; when concentrated, it gradually assumed a light brown tint. In this state it became solid on cooling, of the hardness of cheese, and was very deliquescent. By further heat it melted, then fumed, charred, &c., and gave evidence of the abundant presence of carbonaceous matter.

Some of the impure acid in solution was neutralized by potash, during which no naphthaline or other substance separated. The solution being concentrated until ready to yield a film on its surface, was set aside whilst hot to crystallize: after some hours the solution was filled with minute silky crystals, in tufts, which gave the whole, when stirred, not the appearance of mixed solid salt and liquid, but that of a very strong solution of soap. The agitation also caused the sudden solidification of so much more salt, that the whole became solid, and felt like a piece of soft soap. The salt when dried had no resemblance to sulphate of potash. When heated in the air, it burnt with a dense flame, leaving common sulphate of potash, mixed with some sulphuret of potassium, resulting from the action of the carbon, &c. upon the salt.

Some of the dry salt was digested in alcohol to separate common sulphate of potash. The solution, being filtered and evaporated, gave a white salt soluble in water and alcohol, crystalline, neutral, burning in the air with much flame, and leaving sulphate of potash. It was not precipitated by nitrate of lead, muriate of baryta, or nitrate of silver.

It was now evident that an acid had been formed peculiar in its nature and composition, and producing with bases peculiar salts. In consequence of the solubility of its barytic salt, the following process for the preparation of the pure acid was adopted.

A specimen of native carbonate of baryta was selected, and its purity ascertained. It was then pulverized, and rubbed in successive portions with a quantity of the impure acid in solution, until the latter was perfectly neutralized, during which the slight colour of the acid was entirely removed. The solution was found to contain the peculiar barytic salt. Water added to the solid matter dissolved out more of the salt; and ultimately only carbonate and sulphate of baryta, mixed with a little of another barytic salt, remained. The latter salt being much less soluble in water than the former, was not removed so readily by lixiviation, and was generally found to be almost entirely taken up by the last portions of water applied with heat.

The barytic salt in solution was now very carefully decomposed, by successive additions of sulphuric acid, until all the baryta was separated, no excess of sulphuric acid being permitted. Being filtered, a pure aqueous solution of the peculiar acid was obtained. It powerfully reddened litmus paper, and had a bitter acid taste. Being evaporated to a certain degree, a portion of it was subjected to the continued action of heat; when very concentrated, it began to assume a brown colour, and on cooling became thick, and ultimately solid, and was very deliquescent. By renewed heat it melted, then began to fume, became charred, but did not flame; and ultimately gave sulphuric and sulphurous acid vapours, and left charcoal.

Another portion of the unchanged strong acid solution was

placed over sulphuric acid in an exhausted receiver. In some hours it had by concentration become a soft white solid, apparently dry; and after a longer period was hard and brittle. In this state it was deliquescent in the air, but in close vessels underwent no change in several months. Its taste was bitter, acid, and accompanied by an after metallic flavour, like that of cupreous salts. When heated in a tube at temperatures below $212°$, it melted without any other change, and on being allowed to cool, crystallized from centres, the whole ultimately becoming solid. When more highly heated, water at first passed off, and the acid assumed a slight red tint; but no sulphurous acid was as yet produced, nor any charring occasioned; and a portion being dissolved and tested by muriate of baryta, gave but a very minute trace of free sulphuric acid. In this state it was probably anhydrous. Further heat caused a little naphthaline to rise, the red colour became deep brown, and then a sudden action commenced at the bottom of the tube, which spread over the whole, and the acid became black and opake. Continuing the heat, naphthaline, sulphurous acid, and charcoal were evolved; but even after some time, the residuum examined by water and carbonate of baryta was found to contain a portion of the peculiar acid undecomposed, unless the temperature had been raised to redness.

These facts establish the peculiarity of this acid, and distinguish it from all others. In its solid state it is generally a hydrate containing much combustible matter. It is readily soluble in water and alcohol, and its solution forms neutral salts with bases, all of which are soluble in water, most of them in alcohol, and all combustible, leaving sulphates or sulphurets according to circumstances. It dissolves in naphthaline, oil of turpentine, and olive oil, in greater or smaller quantities, according as it contains less or more water. As a hydrate, when it is almost insoluble in naphthaline, it resembles the *heavier substance* obtained as before described, by the action of sulphuric acid on naphthaline, and which is the solid hydrated acid, containing a little naphthaline and some free sulphuric acid, whilst the *lighter substance* is a solution of the dry acid in naphthaline; the water present in the oil of vitriol originally used being sufficient to cause a separation of a part, but not of the whole.

2. *Salts formed by the peculiar acid with bases.*—These compounds may be formed, either by acting on the bases or their carbonates by the pure acid, obtained as already described ; or the impure acid in solution may be used, the salts resulting being afterwards freed from sulphates, by solution in alcohol. It is however proper to mention that another acid, composed of the same elements, is at the same time formed with the acid in question, in small, but variable proportions. The impure acid used, therefore, should be examined as to the presence of this body, in the way to be directed when speaking of the barytic salts ; and such specimens as contain very little or none of it should be selected.

Potash forms with the acid a neutral salt, soluble in water and alcohol, forming colourless solutions. These yield either transparent or white pearly crystals, which are soft, slightly fragile, feel slippery between the fingers, do not alter by exposure to air, and are bitter and saline to the taste. They are not very soluble in water ; but they undergo no change by repeated solutions and crystallizations, or by long-continued ebullition. The solutions frequently yield the salt in acicular tufts, and they often vegetate, as it were, by spontaneous evaporation, the salt creeping over the sides of the vessel, and running to a great distance in very beautiful forms. The solid salt heated in a tube gave off a little water, then some naphthaline ; after that a little carbonic and sulphurous acid gases arose, and a black ash remained, containing carbon, sulphate of potash, and sulphuret of potassium. When the salt was heated on platinum foil in the air, it burnt with a dense flame, leaving a slightly alkaline sulphate of potash.

Soda yields a salt in most properties resembling that of potash,—crystalline, white, pearly, and unaltered in the air. I thought that, in it, the metallic taste which frequently occurred with this acid and its compounds was very decided. The action of heat was the same as before.

Ammonia formed a neutral salt imperfectly crystalline, not deliquescent, but drying in the atmosphere. Its taste was saline and cooling. It was readily soluble in water and alcohol. When heated on platinum foil it fused, blackened, burnt with flame, and left a carbonaceous acid sulphate of ammonia, which by further heat was entirely dissipated. Its general habits were those of ammoniacal salts. When its solutions, though previously rendered alkaline, were evaporated to dryness at common temperatures, and exposed to air, the salt became strongly acid to litmus paper. This however is a property common to all soluble ammoniacal salts, I believe, without exception.

Baryta.—It is easy by rubbing carbonate of baryta with solution of the impure acid, to obtain a perfectly neutral solution, in which the salt of baryta, containing the acid already described, is very nearly pure. There is in all cases an undissolved portion, which being washed repeatedly in small quantities of hot water, yields to the first portions a salt, the same as that in the solution. As the washings proceed, it is found that the salt obtained does not burn with so much flame on platina foil, as that at first separated; and the fifth or sixth washing will perhaps separate only a little of a salt, which, when heated in the air in small quantities, burns without flame in the manner of tinder. Hence it is evident that there are two compounds of baryta, which as they are both soluble in water, both neutral, and both combustible, leaving sulphate of baryta, differ probably only in the quantity of combustible matter present, or its mode of combination in the acid.

It is this circumstance, of the formation of a second salt in small but variable quantities with the first, which must be guarded against, as before mentioned, in the preparation of salts from the impure acid. It varies in quantity according to the proportions of materials, and the heat employed; and I have thought that when the naphthaline has been in large quantity, and the temperature low, the smallest quantity is produced. When the impure acid is used for the preparation of the salts now under description, a small portion of it should be examined by carbonate of baryta as above, and rejected if it furnish an important quantity of the flameless salt.

These bodies may be distinguished from each other provisionally, as the *flaming* and the *glowing* salts of baryta, from their appearances when heated in the air. The latter is more distinctly crystalline than the former, and much less soluble, which enabled me by careful and repeated crystallizations to obtain both in their pure states.

The *flaming* salt (that corresponding to the acid now under description), when obtained by the slow evaporation of the saturated solution, formed tufts, which were imperfectly crystalline. When drops were allowed to evaporate on a glass plate, the crystalline character was also perceived; but when the salt was deposited rapidly from its hot saturated solution, it appeared in the form of a soft granular mass. When dry, it was white and soft, not changing in the atmosphere. It was readily soluble in water and alcohol, but was not affected by ether. Its taste was decidedly bitter. When heated in the air on platinum foil it burnt with a bright smoky flame, like naphthaline, sending flocculi of carbon into the atmosphere, and leaving a mixture of charcoal, sulphuret of barium, and sulphate of baryta.

After being heated to 212° for some time, the salt appeared to be perfectly dry, and in that state was but very slightly hygrometric. When heated in a tube, naphthaline was evolved; but the substance could be retained for hours at a temperature of 500° F. before a sensible portion of naphthaline had separated: a proof of the strength of the affinity by which the hydrocarbon was held in combination. When a higher temperature was applied, the naphthaline, after being driven off, was followed by a little sulphurous acid, a small portion of tarry matter, and a carbonaceous sulphate and sulphuret were left.

This salt was not affected by moderately strong nitric or nitromuriatic acid, even when boiled with them; and no precipitation of sulphate took place. When the acids were very strong, peculiar and complicated results were obtained. When put into an atmosphere of chlorine, at common temperatures, it was not at all affected by it. Heat being applied, an action between the naphthaline evolved, and chlorine, such as might be expected, took place.

When a strong solution of the pure acid was poured into a strong solution of muriate of baryta, a precipitate was formed, in consequence of the production of this salt. It was re-dissolved by the addition of water. The fact indicates that the affinity of this acid for baryta is stronger than that of muriatic acid.

The *second*, or *glowing* salt of haryta, was obtained in small crystalline groups. The crystals were prismatic, colourless, and transparent : they were almost tasteless, and by no means so soluble either in hot or cold water as the former salts. They were soluble in alcohol, and the solutions were perfectly neutral. When heated on platinum foil, they gave but very little flame, burning more like tinder, and leaving a carbonaceous mixture of sulphuret and sulphate. When heated in a tube, they gave off a small quantity of naphthaline, some empyreumatic fumes, with a little sulphurous acid, and left the usual product.

This salt seemed formed in largest quantity when one volume of naphthaline and two volumes of sulphuric acid were shaken together, at a temperature as high as it could be without charring the substances. The tint, at first red, became olive-green ; some sulphurous acid was evolved, and the whole would ultimately have become black and charred, had it not been cooled before it had proceeded thus far, and immediately dissolved in water. A solution was obtained, which though dark itself, yielded, when rubbed with carbonate of baryta, colourless liquids ; and these when evaporated furnished a barytic salt, burning without much flame, but which was not so crystalline as former specimens. No attempt to form the glowing salt from the flaming salt by solution of caustic baryta, succeeded.

Strontia.—The compound of this earth with the acid already described very much resembled the flaming salt of baryta. When dry it was white, but not distinctly crystalline : it was soluble in water and alcohol; not alterable in the air, but when heated burnt with a bright flame, without any red tinge, and left a result of the usual kind.

Lime gave a white salt of a bitter taste, slightly soluble in water, soluble in alcohol, the solutions yielding imperfect crystalline forms on evaporation : it burnt with flame; and both in the air and in tubes, when heated, gave results similar to those of the former salts.

Magnesia formed a white salt with a moderately bitter taste ; crystallizing under favourable circumstances, burning with flame, and giving such results by the action of heat as might be expected.

Iron.—The metal was acted upon by the acid, hydrogen being evolved. The moist protoxide being dissolved in the acid gave a neutral salt capable of crystallization. This by exposure to air slowly acquired oxygen, and a portion of per-salt was found.

Zinc was readily acted upon by the acid, hydrogen evolved, and a salt formed. The same salt resulted from the action of the acid upon the moist oxide. It was moderately soluble in hot water, the solution on cooling affording an abundant crop of acicular crystals. The salt was white and unchangeable in the air; its taste bitter. It burnt with flame, and gave the usual results by heat.

Lead.—The salt of this metal was white, solid, crystalline, and soluble in water and alcohol. It had a bitter metallic taste, with very little sweetness. The results by heat were such as might be expected.

Manganese.—The protoxide of this metal formed a neutral crystalline salt with the acid. It had a slightly austere taste, was soluble in water and alcohol, and was decomposed by heat, with the general appearances already described.

Copper.—Hydrated peroxide of copper formed an acid salt with the acid, and the solution evaporated in the air left radiated crystalline films. The dry salt when heated fused, burnt with flame, and exhibited the usual appearances.

Nickel.—The salt of this metal was made from the moist carbonate. It was soluble, crystalline, of a green colour, and decomposed by heat in the usual manner. In one instance an insoluble subsalt was formed.

Silver.—Moist carbonate of silver dissolved readily in the acid, and a solution, almost neutral, was quickly obtained. It was of a brown colour, and a powerful metallic taste. By evaporation it gave a splendent, white, crystalline salt; not changing in the air except when heated, but then burning with flame and ultimately leaving pure silver. When the solution of the salt was boiled for some time, a black insoluble matter was thrown down, and a solution obtained, which by evaporation gave abundance of a yellow crystalline salt. The changes which took place during the action of heat in the moist way were not minutely examined.

Mercury.—Moist protocarbonate of mercury dissolved in the acid, forming a salt not quite neutral, crystallizing feebly in the air, white, of a metallic taste, not deliquescent, and decomposed with various phenomena by heat. By re-solution in

o

water or alcohol, and heat, a subsalt of a yellow colour was formed.

The moist hydrated peroxide of mercury also dissolved in the acid, forming an acid solution, which by evaporation gave a yellowish deliquescent salt, decomposed by heat, burning in the air, and entirely volatile.

3. *Analysis of the acid and salts.*—When solution of the pure acid was subjected to the voltaic battery, oxygen and hydrogen gases were evolved in their pure state: no solid matter separated, but the solution became of a deep yellow colour at the positive pole, occasioned by the evolution of free sulphuric acid, which reacted upon the hydrocarbon. A solution of the barytic salts gave similar results.

The analytical experiments upon the composition of this acid and its salts were made principally with the compound of baryta. This was found to be very constant in composition, could be obtained anhydrous at moderate temperatures, and yet sustained a high temperature before it suffered any change.

A portion of the pure salt was prepared and dried for some hours on the sand-bath, at a temperature of about 212°. Known weights were then heated in a platinum crucible to dissipate and burn off the combustible matter; and the residuum being moistened with sulphuric acid to decompose any sulphuret of barium formed, was heated to convert it into a pure sulphate of baryta. The results obtained were very constant, and amounted to 41·714 of sulphate of baryta per cent. of salt used, equivalent to 27·57 baryta per cent.

Other portions of the salt were decomposed by being heated in a flask with strong nitromuriatic acid, so as to liberate the sulphuric acid from the carbon and hydrogen present, and yet retain it in the state of acid. Muriate of baryta was then added, the whole evaporated to dryness, heated red-hot, washed with dilute muriatic acid to remove the baryta uncombined with sulphuric acid, and the sulphate collected, dried and weighed. The results were inconstant; but the sulphate of baryta obtained always much surpassed that furnished by the former method. Judging from this circumstance that the sulphuric acid in the salt was more than an equivalent for the baryta present, many processes were devised for the determination of its quantity, but were rejected in consequence of difficulties and imperfections, arising principally from the presence and action of so much carbonaceous matter. The following was ultimately adopted.

A quantity of peroxide of copper was prepared by heating copper plates in air and scaling them. A sufficient quantity of pure muriatic and nitric acids were provided, and also a specimen of pure native carbonate of baryta. Seven grains of the salt to be examined were then mixed with seven grains of the pulverized carbonate of baryta, and afterwards with 312 grains of the oxide of copper. The mixture being put into a glass tube was successively heated throughout its mass, the gas liberated being passed through a mixture of baryta water and solution of muriate of baryta. It was found that no sulphurous or sulphuric acids came off, or indeed sulphur in any state. The contents of the tube were then dissolved in an excess of nitric and muriatic acids, above that required to take up all that was soluble; and a little solution of muriate of baryta was added for the sake of greater certainty. A portion of sulphate of baryta remained undissolved, equivalent to the sulphuric acid of the salt experimented upon, with that contained accidentally in the oxide of copper acids, &c. This sulphate was collected, washed, dried and weighed. Similar quantities of the carbonate of baryta and oxide of copper were then dissolved in as much of the nitric and muriatic acids as was used in the former experiment; and the washings and other operations being repeated exactly in the same way, the quantity of sulphate of baryta occasioned by the presence of sulphuric acid in the oxide, acids, &c. was determined. This, deducted from the weight afforded in the first experiments, gave the quantity produced from the sulphuric acid actually existing in the salt. Experiments so conducted gave very uniform results. The mean of many indicated 8·9 grains of sulphate of baryta for 10 grains of salt used, or 89 grains per cent., equivalent to 30·17 of sulphuric acid for every 100 of salt decomposed.

In the analytical experiments relative to the quantity of carbon and hydrogen contained in the salt, a given weight of the substance being mixed with peroxide of copper, was heated in a green glass tube. The apparatus used consisted of Mr. Cooper's lamp furnace, with Dr. Prout's mercurial trough; and all the precautions that could be taken, and which are now well known, were adopted for the purpose of obtaining accurate results. When operated upon in this way, the only substances evolved from the salt were carbonic acid and water. As an instance of the results, 3·5 grains of the salt afforded 11·74 cubic inches of carbonic acid gas, and 0·9 of a grain of water. The mean of several experiments gave 32·93 cubic inches of carbonic acid gas, and 2·589 grains of water for every 10 grains of salt decomposed.

On these data, 100 grains of the salt would yield 329·3 cubic inches of carbonic acid, or 153·46 grains, equivalent to 41·9 grains of carbon, and 25·89 grains of water, equivalent to 2·877 grains of hydrogen. Hence 100 grains of the salt yielded—

Baryta	27·57	78
Sulphuric acid	30·17	85·35
Carbon	41·90	118·54
Hydrogen	2·877	8·13
	102·517	

In the second numerical column the experimental results are repeated, but increased, that baryta might be taken in the quantity representing one proportional, hydrogen being unity; and it will be seen that they do not differ far from the following theoretical statement :—

Baryta	1 proportional		78
Sulphuric acid	2	ditto	80
Carbon	20	ditto	120
Hydrogen	8	ditto	8

The quantity of sulphuric acid differs most importantly from the theoretical statement, and it probably is *that* element of the salt in the determination of which most errors are involved. The quantity of oxide of copper and of acids required to be used in that part of the analysis, may have introduced errors, affecting the small quantity of salt employed, which when multiplied, as in the deduction of the numbers above relative to 100 parts, may have created an error of that amount.

As there is no reason to suppose that during the combination of the acid with the baryta any change in its proportions takes place, the results above, *minus* the baryta, will represent its composition: from which it would appear, that one proportional of the acid consists of two proportionals of sulphuric acid, twenty of carbon, and eight of hydrogen; these constituents forming an acid equivalent in saturating power to one proportional of other acids.   Hence it would seem, that half the sulphuric acid present, at least when in combination, is neutralized by the hydrocarbon; or, to speak in more general terms, that the hydrocarbon has diminished the saturating power of the sulphuric acid to one half.   This very curious and interesting fact in chemical affinity was, however, made known to me by Mr. Hennell of Apothecaries' Hall, as occurring in some other compounds of sulphuric acid and hydrocarbon, before I had completed the analysis of the present acid and salts; and a similar circumstance is known with regard to muriatic acid, in the curious compound discovered by M. Kind, which it forms with oil of turpentine.   Mr. Hennell is, I believe, on the point of offering an account of his experiments to the Royal Society, and as regards date they precede mine.

It may be observed, that the existence of sulphuric acid in the new compounds, is assumed, rather than proved; and that the non-appearance of sulphurous acid, when sulphuric acid and naphthaline act on each other, is not conclusive as to the non-reaction of the bodies.   It is possible that part of the hydrogen of the naphthaline may take oxygen from one of the proportions of the sulphuric acid, leaving the hyposulphuric acid of Welter and Gay-Lussac, which with the hydrocarbon may constitute the new acid.   I have not time at present to pursue these refinements of the subject, or to repeat the analyses which have been made of naphthaline, and which would throw light upon the question.   Such a view would account for a part of the overplus in weight, but not for the excess of the sulphuric acid obtained, above two proportionals.

The glowing salt of baryta was now analysed by a process similar to that adopted for the flaming salt.   The specimen operated upon was pure, and in a distinctly crystalline state. It had been heated to about 440° F. for three hours in a metallic bath.   Ten grains of this salt, exposed to air for forty hours, increased only 0·08 of a grain in weight.   These, when converted into sulphate of baryta by heat and sulphuric acid, gave 4·24 grains.   Seven grains by carbonate of baryta, oxide of copper, heat, &c., gave 6·02 grains of sulphate of baryta: hence 10 grains of the salt would have afforded 8·6 grains of the sulphate, equivalent to 2·915 grains of sulphuric acid. Five grains, when heated with oxide of copper, gave 16·68 cubic inches of carbonic acid gas, equal to 7·772 grains, and equivalent to 2·12 grains of carbon. The water-formed amounted to 1·2 grain, equivalent to 0·133 of a grain of hydrogen.

From these data, 100 grains of the salt would appear to furnish—

Baryta . . . 28·03	.	78 or 1 proportional.
Sulphuric acid 29·13	.	81·41 nearly two proportionals.
Carbon . . . 42·40	.	118· approaching to 20 ditto.
Hydrogen . . 2·66	.	7·4 or 7·4 proportionals.

102·22

results not far different from those obtained with the former salt.

I have not yet obtained sufficient quantities of this salt in a decidedly crystalline state to enable me satisfactorily to account for the difference between it and the flaming salt.

Attempts were made to form similar compounds with other acids than the sulphuric. Glacial phosphoric acid was heated and shaken in naphthaline, but without any particular results. A little water was then used with another portion of the materials, to bring the phosphoric acid into solution, but no decided combination could be obtained. Muriatic acid gas was brought into contact with naphthaline in various states, and at various temperatures, but no union could be effected either of the substances or their elements.

Very strong solution of potash was also heated with naphthaline, and then neutralized by sulphuric acid; nothing more, however, than common sulphate of potash resulted.

As the appropriation of a name to this acid will much facilitate future reference and description, I may perhaps be allowed to suggest that of *sulpho-naphthalic acid,* which sufficiently indicates its source and nature without the inconvenience of involving theoretical views.

Royal Institution, January 10, 1826.

On the existence of a Limit to Vaporization*.

[Read June 15, 1826.]

I⊤ is well known that within the limits recognized by experiment, the constitution of vapour† in contact with the body from which it rises, is such, that its tension increases with increased temperature, and diminishes with diminished temperature; and though in the latter case we can, with many substances, so far attenuate the vapour as soon to make its presence inappreciable to our tests, yet an opinion is very prevalent, and I believe general‡, that still small portions are produced; the tension being correspondent to the comparatively low temperature of the substance. Upon this view it has been supposed that every substance *in vacuo* or surrounded by vapour or gas, having no chemical action upon it, has an atmosphere of its own around it; and that our atmosphere must contain, diffused through it, minute portions of the vapours of all those substances with which it is in contact, even down to the earths and metals. I believe that a theory of meteorites has been formed upon this opinion.

Perhaps the point has never been distinctly considered, and it may therefore not be uninteresting to urge two or three reasons, in part dependent upon experimental proof, why this should not be the case. The object, therefore, which I shall hold in view in the following pages, is to show that a *limit* exists to the production of vapour of any tension by bodies placed *in vacuo*, or in elastic media; beneath which limit they are perfectly fixed.

Dr. Wollaston, by a beautiful train of argument and observation, has gone far to prove that our atmosphere is of finite extent, its boundary being dependent upon the opposing powers of elasticity and gravitation§. On passing upwards, from the earth's surface, the air becomes more and more attenuated, in consequence of the gradually diminishing pressure of the superincumbent part, and its tension or elasticity is pro-

* Philosophical Transactions, 1826, p. 484.

† By the term vapour, I mean throughout this paper that state of a body in which it is permanently and indefinitely elastic.

‡ See Sir H. Davy's paper on Electrical Phenomena exhibited *in vacuo.* Phil. Trans. 1822, p. 70.

§ Ibid. p. 89.

portionally diminished : when the diminution is such that the elasticity is a force not more powerful than the attraction of gravity, then a limit to the atmosphere must occur. The particles of the atmosphere there tend to separate with a certain force; but this force is not greater than the attraction of gravity, which tends to make them approach the earth and each other ; and as expansion would necessarily give rise to diminished tension, the force of gravity would then be the strongest, and consequently would cause contraction, until the powers were balanced as before.

Assuming this state of things as proved, the air at the limit of the atmosphere has a certain degree of elasticity or tension; and although it cannot there exist of smaller tension, yet if portions of it were removed to a farther distance from the earth, or if the force of gravity over it could in any other way be diminished, then it would expand, and exist of a lower tension ; upon the renewal of the gravitating force, either by approximation to the earth's surface or otherwise, the particles would approach each other, until the elasticity of the whole was again equal to the force of gravity.

Inasmuch as gases and vapours undergo no change by mere expansion or attenuation, which can at all disturb the analogy existing between them in their permanent state under ordinary circumstances, all the phenomena which have been assumed as occurring with the air at the limit of our atmosphere may, with equal propriety, be admitted with respect to vapour in general in similar circumstances ; for we have no reason for supposing that the particles of one vapour more than another are *free* from the influence of gravity, although the force may, and without doubt does, vary with the weight and elasticity of the particles of each particular substance.

It will be evident also, that similar effects would be produced by the force of gravity upon air or vapour of the extreme tenuity and feeble tension referred to, whatever be the means taken to bring it into that state ; and it is not necessary to imagine the portion of air operated upon, as taken from the extremity of our atmosphere, for a portion of that at the earth's surface, if it could be expanded to the same degree by an air-pump, would undergo the same changes: when of a certain rarity it would just balance the attraction of gravitation and fill the receiver with vapour; but then, if half were taken out of the receiver, the remaining portion, in place of filling the vessel, would submit to the force of gravity, would contract into the lower half of the receiver, until, by the approximation of the particles, the vapour there existing should have an elasticity equal to the force of gravity to which it was subject. This is a necessary consequence of Dr. Wollaston's argument.

There is yet another method of diminishing the elasticity of vapour, namely, by diminution of temperature. With respect to the most elastic substances, as air and many gases, the comparatively small range which we can command beneath common temperatures, does nothing more at the earth's surface than diminish in a slight degree their elasticity, though two or three of them, as sulphurous acid and chlorine, have been in part condensed into liquids. But with respect to innumerable bodies, their tendency to form vapour is so small, that at common temperatures the vapour produced approximates in rarity to the air upon the limits of our atmosphere; and with these, the power we possess of lessening tension by diminution of temperature, may be quite sufficient to render it a smaller force than its opponent, gravity; in which case it will be easy to comprehend that the vapour would give way to the latter, and be entirely condensed. The metal, silver for instance, when violently heated, as on charcoal urged by a jet of oxygen, or by the oxy-hydrogen or oxy-alcohol flame, is converted into vapour; lower the temperature, and before the metal falls beneath a white heat, the tension of the vapour is so far diminished, that its existence becomes inappreciable by the most delicate tests. Suppose, however, that portions are formed, and that vapour of a certain tension is produced at that temperature; it must be astonishingly diminished by the time the metal has sunk to a mere red heat; and we can hardly conceive it possible, I think, that the silver should have descended to common temperatures, before its accompanying vapour will, by its gradual diminution in tension, if uninfluenced by other circumstances, have had an elastic force far inferior to the force of gravity; in which case, that moment at which the two forces had become equal, would be the last moment in which vapour could exist around it; the metal at every lower temperature being perfectly fixed.

I have illustrated this case by silver, because from the high temperature required to make any vapour appreciable, there can be little doubt that the equality of the gravitating and elastic forces must take place much above common temperatures, and therefore within the range which we can command. But there is, I think, reason to believe that the equality in these forces, at or above ordinary temperatures, may take place with bodies far more volatile than silver; with substances indeed which boil under common circumstances at 600° or 700° F.

If, as I have formerly shown*, some clean mercury be put at the bottom of a clean dry bottle, a piece of gold-leaf attached to the under part of the stopper by which it is closed, and the whole left for some months at a temperature of from 60° to 80°, the gold-leaf will be found whitened by amalgamation, in consequence of the vapour which rises from the mercury beneath; but upon making the experiment in the winter of 1824–25, I was unable to obtain the effect, however near the gold-leaf was brought to the surface of the mercury; and I am now inclined to believe it was so because the elastic force of any vapour which the mercury could have produced at that temperature, was less than the force of gravity upon it, and that consequently the mercury was then perfectly fixed.

Sir Humphry Davy, in his experiments on the electrical phenomena exhibited *in vacuo*, found, that when the temperature of the vacuum above mercury was lowered to 20° F. no further diminution, even down to − 20° F., was able to effect any change, as to the power of transmitting electricity, or in the luminous appearances; and that these phenomena were then nearly of the same intensity as in the vacuum made over tin†. Hence, in conjunction with the preceding reasoning, I am led to conclude that they were then produced independent of any vapour of the metals, and that under the circumstances described; no vapour of mercury existed at temperatures beneath 20° F.

Concentrated sulphuric acid boils at about 600° F., but as the temperature is lowered the tension of its vapour is rapidly diminished. Signor Bellani‡ placed a thin plate of zinc at the upper part of a closed bottle, at the bottom of which was some

---

* Quarterly Journal of Science, x. 354. See page 57.
† Phil. Trans. 1822, p. 71.   ‡ Giornale di Fisica, v. 197.

concentrated sulphuric acid. No action had taken place at the end of two years, the zinc then remaining as bright as at first; and this fact is very properly adduced in illustration of the fixedness of sulphuric acid at common temperatures. Here I should again presume, that the elastic force which tended to form vapour was surpassed by the force of gravity.

Whether it be admitted or not, that in these experiments the limit of volatilization, according to the principle of the balance of forces before stated, had been obtained, I think we can hardly doubt that such is the case at common temperatures with respect to the silver and with all bodies which bear a high temperature without appreciable loss by volatilization, as platina, gold, iron, nickel, silica, alumina, charcoal, &c.; and consequently, that, at common temperatures, no portion of vapour rises from these bodies or surrounds them; that they are really and truly fixed; and that none of them can exist in the atmosphere in the state of vapour.

But there is another force, independent of that of gravity, at least of the general gravity of the earth, which appears to me sufficient to overcome a certain degree of vaporous elasticity, and consequently competent to the condensation of vapour of inferior tension, even though gravity should be suspended; I mean the force of homogeneous attraction.

Into a clean glass tube, about half an inch in diameter, introduce a piece of camphor; contract the tube at the lamp about 4 inches from the extremity; then exhaust it, and seal it hermetically at the contracted part; collect the camphor to one end of the tube; and then, having placed the tube in a convenient position, cool the other end slightly, as by covering it with a piece of bibulous paper preserved in a moist state by a basin of water and thread of cotton; in this way, a difference in temperature of a few degrees will be occasioned between the ends of the tube, and after some days, or a week or two, crystals of camphor will be deposited in the cooled part; there will not, however, be more than three or four of them, and these will continue to increase in size as long as the experiment is undisturbed, without the formation of any new crystals, unless the difference of temperature be considerable.

A little consideration will, I think, satisfy us that, after the first formation of the crystals in the cooled part, they have the power of diminishing the tension of the vapour of camphor below that point at which it could have remained unchanged in contact with the glass, or in space; for the vapour of the camphor is of a certain tension in the cooled end of the tube, which it can retain in contact with the glass, and therefore it remains unchanged; but which it cannot retain in contact with the crystal of camphor, for there it is condensed, and continually adds to its mass. Now this can only be in consequence of a positive power in the crystal of camphor of attracting other particles to it; and the phenomena of the experiment are such as to show that the force is able to overcome a certain degree of elasticity in the surrounding vapour. There is therefore no difficulty in conceiving, that, by diminishing the temperature of a body and its atmosphere of vapour, the tension of the latter may be so far decreased, as at last to be inferior to the force with which the solid portion, by the attraction of aggregation, draws the particles to it; in which case it would immediately cause the entire condensation of the vapour.

The preceding experiment may be made with iodine, and many other substances; and indeed there is no case of distinct crystallization by sublimation* which does not equally afford evidence of the power of the solid matter to overcome a positive degree of tension in the vapour from which the crystals are formed. The same power, or the force of aggregation, is also illustrated in crystallizing solutions; where the solution has a tendency to deposit upon a crystal, when it has not the same tendency to deposit elsewhere.

It may be imagined that crystallization would scarcely go on from these attenuated vapours, as it does in the denser states of the vapours experimented upon. There is, however, no good reason for supposing any difference in the force of aggregation of a solid body, dependent upon changes in the tension of the vapour about it; and indeed, generally speaking, the method I have assumed for diminishing the tension of the vapour, namely, by diminishing temperature, would cause increase in the force of aggregation.

---

* Calomel, corrosive sublimate, oxide of antimony, naphthaline, oxalic acid, &c. &c.

Such are the principal reasons which have induced me to
believe in the existence of a limit to the tension of vapour.
If I am correct, then there are at least two causes, each of
which is sufficient to overcome and destroy vapour when reduced to a certain tension; and both of which are acting
effectually with numerous substances upon the surface of the
earth, and retaining them in a state of perfect fixity. I have
given reasons for supposing that the two bodies named, which
boil at about 600° F., are perfectly fixed within limits of low
temperature which we can command; and I have no doubt,
that nearly all the present recognized metals, the earths, carbon, and many of the metallic oxides, besides the greater
number of their compounds, are perfectly fixed bodies at common temperatures. The smell emitted by various metals when
rubbed may be objected to these conclusions, but the circumstances under which these odours are produced, are such as
not to leave any serious objections on my mind to the opinions
above advanced.

I refrain from extending these views, as might easily be
done, to the atomic theory, being rather desirous that they
should first obtain the sanction or correction of scientific men.
I should have been glad to have quoted more experiments
upon the subject, and especially relative to such bodies as
acquire their fixed point at, or somewhat below common temperatures. Captain Franklin has kindly undertaken to make
certain experiments for me in the cold regions to which he has
gone, and probably when he returns from his arduous under-
taking, he may have some contributions towards this subject.

Royal Institution, May 4, 1826.

On the Limits of Vaporization*.

I was induced some time since to put together a few remarks
and experiments on the existence of a limit to vaporization,
which were favoured with a place in the Philosophical Transactions for the year 1826†. When the experiments there mentioned were published, I arranged some others bearing upon
the same subject, but which required great length of time for

* Royal Institution Journal, 1830, i. 70.   † See page 199.

the development of their result. Four years have since elapsed, during which, the effects, if any, have been accumulating, and it is the object of this brief paper to give an account of them.

The point under consideration originally was, whether there existed any definite limit to the force of vaporization. Water at 220° sends off vapour so powerfully, and in such abundance as to impel the steam-engine; at 120° it sends off much less; at 40°, though cold, still vapour rises; below 32°, when the water becomes ice, yet the ice evaporates; and there is no cold, either natural or artificial, so intense as entirely to stop the evaporation of water, or in the open air prevent a wet thing from becoming dry.

The opinion of many, among whom were the eminent names of Sir H. Davy and Mr. Dalton, was, that though the power of evaporating became continually less with diminution of temperature, it never entirely ceased, and that therefore every solid or fluid substance had an atmosphere of its own nature about it and diffused in its neighbourhood; but which being less powerful as the body was more fixed, and the existing temperature lower, was, with innumerable substances, as the earths, metals, &c., so feeble as to be quite insensible to ordinary or even extraordinary examination, though in certain cases they might affect the transmission of electricity; or, rising into the atmosphere, produce there peculiar and strange results.

The object of my former paper was to show that a real and distinct limit to the power of vaporization existed, and that at common temperatures we possess a great number of substances which are perfectly fixed. The arguments adduced, were drawn first from the power of gravity, as applied by Dr. Wollaston, to show that the atmosphere around our globe had an external limit, and then from the power of cohesion; either of these seemed to me alone sufficient to put a limit to vaporization, and experiments upon the sufficiency of the latter force were detailed in the paper.

The conclusion was, that although such substances as ether, alcohol, water, iodine, &c. could not as such be entirely deprived of their vaporizing force, by any means we could apply to them, but, if in free space or in air, would send off a little vapour, yet there were other bodies, as iron, silver, copper,

&c., most of the metals, and also the earths, which were absolutely fixed under common circumstances, the limit of their
vaporization being passed; and further, that there were a few
bodies, the limits of whose vaporization occurred at such temperatures as to be within our command, and therefore passable
in either direction. Thus mercury is volatile at temperatures
above 30°, but fixed at temperatures below 20°, and concentrated
sulphuric acid, which boils at temperatures about 600°, is fixed
at the ordinary temperature of the atmosphere.

It is well known in the practical laboratory that vaporization
may be very importantly assisted so as to make certain processes of distillation effectual, which otherwise would fail.
Thus with the essential oils, many of them which would require a high temperature for their distillation if alone, and be
seriously injured in consequence, will, when distilled with water,
pass over in vapour with the vapour of the water at a much
lower temperature, and, being condensed, may be obtained
in their unaltered state.

It has been supposed that the vapour of the water, either by
affinity for the vapour of the essential oil or in some other way,
has increased the vaporizing force of the latter at the temperature applied, and so enabled it to distil over; but there is
no doubt that if air or any other similar elastic medium were
made to come in contact with the mass of essential oil at 212°
in equal quantity, and in a manner to represent the vapour of
water, it would, according to well-known laws, carry up the
vapour of the essential oil perhaps to an equal extent, and pass
it forward; only the facility with which the carrying agent
is condensed when it consists of steam, allows of the condensation of every particle of the essential oil vapour, whereas the
permanency of the elastic state of the air would cause it to
retain a large proportion of the vapour of the oil when cold,
and consequently a diminished result would be obtained.

There are, nevertheless, some appearances which seem to
favour the idea that water occasionally favours vaporization,
not merely in the manner referred to above, but by some peculiar process; and it was to ascertain whether substances which,
from a consideration of the general reasoning already referred
to, and the high temperature at which they sensibly volatilized,
might be considered as *fixed* at common temperatures, could,

by reason of this peculiarity, have any sensible degree of volatility, in conjunction with water or its vapour, conferred upon
them at ordinary temperatures, that the following experiments
were made.   It is well known that a theory of meteoric stones
has been founded on the supposition that the earthy and metallic
matter found in them had been raised in vapour from similar
matter upon the earth's surface; which vapours, though extremely attenuated and dilute at first, gradually accumulated,
and by some natural operation in the upper regions of the atmosphere became condensed, forming those extraordinary masses
of matter which occasionally fall to us from above.   The theory
has in its favour the remarkable circumstance, that, notwithstanding many substances occur in meteoric iron and stones,
yet there is none but what also occur on this our earth*; and
it also has a right to the favouring action of water, if there be
such an action; because vaporization is one of the most important, continual, and extensive operations that go on between the
surface of the globe and the atmosphere around it.

   In September 1826, several stoppered bottles were made
perfectly clean, and several wide tubes closed at one extremity,
so as to form smaller vessels capable of being placed within the
bottles, were prepared.   Then selected substances were put
into the tubes, and solutions of other selected substances into
the bottles: the tubes were placed in the bottles so that nothing
could pass from the one substance to the other, except by way
of vaporization.   The stoppers were introduced, the bottles tied
over carefully and put away in a dark safe cupboard, where,
except for an occasional examination, they have been left for
nearly four years, during which time such portion of the substances as could vaporize have been free to act and produce
accumulation of their specific effects.

   No. 1. The bottle contained a clear solution of sulphate of
soda with a drop of nitric acid,—the tube, crystals of muriate
of baryta.   One half or more of the water has passed by evaporation into the tube, and formed a solution of muriate of

---

* This very striking circumstance does not *prove* that aërolites in any way
originate from our planet; but then, if we could by other arguments deduce
that they were extraneous, it would lead to the conclusion that the substances
which have been used in the construction of this our globe, are the same with
those which have been used extensively elsewhere in the material creation.

baryta above crystals, but both that and the remaining solution of sulphate of soda is perfectly clear; there is not the slightest trace of sulphate of baryta in either the one or the other, so that neither muriate of baryta nor sulphate of soda appear to have volatilized with the water.

No. 2. Bottle, solution of nitrate of silver; tube, fused chloride of sodium. All the water has passed from the nitrate of silver to the salt; but there is no trace of chloride of silver either in one or the other. No nitrate of silver has sublimed with the water, nor has any chloride of sodium passed over to the nitrate.

No. 3. Bottle, solution of muriate of lime; tube, crystals of oxalic acid. The water here remained with the muriate of lime. In the tube, the oxalic acid when put in had formed a loose aggregation, with numerous vacancies, and with a very irregular upper surface about an inch below the upper edge of the tube. No particular appearances occur in the vacancies; but at the top there has evidently been a sublimation of the oxalic acid, for upon the crystals and glass new crystals in exceedingly thin plates and reflecting colour have been formed; these rise no higher in the tube than to the level of the most projecting part of the original portion of oxalic acid; no appearance of sublimation is evident above this, and it seems as if the most elevated parts of the salt have given off vapour, which has sunk and formed crystals on the neighbouring lower surfaces, but that no vapour has risen to the upper part of the tube. On examining the solution by a drop or two of pure ammonia, it was found that a slight precipitate of oxalate of ammonia occurred. The experiment shows, therefore, that oxalic acid is volatile at common temperatures, and had not only formed crystals in the tube, but had passed over to the solution of lime.

No. 4. Bottle, solution half sulphuric acid, half water; tube, crystallized common salt. No water has passed to the salt. On opening the bottle, the clear diluted sulphuric acid was examined for muriatic acid, but no trace could be found. Hence chloride of sodium has not been volatilized under these circumstances.

No. 5. Bottle, solution of muriate of lime; tube, crystals of oxalate of ammonia. The oxalate of ammonia appeared quite unchanged. The solution of muriate of lime was perfectly clear;

but when a little pure ammonia was added to it, a very faint precipitate of oxalate of lime was produced.

No. 6. Bottle, little solution of potash; tube, white arsenic in pieces and powder. This bottle was opened because of the appearances, in October 1829, having then remained three years undisturbed. The arsenious acid was to all appearance unchanged. The solution of potash was turbid and foul. On chemical examination, it proved to have acted powerfully on the glass. It had dissolved so much silica as to become a soft solid, by the action of an acid, and it had also dissolved a considerable quantity of lead; but there was no trace of arsenious acid in it; so that this substance, although abundantly volatile at 600°, had not risen in vapour when aqueous vapour and air were present at common temperatures.

No. 7 was some of the sulphuric acid used in these experiments, preserved for comparison.

No. 8. Bottle, solution half sulphuric acid, half water; tube, pieces of muriate of ammonia. When this bottle was opened, the pieces of muriate of ammonia presented no appearance of change; there was no moisture about them, nor any appearances of dissection that I could distinguish. The diluted sulphuric acid being examined by sulphate of silver, gave no evidence of muriatic acid; so that muriate of ammonia appears fixed under these circumstances.

No. 9. Bottle, a little solution of persulphate of iron; tube, crystals of the ferro-prussiate of potash. Both were unchanged; there was no appearance of prussian blue about either the crystals or solution; neither of the salts had been volatilized.

No. 10. Bottle, a little solution of potash; tube, fragments of calomel. Here the potash had acted upon the glass, as in No. 6; but, with respect to the calomel, the volatility of which was in question, there was not the slightest trace of such an effect. No black oxide nor other substance existed in the potash solution, which could allow the presumption that any calomel had passed.

No. 11. Bottle, solution of potash; tube, fragments of corrosive sublimate. Here the potash had acted on the glass as before; carbonic acid had also gained access by the stopper; so that no caustic potash was present; but there were distinct appearances of the sublimation of corrosive sublimate, and minute crystals of the substance were even attached to the under part of the stopper in the bottle. Hence corrosive sublimate is volatile at common temperatures.

No. 12 and 13. Bottles, solution of chromate of potash; tubes, in one, chloride of lead in powder, in the other nitrate of lead in crystals. In both these experiments the chromate of potash had acted upon the lead of the glass, and rendered it yellow and dim; so that no indication could be gathered relating to the volatility of the compounds of lead.

No. 14. Bottle, solution of iodide of potassium; tube, chloride of lead. Both remained unaltered; the solution of iodide was perfectly clear and colourless; no trace of the chloride of lead had passed over in vapour.

No. 15. Bottle, solution of muriate of lime; tube, crystals of carbonate of soda. A part of the water has passed to the carbonate of soda; but both it and the remaining solution of muriate of lime are perfectly clear. No portion of either salt has volatilized from one place to another.

No. 16. Bottle, dilute sulphuric acid; tube, nitrate of ammonia in fragments. The nitrate was slightly moist. The acid being examined was found to contain nitric acid, whilst the test acid, No. 7, was perfectly free from it. It would therefore appear that nitrate of ammonia is a salt volatile at common temperatures; although it is just possible that slow decomposition may take place in it, and so nitric acid or its elements pass over.

No. 17. Bottle, solution of persulphate of copper; tube, crystals of ferro-prussiate of potash. The crystals had attracted most of the water from the cupreous salt; but the solution of ferro-prussiate and that of the copper had their proper colour; neither were rendered brown; no salts had been volatilized.

No. 18. Bottle, solution of acetate of lead; tube, iodide of potassium. The acetate of lead is now dry; the iodide of potassium has taken all the water and formed a brown solution, in which there is free iodine; probably a little acetic acid has passed over and caused the change in the iodide of potassium. There is no appearance of iodide of lead in the tube, but there is in the bottle, and most probably in consequence of the vaporization of the free iodine from the solution in the tube.

From these experiments it would appear that there is no

reason to believe that water or its vapour confer volatility, even in the slightest degree, upon those substances which when alone have their limits of vaporization at temperatures above ordinary occurrence, and that consequently natural evaporation can produce no effects of this kind in the atmosphere.

It would also appear that nitrate of ammonia, corrosive sublimate, oxalic acid, and perhaps oxalate of ammonia, are substances which evolve vapour at common temperatures.

Royal Institution, Aug. 30, 1830.

Fluidity of Sulphur at Common Temperatures.*

HAVING placed a Florence flask containing sulphur upon a hot sand-bath, it was left to itself. Next morning, the bath being cold, it was found that the flask had broken, and in consequence of the sulphur running out, nearly the whole of it had disappeared. The flask being broken open, was examined, and was found lined with a sulphur dew, consisting of large and small globules intermixed. The greater number of these, perhaps two-thirds, were in the usual opake solid state; the remainder were fluid, although the temperature had been for some hours that of the atmosphere. On touching one of these drops, it immediately became solid, crystalline, and opake, assuming the ordinary state of sulphur, and perfectly resembling the others in appearance. This took place very rapidly, so that it was hardly possible to apply a wire or other body to the drops quick enough to derange the form before solidity had been acquired; by quick motion, however, it might be effected, and by passing the finger over them, a sort of smear could be produced. Whether touched by metal, glass, wood, or the skin, the change seemed equally rapid; but it appeared to require actual contact; no vibration of the glass on which the globules lay rendered them solid, and many of them were retained for a week in their fluid state. This state of the sulphur appears evidently to be analogous to that of water cooled whilst quiescent below its freezing-point. The same property is also exhibited by some other bodies, but I believe no instance is known where the difference between the

* Quarterly Journal of Science, xxi. 392.

usual point of fluidity and that which could thus be obtained is so great: it, in the present instance, amounts to 130°, and it might probably have been rendered greater if artificial cold had been applied.

On the Fluidity of Sulphur and Phosphorus at Common Temperatures*.

I PUBLISHED some time ago a short account of an instance of the existence of fluid sulphur at common temperatures†; and though I thought the fact curious, I did not esteem it of such importance as to put more than my initials to the account. I have just learned, through the 'Bulletin Universel' for September, p. 178, that Signor Bellani had observed the same fact in 1813, and published it in the 'Giornale di Fisica‡,' vol. vi. (old series). I also learn, by the same means, that M. Bellani complains of the manner in which facts and theories, which have been published by him, are afterwards given by others as new discoveries; and though I find myself classed with Gay-Lussac, Sir H. Davy, Daniell, Bostock, &c., in having thus erred, I shall not rest satisfied without making restitution, for M. Bellani in this instance certainly deserves it at my hand.

Not being able to obtain access to the original journal, I shall quote M. Bellani's very curious experiments from the 'Bulletin,' in which they appear to be fully described. "The property which water possesses of retaining its fluid state, when in tranquillity, at temperatures 10° or 15° below its freezing-point, is well known; phosphorus behaves in the same manner; sometimes its fluidity may be retained at 13° (Centigrade?) for a minute, an hour, or even many days. What is singular is, that, though water cooled below its freezing-point congeals easily upon slight internal movement, however communicated, phosphorus, on the contrary, sometimes retains its liquid state at 3°, even though it be shaken in a tube, or poured upon cold water. But, as soon as it has acquired the lowest temperature which it can bear without solidifying, the moment it is touched with a body at the same temperature, it solidifies so

---

* Quarterly Journal of Science, xxiv. 469.   † Ibid, xxi. 392; or page 212.
‡ The Italian Journal has not yet arrived in this country.

quickly, that the touching body cannot penetrate its mass. If the smallest morsel of phosphorus is put into contact with a liquefied portion, the latter infallibly solidifies, though it be only a single degree below the limit of temperature necessary; this does not always happen when the body touching it is heterogeneous.

"Sulphur presented the same phenomena as phosphorus; fragments of sulphur always produced the crystallization of cold fluid portions. Having withdrawn the bulb of a thermometer which had been plunged into sulphur at 120°, it came out covered with small globules of sulphur, which remained fluid at 60°; and having touched these one after another with a thread of glass, they became solid: although several seemed in contact, yet it required that each should be touched separately. A drop of sulphur, which was made to move on the bulb of the thermometer by turning the instrument in a horizontal position, did not congeal until nearly at 30°; and some drops were retained fluid at 15°, *i. e.* 75° of Reaumur below the ordinary point of liquefaction."

The 'Bulletin Universel' then proceeds to describe some late and new experiments of M. Bellani, on the expansion in volume of a cold dense solution of sulphate of soda during the solidification of part of the salt in it. The general fact has, however, been long and well known in this country and in France; and the particular form of experiment described is with us a common lecture illustration. The expansion, as ascertained by M. Bellani, is $\frac{2}{87}$ of the original volume of fluid.

According to the 'Bulletin,' M. Bellani also claims, though certainly in a much less decided manner than the above, the principal ideas in a paper which I have published on the existence of a limit to vaporization; and I referred back to the 'Giornale di Fisica' for 1822 (published prior to my paper), for the purpose of rendering justice in this case also. Here, however, the contact of our ideas is so slight, and for so brief a time, that I shall leave the papers in the hands of the public without further remarks. It is rather curious to observe how our thoughts had been simultaneously engaged upon the same subject. Being charged in the 'Bulletin' with quoting an experiment from a particular page in M. Bellani's memoir (which

I did from another journal, in which the experiment only was described), I turned to the original place, and there, though I found the experiment I had transferred, I also found another which I had previously made on the same subject, and which M. Bellani had quoted.

I very fully join in the regret which the 'Bulletin Universel' expresses, that scientific men do not know more perfectly what has been done, or what their companions are doing; but I am afraid the misfortune is inevitable. It is certainly impossible for any person who wishes to devote a portion of his time to chemical experiment, to read all the books and papers that are published in connexion with his pursuit; their number is immense, and the labour of winnowing out the few experimental and theoretical truths which in many of them are embarrassed by a very large proportion of uninteresting matter, of imagination, and of error, is such, that most persons who try the experiment are quickly induced to make a selection in their reading, and thus inadvertently, at times, pass by what is really good.

On a peculiar Perspective Appearance of Aërial Light and Shade*.

ONE evening last month (Aug. 19, 1826), a curious aërial phenomenon was observed from the undercliff at the back of the Isle of Wight, just above Puckaster Cove. The sky was clear, the sun had just set to those who were standing where the appearance was observed, when several enormous rays of light and shade were remarked towards the E., N.E., and S.E., all radiating in strait lines from a spot rather south of east, and just upon the horizon. They were ten or twelve in number, did not join at the place from whence they appeared to originate, but seemed to emerge from an obscure portion of surface of a convex form 8° or 9° in horizontal extent, and about the third of that in height. The rays extended from 30° to 40° on the right and left from the centre, but were of less extent as they became more vertical. They diminished gradually in intensity at the extremities until they could be traced no further. The appearance slowly faded away, some of the rays disappearing

* Quarterly Journal of Science, xxii. 81.

before others, but was observed upon the whole for about half an hour.

At first the phenomenon seemed inexplicable, but after a little consideration, was referred. (and as it appeared from after observations correctly) to an effect of aërial perspective. The rays which seemed to originate from a common centre on the east, were really only the intervals between long shadows caused by the occurrence of clouds far to the west, and were in fact passing to the place from whence they *seemed* to originate, and the circumstances of the case seem to have been as follow:—the atmosphere contained a slight haze, which allowed the sun's beams to pass forward with but little interruption, but was yet in sufficient quantity to reflect a considerable portion of light to the eye. The sun was just setting;—clouds, very far to the west and out of sight from the place where the observer stood, stopped the light wherever they interfered, and cast immense horizontal or nearly horizontal shadows along the sky, parallel to each other, and over the head of the observer. The difference between these shadows and the intervening illuminated parts, could not be observed over head or on the right or left hand, *i. e.* perpendicular to their direction, because of the want of sufficient depth, as it were, in the parts thus circumstanced, to make them visible; but as they receded from the observer in the direction from the sun, they became fore-shortened, and then, from the greater depth of mass, and consequently greater number of particles looked at, became visible. This is at least one reason why they were so visible towards the east; but another is the probable existence of more haze in that direction than towards the west, or to the right or left of the observer's situation: the rays could not be seen between the sun and the observer, though the sun was out of sight, and consequently the general light, it may be supposed, not too great; which seems to imply that less haze existed in that direction; and its presence was fully proved towards the east by the dull red colour which the moon assumed upon rising a short time after the appearance had ceased. The convergence of the rays to one spot, and that opposite the sun, was merely an effect of perspective, and requires no explanation here. (See Plate I. fig. 6.)

Although the appearance on this evening was exceedingly

beautiful and rare, and the more striking from the absence, to the observer, of the sun or clouds, and the complete insulation of the phenomenon, yet by close observation upon other evenings, it was found that partial effects of the same kind were very common, and, from the manner in which these could be observed, the explanation above given was fully confirmed. On several evenings after, when observing the sun-set from the neighbouring hill of St. Catherine, it was found that if the atmosphere was generally clear, but with compact and distinctly-formed clouds floating in it, the effect was produced. The usual appearance of rays at sun-set, diverging amongst the clouds in the west, from the sun, is well known; but even when these were not visible, upon looking to the eastern half of the hemisphere, and especially to the north or south of east towards the horizon, it was rare that some clouds could not be distinguished with long shadowy projections behind them, always converging to the spot opposite to the sun. Frequently clouds could be selected moving more immediately in the neighbourhood of the observer: of those which passed overhead, the shadows could not be observed close to the clouds; but carrying the eye onwards towards the east, the same shadows became visible, when considerably fore-shortened, and could be observed moving on and changing with the clouds themselves. All these phenomena, with their variations, were easily referable to their causes, and may be observed at almost any sun-set in fine weather; but the effect of the first evening, so similar in kind, though so different in appearance, was not again remarked. It is with a view of guarding persons who may observe the same effect, against any mistake as to its origin, that the appearance, with its nature, has been thus particularly described.

On the confinement of Dry Gases over Mercury [*].

THE results of an experiment made by myself, and quoted as such, having been deemed of sufficient interest to be doubted, I have been induced to repeat it; and though the original experiment was not published by me, I am inclined to put the latter and more careful one upon record, because of the strong

[*] Quarterly Journal of Science, xxii. 220.

illustration it affords of the difficulty of confining dry gases over mercury alone. Two volumes of hydrogen gas were mixed with one volume of oxygen gas, in a jar over the mercurial trough, and fused chloride of lime introduced, for the purpose of removing hygrometric water. Three glass bottles, of about three ounces capacity each, were selected for the accuracy with which their glass stoppers had been ground into them; they were well cleaned and dried, no grease being allowed upon the stopper. The mixture of gases was transferred into these bottles over the mercurial trough, until they were about four-fifths full, the rest of the space being occupied by the mercury. The stoppers were then replaced as tightly as could be, the bottles put into glasses in an inverted position, and mercury poured round the stoppers and necks, until it rose considerably above them, though not quite so high as the level of the mercury within. Thus arranged they were put into a cupboard, which happened to be dark, and were sealed up. This was done on June 28, 1825, and on September the 15th, 1826, after a lapse of fifteen months, they were examined. The seals were unbroken, and the bottles found exactly as they were left, the mercury still being higher on the inside than the outside. One of them was taken to the mercurial trough, and part of its gaseous contents transferred; upon examination it proved to be common air, no traces of the original mixture of oxygen and hydrogen remaining in the bottle. A second was examined in the same manner; it proved to contain an explosive mixture. A portion of the gas introduced into a tube, with a piece of spongy platina, caused dull ignition of the platina; no explosion took place, but a diminution to rather less than one-half. The residue supported combustion a little better than common air. It would appear, therefore, that nearly a half of the mixture of oxygen and hydrogen had escaped from it, and been replaced by common air. The third bottle, examined in a similar manner, yielded also an explosive mixture, and upon trial was found to contain nearly two-fifths of a mixture of oxygen and hydrogen, the rest being very little better in oxygen than common air.

There is no good reason for supposing that this capability of escape between glass and mercury is confined to the mixture here experimented with; probably every other gas, having no action on the mercury or the glass, would have made its way out in the same manner.   There is every reason for believing that a small quantity of grease round the stoppers would have made them perfectly tight.

On the probable Decomposition of certain Gaseous Compounds of Carbon and Hydrogen during sudden expansion* .

SOME very singular appearances have been observed by Mr. Gordon, of the Portable-Gas Works, which have led him to believe that chemical changes are occasioned by the sudden expansion of oil-gas, which do not happen when the expansion is gradual; a striking result of the change being the separation of carbon from the gas.   The effect referred to is exhibited when oil-gas, compressed into vessels by a power equal to that of thirty atmospheres, is suddenly allowed to escape through a small aperture into the air.   It was first observed accidentally, in consequence of the derangement of the valve of a large apparatus, into which the gas had been compressed to twenty-seven atmospheres.   The gas escaped with immense velocity, and when an examination took place of what had happened, it was found that all the metallic part of the valve upon which the gas had rushed was covered with a black, moist carbonaceous substance, and the contiguous brick wall with dry, black carbon, the moisture in this case having been absorbed by the brick.   Since that time, Mr. Gordon has repeatedly shown the effect, by allowing the gas to rush out with very great violence from a portable lamp against a piece of white paper, which becomes immediately covered with black carbonaceous deposit.

The general conclusion is, that as the gas thus rapidly expands, a partial decomposition takes place and carbon is separated.   If this explanation should ultimately prove by further experiments to be true, it will be highly important, as affording an instance of the exertion of mechanical and chemical powers in those circumstances where they most closely verge upon each other.   At present, we have but little knowledge of such phenomena, though the announcement in France of the production of several new compound bodies, possessed of peculiar

* Quarterly Journal of Science, xxiii. 204.

properties, solely by the exertion of physical powers, may lead us to hope for an accession of information on the subject; that which we thought we had, was in part rendered uncertain by the contrary conclusions arrived at by Mr. Perkins and Dr. Brewster, the one believing that in a case of crystallization the effect was produced entirely in consequence of pressure*; the other, that pressure had been the only cause why bodies, otherwise ready to crystallize, had retained the fluid state.

A natural suspicion, upon first hearing of and seeing the results obtained by Mr. Gordon, was, that the rapidity of the current of gas had carried away a minute portion of the metal from the surface of the valve past which it rushed, or of the interior of the air-way against which it was thrown, and that that metal had caused the stain upon the paper; but upon examination this proved not to be the case; for the black deposit upon a card, when subjected to acids, remained insoluble, and when burnt and tested chemically, gave no traces of copper.

Further examination of the substance showed that it was not a pure carbon, but one of those compounds, containing a very large proportion of carbon, combined with a small quantity of hydrogen; being analogous to tar, pitch, or asphaltum. It dissolved readily in the fluid hydrocarbons obtained by the compression of oil-gas. As these black carbonaceous compounds are formed in the process of making oil-gas, a suspicion cannot but arise, that the effect observed *may* have been produced by the current of gas having swept off small portions of such substances previously deposited, or slowly formed in the interior of the vessels at former periods; and have left them upon the wall in the accidental result, or upon the paper placed in the current of the gas, when the effect has been purposely shown.

It may, however, be remarked, that in experiments made in the laboratory of the Royal Institution† upon the fluid product obtained by condensing oil-gas at high pressures, it was observed, after rectifying the products and separating the more fixed from the more volatile, that although they were perfectly clear and transparent at first, yet by spontaneous evaporation through the corks which closed the vessels, and after a lapse of time, chemical changes were produced; for

* Philosophical Transactions.   † page 154.

ultimately there remained nothing in several of the receivers but a brown substance, heavy, adhesive like honey or treacle, and in certain cases even almost solid.   From the circumstances of the experiments, no hesitation could arise in concluding that a spontaneous chemical change had taken place; and it does not seem at all unlikely that a similar change, or one to a much greater extent, may have occurred *suddenly* during the rapid alteration in the mechanical condition of the gas in Mr. Gordon's experiment; the most condensable of the substances in the mixture of elastic matters which constitute oil-gas, being perhaps those which are most altered, and in that case Mr. Gordon's account of the phenomena would be correct.

Transference of Heat by change of Capacity in Gas.*

MANY of the copper vessels in which gas is compressed at the Portable-Gas Works are cylinders from two to three feet in length, terminated by hemispherical ends.   These are attached at one end to the system of pipes by which the gas is thrown in, and being so fixed, the communication is opened; it frequently happens, that gas previously at the pressure of thirty atmospheres in the pipes and attached recipients, is suddenly allowed to enter these long gas-vessels, at which time a curious effect is observed.   That end of the cylinder at which the gas enters becomes very much cooled, whilst, on the contrary, the other end acquires a considerable rise of temperature. This effect is produced by change of capacity in the gas; for, as it enters the vessel from the parts in which it was previously confined, at a pressure of thirty atmospheres it suddenly expands, has its capacity for heat increased, falls in temperature, and consequently cools that part of the vessel with which it first comes in contact; but the part which has thus taken heat from the vessel being thrust forward to the further extremity of the cylinder by the successive portions which enter is there compressed by them, has its capacity diminished, and now gives out that heat, or a part of it, which it had the moment before absorbed; this it communicates to the metal of that part of the gas-vessel in which it is so compressed, and raises

* Quarterly Journal of Science, xxiii. 474.

its temperature. Thus the heat is actually taken up by the gas from one end of the cylinder, and conveyed to the other, occasioning the difference of temperature spoken of. The effect is best observed when, as before stated, the gas, at a pressure of thirty atmospheres, is suddenly let into the vessels: the capacity of the recipients is such, that the pressure usually sinks to about ten atmospheres.

Experiments on the nature of Labarraque's *Disinfecting Soda Liquid* *.

1. THE following experimental investigations relate to the nature of that medicinal preparation which M. Labarraque has lately introduced to the world, and named *Chloride of oxide of Sodium.* They were occasioned by the accounts which were given of this and other substances of similar power to the Members of the Royal Institution, at two of their Friday evening meetings; the value of the preparation, the uncertainty of its nature, and the inaccuracy of its name, all urging the inquiry.

2. In the first instance the inquiry was directed to the nature of the action exerted by chlorine gas upon a solution of carbonate of soda, questions having arisen in the minds of many, whether it was or was not identical with the action exerted by the same gas on a solution of the caustic alkali, and whether carbonic acid was evolved during the operation or not. Chlorine gas was therefore carefully prepared, and after being washed was sent into a solution of carbonate of soda, in the proportions directed by M. Labarraque: *i. e.* 2800 grains of crystallized carbonate of soda were dissolved in 1·28 pint of water; and being put into a Woulfe's apparatus, two-thirds of the chlorine evolved from a mixture of 967 grains of salt with 750 grains of oxide of manganese, when acted upon by 967 grains of oil of vitriol, previously diluted with 750 grains of water, were passed into it; the remaining third being partly dissolved in the washing water, and partly retained in the open space of the retort and washing vessel. The operation was conducted slowly, that as little muriatic acid as possible might

* Quarterly Journal of Science, xxiv. 84.

be carried over into the alkali. The common air ejected from the bottle containing the solution was collected and examined; but from the beginning to the end of the operation not a particle of carbonic acid was disengaged from the solution, although the chlorine was readily absorbed. Ultimately a liquid of a very pale yellow colour was obtained, being the same as M. Labarraque's soda liquor, and with which the investigations were made that will hereafter be described.

3. An experiment was then instituted, in which the effect of excess of chlorine, upon a solution of carbonate of soda of the same strength as the former, was rendered evident. The solution was put into two Woulfe's bottles, the chlorine well washed and passed through, until ultimately it bubbled through both portions without absorption of any appreciable quantity. As soon as the common air was expelled, the absorption of the chlorine was so complete in the first bottle, that no air or gas of any kind passed into the second, a proof that carbonic acid was not liberated in that stage of the experiment. Continuing the introduction of the chlorine, the solution in the first bottle gradually became yellow, the gas not being yet visible by its colour in the atmosphere above the solution, although chlorine could be detected there by litmus paper. Up to this time no carbonic acid gas had been evolved; but the first alkaline solution soon acquired a brighter colour, and now carbonic acid gas began to separate from all parts of it, and passing over into the second bottle, carried a little chlorine with it. The soda solution in the first bottle still continued to absorb chlorine, whilst the evolution of carbonic acid increased, and the colour became heightened. After some time the evolution of carbonic acid diminished, smaller quantities of the chlorine were absorbed by the solution, and the rest passing into the atmosphere in the bottle, went from thence into the second vessel, and there caused the same series of changes and actions that had occurred in the first. The solution in the first bottle was now of a bright chlorine yellow colour, and the gas bubbled up through it as it would through saturated water.

4. When the chlorine had saturated the soda solution in the second bottle, and an excess of gas sufficient to fill several large jars had been passed through the whole apparatus, the latter was dismounted, the solutions put into bottles and distinguished

as the saturated solutions of carbonated soda ; they were of a bright greenish yellow colour, and had an insupportable odour of chlorine.

5. The saturated solution (4) was then examined as to the change which had been occasioned by the action of the chlorine. It bleached powerfully, and apparently contained no carbonated alkali; but when a glass rod was dipped into it and dried in a warm current of air, the saline matter left, when applied to moistened turmeric paper, reddened it considerably at first, and then bleached it; and this piece of paper being dried and afterwards moistened upon the bleached part, gave indications of alkali to fresh turmeric paper.

6. A portion of the saturated solution (4) being warmed, instantly evolved chlorine gas, then assumed a dingy appearance, and ultimately became nearly colourless; after which it had an astringent and saline taste. Being evaporated to dryness at a very moderate temperature, it left a saline mass consisting of much common salt, a considerable quantity of chlorate of soda, and a trace of carbonate of soda. This mixture had no bleaching powers. The dingy appearance, assumed in the first instance, was found to be occasioned by a little manganese which had passed over into the solutions, notwithstanding the care taken in evolving and washing the gas.

7. From these experiments it was evident that when chlorine was passed *in excess* into a solution of carbonate of soda (3), the carbonic acid was expelled, and the soda acted upon as if it were caustic, a mixture of chloride of sodium and chlorate of soda being produced; with the exception of the small portion of carbonate of soda which, it appears, may remain for some time in the solution in contact with the excess of chlorine at common temperatures, without undergoing this change. The quantities of chloride of sodium and chlorate of soda were not ascertained, no doubt being entertained that they were in the well-known proportions which occur when caustic soda is used.

8. The Labarraque's soda liquor which had been prepared as described (2), was now examined relative to the part the chlorine played in it, or the change the alkali had undergone, and was soon found to be very different to that which has just been described, as indeed the experiments I had seen made

by Mr. Phillips* led me to expect. The solution had but little odour of chlorine; its taste was at first sharp, saline, scarcely at all alkaline, but with a persisting astringent biting effect upon the tongue. When applied to turmeric paper, it first reddened and then bleached it.

9. A portion of the solution (2) being boiled, gave out no chlorine; it seemed but little changed by the operation, having the same peculiar taste, and nearly the same bleaching power as before. This is a sufficient proof that the chlorine, though in a state ready to bleach or disinfect, must not be considered as in the ordinary state of solution, either in water or a saline fluid; for ebullition will freely carry off the chlorine under the latter circumstances.

10. A portion evaporated on the sand-bath rather hastily, gave a dry saline mass, quite unlike that left by the *saturated solution* already described (6); and which, when dissolved, had the same astringent taste as before, and bleached solution of indigo very powerfully: when compared with an equal portion of the unevaporated solution which had been placed in the mean time in the dark, its bleaching power upon diluted sulphate of indigo was 30, that of the former being 76. Another portion evaporated in a still more careful manner, gave a mass of damp crystals, which, when dissolved, had the taste, smell, and bleaching power of the original solution, with almost equal strength.

11. These experiments showed sufficiently that the whole of the chlorine had not acted upon the carbonate of soda to produce chloride of sodium and chlorate of soda; that much was in a peculiar state of solution or union which enabled it to withstand ebullition, and yet to act freely as a bleaching or disinfecting agent; and that probably little or none had combined with the sodium, or been converted into chloric acid. To put these ideas to the test, two equal portions of the Labarraque solution were taken; one was placed in a large tube closed at one extremity, diluted sulphuric acid was added till in excess, and then air blown through the mixture by a long small open tube, proceeding from the mouth, for the purpose of carrying off the chlorine; the contents of the tube

* Phil. Mag. N. S. i. 376.

Q

were then heated nearly to the boiling-point, air being continually passed through. In this way all the chlorine which had combined with the carbonated alkali without decomposing it, was set free by the sulphuric acid, and carried off by the current of air and vapour, whilst any which had acted chemically upon the alkali would, after the action of the sulphuric acid, be contained in solution as muriatic and chloric acids, and from the diluted state of the whole, would not be removed by the after process, but remain to be rendered evident by tests. The other portion being diluted, had sulphuric acid added also in excess, but no attempt was made to remove the chlorine. Equal quantities of these two portions in the same state of dilution were then examined by nitrate of silver for the quantities of chlorine sensible in them, and it was found that the latter portion, or that which retained the whole of the chlorine thrown into it, contained above sixty times as much as the former.

12. Now although it may be supposed that in the former portion that part of the chlorine which, in acting energetically, had produced chloric acid, could not be detected by the nitrate of silver, yet more than a sixth of the small portion which remains cannot be thus hidden; and even that quantity is diminished by the sulphuric acid present in excess, which tends to make the chlorine in the chlorate sensible to nitrate of silver: so that the experiment shows that nearly 59 parts out of 60 of the chlorine in M. Labarraque's liquid are in a state of weak combination with the carbonated alkali, and may be separated by acids in its original condition; that this quantity is probably wholly available in the liquid when used as a bleaching or disinfecting agent; that little, if any, of the chlorine forms chloride of sodium and chlorate of soda with the alkali of the solution; and that the portion of chlorine used in preparing the substance which is brought into an inactive state, is almost insensible in quantity.

13. The peculiar nature of this compound or solution, with the results Mr. Phillips had shown me (8), obtained by evaporation of a similar preparation to dryness, induced me to try the effects of slow evaporation, crystallization, heat, and air upon it. In the first place, five equal portions of the solution prepared by myself were measured out: two were put into

stoppered bottles, two were put into basins and covered over with bibulous paper, and one was put into a basin and left open; all were set aside in an obscure place, and remained from July 16th to August 28th. Being then examined, the portions in the basins were found crystallized and dry; the crystals were large and flat, striated and imperfect, resembling those formed in a similar way from carbonate of soda. They were not small and acicular, were nearly alike in the three basins, and had effloresced only on a few minute points. A part of one portion, when dissolved, gave a solution having an alkaline taste, without any of the pungency of Labarraque's liquid; and when tested by turmeric paper, it reddened, but did not bleach it.

14. One of these portions, that had effloresced least, was selected, and being dissolved, was compared in bleaching power upon diluted sulphate of indigo, with one of the portions of solution that had been preserved in bottles. The former had scarcely any visible effect, though sulphuric acid was added to assist the action; a single measure of the indigo liquor coloured the solution permanently blue, whereas seventy-seven such measures were bleached by the portion from the bottle. Hence the process of slow crystallization had either almost entirely expelled the chlorine, or else had caused it to react upon the alkali, and by entering into strong chemical combination as chloride and chlorate, had rendered it inert as a bleaching or disinfecting agent.

15. From the appearance of the crystals there was no reason to expect the latter effect; but to put the question to the proof, one of the evaporated portions, and one of the fluid portions contained in the bottles, were acted upon by sulphuric acid, heat, and a current of air, in the manner already described (11), to separate the chlorine that had not combined as chloride or chlorate. They were then compared with an equal portion of the solution, which retained all its chlorine, nitrate of silver being used as before; the quantity of chloride indicated for the latter portion was 60 parts; whilst that of the fluid portion deprived of as much free chlorine as could be, by sulphuric acid and blowing, was 6 parts; and for the evaporated and crystallized portion, similarly cleared of free chlorine, only 1·5 part.

16. This result, as compared with the former experiment of a similar kind (11), showed, that though reaction of the chlorine on the carbonate had taken place in the evaporated portion, it was only to a very slight extent, since the chlorine was almost as much separated from it by the process altogether, as it had been from the recent preparation by sulphuric acid, blowing and heat. The experiment showed also that there was a gradual reaction of the chlorine and alkali in the fluid preparation, proceeding to a greater extent than in the evaporated portion; for chlorine, equal to five parts, was found by the nitrate of silver to remain. Hence this preparation is one which deteriorates even in the small space of forty-three days. Whether the effect will proceed to any great extent, prolonged experiments only can show.

17. From an experiment made upon larger quantities of the Labarraque liquor, it would appear that the force of crystallization alone is sufficient to exclude the chlorine. A quantity was put into an evaporating basin, and left covered over with paper from July 16th to August 28th. Being then examined, a few large crystals were found covered over with a dense solution; the whole had the innocuous odour of Labarraque's fluid, and the fluid the usual acrid, biting taste. The crystals being separated, one of the largest and most perfect was chosen, and being well wiped on the exterior, and pressed between folds of bibulous paper, was rubbed down in water, so as to make a saturated solution. This had no astringent taste like that of Labarraque's fluid, or the mother-liquor, but one purely alkaline; and when applied to turmeric paper, reddened, but did not bleach it. Equal portions of this saturated solution and of the mother-liquor were then compared in bleaching power, acid being added to the former to assist the effect : it was found, notwithstanding that portions of mother-liquor must have adhered to the crystal, that its solution had not $\frac{1}{21}$st part the power of the mother-liquor. This, in conjunction with the other experiments, is a striking instance of the manner in which the carbonate of soda acts, as a simple substance, with the chlorine in the solution. The crystal itself had never been in contact with the air: but whether it should be considered as the excess of carbonate of soda only which crystallized; or whether it is essential to the formation of these crystals that

chlorine should simultaneously be given off into the air; or what would take place, if the water were abstracted without the evolution of chlorine, I have not determined.

18. Notwithstanding the perfect manner in which the chlorine may be thus separated by crystallization and slow evaporation to dryness, yet it is certain that by quick evaporation a substance apparently quite dry may be obtained, which yet possesses strong bleaching power. In one experiment, where, of two equal portions, one had been evaporated in the course of twenty-four hours to dryness upon the warm part of a sand-bath, it, when compared with the former, had not lost more than one-third of its bleaching power.

19. With the desire of knowing what effect carbonic acid would have on Labarraque's fluid, and whether it possessed in a greater or smaller degree the power of ordinary acids to expel the chlorine, portions of the solution were put into two Woulfe's bottles, and a current of carbonic acid gas passed through them. The gas was obtained from sulphuric acid and whitening in a soda-water apparatus, and was well washed in water. The stream of gas brought away small portions of chlorine with it, but they were not sensible to the smell, and could only be detected by putting litmus paper into the current. An immense quantity of gas, equal to nearly 1300 times the volume of the fluid, was sent through; but yet very little chlorine was removed, and the bleaching powers of the fluid were but little diminished, though it no longer appeared alkaline to turmeric paper. Air was then passed through the solution in large quantity; it also removed chlorine, but apparently not quite so much as carbonic acid.

20. One other experiment was made upon the degree in which the carbonate of soda in Labarraque's liquor resisted decomposition by the chlorine, even at high temperature. Two equal portions of the fluid were taken, and one of them boiled rapidly for fifteen minutes; both were then acted upon by sulphuric acid, blowing and heat, as described (11), and the two were then tested by nitrate of silver, to ascertain the quantity of chlorine remaining: it was nearly three times as much in the boiled as in the unboiled portion; and by comparing this with the results before obtained (11), it will be seen, that after boiling for a quarter of an hour, not more than a twentieth

part of the chlorine had acted upon the alkali to form chloride and chlorate.

21. It would seem as if I were unacquainted with Dr. Granville's paper upon this subject, in the Quarterly Journal of Science, p. 371, were I to close my remarks without taking any notice of it. Unfortunately Dr. Granville has mistaken M. Labarraque's direction, and by passing chlorine, to "complete saturation," through the carbonate, instead of using the quantities directed, has failed in obtaining Labarraque's really curious and very important liquid; to which, in consequence, not one of his observations or experiments applies, although the latter are quite correct in themselves.

Royal Institution, Sept. 3, 1827.

Anhydrous Crystals of Sulphate of Soda*.

IF a drop of a solution of sulphate of soda be placed upon a glass plate and allowed to evaporate spontaneously, it will leave crystals which may be distinguished by their form and ultimate efflorescence, as being the salt in question. Most of the potash and soda salts may be distinguished as to their base by such an experiment. They are easily converted into sulphates by a drop or two of sulphuric acid and ignition, and then, being dissolved and tried as above, will yield crystals which may be known by their forms, and more especially by their efflorescence if of soda, and their unchangeable state if of potash. This test is, however, liable in certain circumstances to uncertainty, arising from a curious cause. If the drop of solution on the glass be allowed to evaporate at common temperatures, then the efflorescence takes place and the distinction is so far perfect; but if the glass plate with the drop upon it be placed upon a warm part of a sand-bath or hot iron plate, or in any other situation of a certain temperature, considerably beneath the boiling-point of the solution, the crystals which are left upon evaporation of the fluid are smaller in quantity, more similar in appearance to sulphate of potash, and finally *do not* effloresce. Upon examining the cause of this

* Quarterly Journal of Science, xxv. 223.

difference, I found they were *anhydrous*; consequently incapable of efflorescing, and indeed exactly of the same nature as the crystals obtained by Dr. Thompson from certain hot saturated leys*.

Hence it would appear that a mere difference in the temperature at which a solution of sulphate of soda is evaporated, will cause the formation of hydrated or anhydrous crystals at pleasure, and that whether the quantity of the solution be large or small. This, indeed, might have been expected from that which takes place when hydrated crystals of sulphate of soda are carefully melted; a portion dissolves, and a portion separates, the latter in an anhydrous state*. I find that, if it were desirable, crystallized anhydrous sulphate of soda might easily be prepared for the market; though, as the pure salt is now but little used, it is not likely this condensed form will be required. Whenever a soda salt is to be distinguished from one of potash, in the manner above described, this effect of temperature must be carefully guarded against.

The Bakerian Lecture.—*On the Manufacture of Glass for Optical Purposes*†.

[Read November 19, December 3 and 10, 1829.]

Introduction.

Perfect as is the manufacture of glass for all ordinary purposes, and extensive the scale upon which its production is carried on, yet there is scarcely any artificial substance in which it is so difficult to unite what is required to satisfy the wants of science. Its general transparency, hardness, unchangeable nature and varied refractive and dispersive powers, render glass a most important agent in the hands of the philosopher engaged in investigating the nature and properties of light; but when he desires to apply it, according to the laws he has discovered, in the construction of perfect instruments, and especially of the achromatic telescope, it is

---

* Quarterly Journal of Science, xxii. 399, or Ann. Phil. N. S. xx. 401.

† Philosophical Transactions, 1830, p. 1. The use of the glass manufactured as described has since become so important in diamagnetic and magneto-optical researches, that I deem the paper worthy of insertion at full length in this collection.—M. F. 1858.

found liable to certain imperfections, not essentially existing, but almost always involved during its preparation, and fatal to its use. These are so important and so difficult to avoid, that science is frequently stopped in her progress by them; a fact fully proved by the circumstance that Mr. Dollond, one of our first opticians, has not been able to obtain a disc of flint-glass four inches and a half in diameter, fit for a telescope, within the last five years, or a similar disc of five inches in diameter within the last ten years.

It must be well known to the scientific world, that these difficulties have induced some persons to labour hard and earnestly for years together, in hopes of surmounting them. Guinand was one of these: his means were small, but he deserves the more honour for his perseverance and his success. He commenced the investigation about the year 1784, and died engaged in it in the year 1823. Fraunhofer laboured hard at the solution of the same practical problem. He was a man of profound science, and had all the advantages arising from extensive means and information, both in himself and others. He laboured in the glass house, the workshop, and the study, pursuing without deviation the great object he had in view, until science was deprived of him also by death. Both these men, according to the best evidence we can obtain, have produced and left some perfect glass in large pieces: but whether it is that the knowledge they acquired was altogether practical and personal, a matter of minute experience, and not of a nature to be communicated; or whether other circumstances were connected with it,—it is certain that the public are not in possession of any instruction, relative to the method of making a homogeneous glass fit for optical purposes, beyond what was possessed before their time; and in this country it seems doubtful whether they ever attained a method of making such glass with certainty and at pleasure, or have left any satisfactory instructions on the subject behind them.

The philosophical deficiencies referred to above, induced the President and Council of the Royal Society in 1824, to appoint a Committee for the improvement of glass for optical purposes, consisting of Fellows of the Royal Society and Members of the then Board of Longitude. The Government on being applied to, not only removed the restrictions to experiments on glass, occasioned by the Excise laws and regulations, but undertook to bear all the expenses of furnaces, materials, and labour, as long as the investigations offered a reasonable hope of success. In consequence of these facilities, a small glass furnace was erected in 1825, and many experiments both upon a large and small scale were made with flint and other glasses. During their continuance, Messrs. Green and Pellatt gave every instruction and assistance in their power, and evinced the most earnest desire for success. The researches, however, soon showed themselves to be a work of labour, which, to be successful, would require to be pursued unremittingly for a long period; and on May 5, 1825, a sub-committee was appointed, to whom the direct superintendence and performance of experiments were entrusted. This committee consisted of Mr. Herschel, Mr. Dollond, and myself; but in March 1829 was reduced to two, by the retirement of Mr. Herschel, who about that period went to the continent. From the respective pursuits of the three persons appointed upon this committee it may be easily gathered, that though all were to do what they could in every way for the general good of the cause in which they were jointly engaged, yet a distinction in the duties of each existed. It was my business to investigate particularly the chemical part of the inquiry; Mr. Dollond was to work and try the glass, and ascertain practically its good or bad qualities; whilst Mr. Herschel was to examine its physical properties, reason respecting their influence and utility, and make his competent mind bear upon every part of the inquiry.

The experimental glass house was erected on a part of the premises of Messrs. Green and Pellatt, at the Falcon Glassworks; whilst my duties as Director of the Laboratory of the Royal Institution, required my presence almost constantly at the latter place, nearly three miles from the former. As I found it impossible under these circumstances to make the numerous experiments and pay that close attention which appeared essentially necessary to produce any degree of success, the President and Council of the Royal Society applied to the President and Managers of the Royal Institution, for leave to erect on their premises an experimental room, with a furnace, for the purpose of continuing the investigation. They were guided in this by the desire which the Royal Institution has always evinced to assist in the advancement of science; and the readiness with which the application was granted, showed that no mistaken notion had been formed in this respect. As a member of both bodies, I felt much anxiety that the investigation should be successful. A room and furnaces were built at the Royal Institution in September 1827, and an assistant was engaged, Sergeant Anderson of the Royal Artillery, whose steady and intelligent care has been of the greatest service to me in the experiments that have been proceeding constantly from that time to the present. At first, the inquiry was pursued principally as related to flint and ground glass; but in September 1828 it was directed exclusively to the preparation and perfection of peculiar heavy and fusible glasses, from which time to the present continual progress has been made.

I have thought it right to give this brief explanatory statement of the manner in which it has happened to become my duty, on the present occasion, to give an account of what has been done in the improvement of glass for optical purposes by the Committee of the Royal Society, working at the Royal Institution. I would willingly have deferred this account until the inquiry were more complete than at present; for though glass has been made, and telescopes manufactured, yet I have no doubt that much more of improvement will be effected. It may be said that a long time has elapsed since the experiments were first instituted; and that if anything could be done, it should have been effected in so long a period. But be it remembered, that it is not a mere analysis, or even the development of philosophical reasoning, that is required: it is the solution of difficulties, which, as in the cases of Guinand and Fraunhofer, required many years of a practical life to effect, if it was ever effected. It is the foundation and development of a manufacturing process, not in principle only, but through all the difficulties of practice, until it is competent to give constant success: and I may be allowed to plead the acknowledged difficulty and importance of the subject as a reason, both why it may not yet have obtained perfection, and why it should still be pursued.

My wish, however, to delay the account of the researches until I could have carried the experiments further, is overcome by the conviction that much more time must be expected to elapse before I shall consider the investigation finished; by the consideration that a decided step has been made in the manufacture of glass for optical purposes; and by the feeling that the Royal Society which instituted, and the Government which defrays the expenses of the experiments, have a right to an official account of the present state of the investigation. Although much useful information has been obtained respecting flint and other glasses, yet as that train of research is very imperfect, uncertain, and will probably be resumed, I shall confine my present statement altogether to the heavy optical glass already referred to. It will be impossible for me to describe all that has been done on this subject; but I shall endeavour to give such an account of the glass, and the process by which it is obtained in a homogeneous state, as shall enable other persons to do what has been done at the Royal Institution, without incurring the laborious prefatory experiments and investigations which we have had to undertake; only introducing so much of the latter, and the principles of the process, as are necessary to make the descriptions clear to a practical man, and enable him to avoid those circumstances which might otherwise occasion failure. That the paper may appear long and tedious I am aware; but it should be remembered, that it can have no other utility than as containing efficient instructions to the few who may desire to manufacture optical glass; and that to render whatever of this character it may have, imperfect, for the sake of giving to it a more abbreviated and popular form, would have been doing injustice to the objects and motives of those who have instituted and supported the experiments.

§ 1. *Process of Manufacture, &c.*

1. The general properties of transparency, hardness, and a certain degree of refractive and dispersive power, which render glass so valuable as an optical agent, are easily obtained : but there is one condition essential in all delicate cases of its application, which is not so readily fulfilled ; this is, a perfectly homogeneous composition and structure. Although every part of the glass may in itself be as good as possible, yet without this condition they do not act in uniformity with each other; the rays of light are deflected from the course which they ought to pursue, and the piece of glass becomes useless. The streaks,

striæ, veins or tails, which are seen within glass otherwise perfectly good, result from a want of this equality; they are visible only because they bend the rays of light which pass through them from their rectilinear course, and are constituted of a glass having either a greater or a smaller refractive power than the neighbouring parts.

2. When these irregularities are so powerful as to render their effects observable by the naked eye, it may easily be supposed to what an injurious extent their influence must extend in the construction of telescopes and other instruments of a similar nature, where these faults are not only magnified many times, but where the effect is to give an equally magnified erroneous representation of the object looked at, when the very point to be attained is to examine that object with the utmost accuracy; and it is accordingly found that these striæ are the most fatal faults of glass intended for optical purposes. Besides this, not only do the striæ themselves occasion harm, but there is every reason to believe that they rarely occur in glass otherwise homogeneous. Sometimes, it is true, a grain of sand, in passing through and at the same time dissolving in glass, will give a streak of different composition to the rest of the substance; and at others, a bubble ascending may lift a line of heavy or more refractive matter into a lighter and less refractive portion above. But very often, and especially as glass is usually manufactured and collected for use, striæ are merely the lines or planes where two different kinds of glass approximate; and even if the striæ could be covered so as to produce no bad effect, yet the other parts, not being in every respect alike, would exert an unequal action on light, and the piece be therefore improper for the construction of a telescope. Many a disc, which upon the most careful examination has appeared perfectly free from striæ and quite uniform, has, when worked into an object-glass, been found incapable of giving a good image, on account of the existence of irregularities in the mass, which, though not sudden or strong enough to occasion striæ, still produce a confused effect; and if this happens with glass approaching so near to perfection, it happens still more frequently and to a much stronger degree with such as contain visible irregularities.

3. It must not be imagined that striæ, or those fainter differ-

ences, are, according to an expression sometimes used, due to impurity. The glass, either of the streak or of the neighbouring parts, would be equally good for optical purposes were it all alike. It is the irregularity that constitutes the fault; and hence, in this respect, a particular composition is of very little importance. As glass is always the result of a mixture of materials having different refractive and dispersive powers, it is evident that striæ must exist at one period during its preparation; and the point required is not so much to seek for a difference of composition, or for those proportions which are found by analysis to exist in specimens of tried and acknowledged good glass, as to devise and perfect a process by which the striæ period should be passed over before the glass is finished and the formation of fresh striæ be prevented.

4. Besides these, there are other faults in glass. Sometimes it is said to be wavy, when it has the appearance of waves within its mass; but this is only a variety of that irregularity which has just been explained as constituting, when in a stronger degree, streaks and striæ. Occasionally appearances are observed in it, which seem to indicate a peculiar structure of crystallization, or an irregular tension of its parts: these, there is every reason to believe, may be avoided by careful annealing. Again: the glass sometimes includes bubbles, which, when small and numerous, render it what is called seedy. Bubbles are not usually considered as of much consequence to the performance of the glass, but objectionable only because of their appearance when the glass is looked at, rather than when looked through. They each act like a very powerful but very small double convex lens of a rare substance in a very dense medium, or as equally deep double concave lenses of glass would do in air; they rapidly, therefore, turn the rays impinging on them on one side, and occasion a loss of light, just as so many opake spots would do. But as even when numerous their united area may amount to only a very small proportion of the area of the plate of glass required for a telescope, this loss of light is usually of but little consequence. In practice, it is said that no other real evil than such loss of light is dependent on them.

5. Of all these faults, that of the irregularity constituting streaks, striæ, and waves, is the most difficult to avoid, and the

most injurious in its effect. It is not an improvement only beyond what is ordinarily done in this respect that is required, but absolute perfection, a homogeneity equal to that of pure water. In the two kinds of glass required to render a telescope achromatic, namely, crown or plate glass, and flint glass, it is the latter which is obtained perfect with the greatest difficulty, and to which therefore the greatest attention has been paid. The reason of this will be evident, if the general composition of the two glasses be taken into account. The required difference between them in refractive and dispersive power is found to be at command, by attention to composition; and it has been also ascertained, that crown and plate glass answer exceedingly well for the one variety, and flint glass for the other. Crown glass consists of silica, lime, oxide of iron, sometimes a little alkali, and small quantities of other matters: these substances are not very different in their refractive powers, and when fused do not produce very strong streaks, even though a little difference in the composition of different parts of the glass may exist. The glass also is not a very powerful fluxing agent upon the crucible in which it is melted; so that although it is in contact with it in a fluid and heated state for many hours, it does not dissolve much from it; and what it does dissolve having a refractive power little different from that of the glass itself, proportionately less harm is occasioned. Again: the specific gravity of the different materials used is not very different; so that the mixing agencies which affect the contents of the pot,—such as the ascent of bubbles, the ascending and descending currents from difference of temperature,—are more energetically exerted, and the whole mass approaches nearer to uniformity in a given time, or acquires it sooner than would happen were greater differences to exist.

6. With plate glass the same circumstances hold nearly in an equal degree. This substance is composed of silica and alkali essentially, other elements being only in small quantities. Its action upon the crucible is greater than crown glass, but then it has a second application of heat in such circumstances as are calculated to give a very uniform temperature to the contents of a whole pot, and it is delivered into its final form in the manner least likely to cause mixture of the different parts.

7. With flint glass many circumstances are altogether differ-

ent. Oxide of lead enters into its composition to the amount
of one third of its weight, or more, and by its presence gives
that proportion of refractive and dispersive power, which makes
the glass valuable in conjunction with crown or plate: this it
does in consequence of its own powerful action on light; and
it makes the glass heavy also, because of its own great specific
gravity. A third property belonging to it, namely, its high
fluxing or dissolvent powers, it also confers upon the glass.
Now these three properties are unfortunately very conducive
to the formation of striæ. If the least difference in composition exists between one part and another it becomes evident,
because of the great difference between the qualities of the
oxide of lead and the other ingredients; and a variation in
proportions which in crown or plate glass would produce no
sensible effect on the naked eye, would, in flint glass, form
strong striæ. Hence it is required that the mixture be in this
case far more perfect than in the other glasses; and yet it
unfortunately happens that every thing tends to make it much
less so. The oxide of lead is so heavy a material, and at the
same time so fusible, that it melts and sinks to the bottom,
leaving the lighter materials to accumulate at the top: and so
imperfect are the means of mixture, under ordinary circumstances, that glass of very different specific gravity is procured
from the bottom and top of the same crucible. The following
are some cases of this kind, from pots containing glass not more
than six inches in depth, made from the usual materials, and
retained at a full heat for twenty-four hours:—

Top ......3·38 3·30 3·28 3·21 3·15 3·73 3·85 3·81 3·31 3·30
Bottom...4·04 3·77 3·85 3·52 3·80 4·63 4·74 4·75 3·99 3·74

These differences are great, and selected for illustration;
but from appearances there is little reason to doubt that the
same state of things, though not to such an extent, occurs in
every pot of flint glass made in the ordinary way.

8. Another curious illustration of the predominance of oxide
of lead at the bottom is shown in many of our specimens, which
have been broken through vertically: they have been affected
by sulphuretted vapours and tarnished; but the tarnish has
occurred only at the bottom, where the lead is abundant, and
is there very strong, whilst there is no appearance of it towards
the top.

9. Whilst the crucible is in the condition described, it is clear that all those circumstances, as currents, bubbles, &c., which tend to mix the glass, form abundant striæ and veins of enormous strength, and do harm unless they are continued in activity until the mixture is nearly complete; a state rarely, if ever, acquired in the ordinary flint glass pot. But even if this could be the case, there is a constant cause of deterioration arising from the highly fluxing and dissolving quality given to the glass by the oxide of lead. In this respect, flint glass far surpasses crown or plate glass, and it is also during one stage of its preparation more fluid: it consequently is continually exerting a solvent power upon the crucible to a considerable extent, occasioning that very irregularity in composition which produces striæ, whilst the comparative levity of the matter dissolved at the sides and bottom, and the ascending currents at the hottest parts of the crucible, are constantly mixing this deteriorating portion with the general mass.

10. The difficulties which are thus introduced into the manufacture of flint glass fit for optical uses appeared to the committee, who, however, were none of them practical glass-makers, to increase, as the scale upon which the inquiries were carried on diminished: and the enormous expense of large experiments, —the time required for each,—the number necessary to give that experience which should render any one who undertook the charge of this part of the inquiry an ordinary practical workman,—and the uselessness of the resulting glass for any other purpose than the one directly contemplated,—compelled the sub-committee to consider seriously on the possibility of making other glasses than those ordinarily in use, which, at the same time that they had the high dispersive power enabling them to replace flint glass, might have also such fusibility as would allow of their being perfectly stirred and mixed, and might be retained, without alteration, in such vessels as could be procured of any desired size.

11. The borate of lead, and the borate of lead with silica, were the substances which, after some trials, were found to offer such reasonable hopes of success as to justify perseverance in a series of experiments; and the metal platina was looked to as the material out of which to form the vessels intended to be used. It was soon ascertained that the borate of lead could

be readily formed from dry materials, and that silica might be added with great advantage to the resulting glass; a range of proportions between the three ingredients being permissible, which gave much command over the properties of hardness, colour, weight, refractive and dispersive power, &c., and yet remained within the required range of fusibility. Platinum also was ultimately found to answer perfectly the purpose of retaining the glass; for though at first it was continually liable to failure, yet it was ultimately ascertained that neither the glass nor any of the substances entering into its composition, separate or mixed, had the slightest action upon it. Finally, it was found that several kinds of glass formed of these materials, were in their physical properties fitted to replace flint glass in the construction of telescopes, in some cases apparently even with advantage; since which time the experiments have been unremittingly pursued.

12. The great proportion of oxide of lead in these glasses rendered attention to very minute points essential; for otherwise striæ were inevitably formed, and even the destruction of the apparatus involved. For this reason, after a certain number of trials upon composition had been made, one unvarying set of proportions were adopted, and the attention given altogether to the discovery and establishment of a process which should yield constantly good results. This, as far as it has been carried into effect and proved, it is now my object to describe.

13. The glass with which I have principally worked is a silicated borate of lead, consisting of single proportionals of silica, boracic acid, and oxide of lead. The materials are first purified, then mixed, fused, and made into a rough glass, which is afterwards finished and annealed in a platina tray.

14. *Purification of Materials. Oxide of Lead.*—The oxide of lead at first used was litharge; but this source occasioned frequent destruction of the platinum trays, in consequence of the existence of particles of metallic lead, which alloying with the platina, rendered it fusible. When red lead was substituted for litharge, the same effect took place, due to the presence of particles of carbonaceous and reducing matter. Both these substances also contained so much iron and other impurities, as to give a deep colour to the glass, far beyond what was expected

R

from the quantity of impurity present; this was afterwards explained. Carbonate of lead was also found to be too impure. Finally, all the oxide of lead necessary was purified, by being converted into a nitrate, and crystallized once or twice, as occasion might require.

15. For this purpose litharge is first washed, by which many black carbonaceous and ferruginous particles are separated; it is then dissolved in diluted nitric acid, so as to form a hot saturated solution, the operation being performed in clean earthenware vessels. Both the perfectly pure and the moderately pure acid have been tried without any sensible difference in the results: a little sulphuric acid does not seem injurious; and I find that sulphate of lead will dissolve perfectly in the glass; but muriatic acid has been always avoided. As the acid, water, and litharge are made to act on each other by heat, either purposely applied or resulting from the chemical action going on, it will be found that, when approaching towards neutrality, the liquid will become very turbid. The hot saturated solution is then to be poured from the remaining litharge and undissolved nitrate of lead, and, after standing a few moments, again poured from the sediment, and set aside to crystallize in a cool place. Before it is left, however, it is to be examined as to its acidity: if strongly acid to litmus paper, it is in a right state; if not, a little nitric acid should be added, for the crystals of nitrate have always been compact and pure under such circumstances, and more readily separable from insoluble matter.

16. After eighteen or twenty-four hours, the basins of crystals are to be examined; the clear mother-liquor carefully poured off; the crystals broken up in the basins, and then repeatedly washed in fresh clear portions of the mother-liquor, that any insoluble deposited matter may be removed. There will generally be a portion of this deposit; but if the process has been well performed, the crystals will be quite free. If they appear perfectly white or bluish white, they need not be recrystallized; but if yellow, they must be dissolved in water, a little nitric acid added, and the crystallization repeated. The nitrate in the mother-liquors and washings should be purified by repeated processes.

17. The good crystals are to be washed in three or four waters, to remove the last portion of deposit and adhering so-

luble impurities: but to prevent excessive solution of the nitrate, the same portions of water may be used for several basins of crystals washed at the same time, by making it pass from one to another in succession. Being thus cleansed, they are to be drained, put over the sand-bath, stirred and dried, and finally preserved in glass bottles. By this process much iron and sulphate of lead are excluded; and the purified nitrate is found to yield a glass very far superior in colour to that prepared with the ordinary oxides of lead, and to exert not the slightest action on the platina: its use put an end to all the accidents and failures which resulted from the presence of metallic lead in the oxide. 166 parts by weight are to be considered as equivalent to one proportional or 112 parts of protoxide of lead.

18. *Boracic acid.*—The boracic acid for these experiments was obtained pure from the manufacturer, but before being used was carefully examined. It was rejected unless it was in white or bluish-white crystals, clean and entirely soluble in water. Its solution was tested for iron by the ferro-prussiate of potash and a drop of sulphuric acid, and also for other metallic impurities by a little solution of sulphuretted hydrogen. An ounce or two were heated and dissolved in a little water; and when cold, the soluble part separated and examined for sulphuric acid, by a few drops of nitrate of baryta and a little nitric acid. It was also examined for soda by dissolving three or four ounces in hot water, adding ten or fifteen drops of sulphuric acid, and allowing the whole to cool and crystallize; expressing the mother-water from the crystals; concentrating it; again crystallizing, and then acting upon the mother-liquor, obtained at the second time, by strong alcohol; continuing to wash with the latter fluid until all was dissolved or an insoluble part left. If the latter circumstance occurred, the insoluble substance was examined for sulphate of soda, which, if in any sensible quantity, occasioned the condemnation of the boracic acid. This care respecting alkali in boracic acid was taken in consequence of observing certain bad effects produced in glasses, which appeared referable to its presence.

19. When the boracic acid was acknowledged as pure, 36 parts by weight of the crystals were considered as equivalent to 24 parts or one proportional of the dry substance.

20. *Silica.*—This material is in its most convenient state

R 2

when it forms part of a combination consisting of two proportions silica, and one oxide of lead. As yet, the silica I have
used has been the flint-glass-maker's sand, obtained from the
coast of Norfolk, well washed and calcined. The silicate has
been prepared by mixing two by weight of this sand with one
of litharge, or with such quantity of nitrate of lead as is equivalent to one of litharge (16); the mixture is put into a large
Hessian or Cornish crucible, which being covered over, has
been put into a furnace and raised to a bright red heat for
eighteen or twenty-four hours. On taking out the crucible,
the charge has been found diminished somewhat in bulk, and
of a porous structure and appearance like loaf-sugar. It
has been freed from the crucible, the outside portions removed,
and the pure parts carefully pulverized in a clean Wedgwood
mortar. The powder has then been washed over in water, so
as to obtain the whole in a fine state of division; after which
it has been dried and preserved in bottles. No sieve should
be used in these comminuting operations, nor any reducing
or metallic matter brought in contact with the substance.
Every care should be taken to avoid contamination. 24 parts
by weight of the silicate are equivalent to 16 parts, or one proportional of silica, and 8 parts of protoxide of lead.

21. The advantage of the silica in this combined state depends upon the known composition of the substance, its comparatively easy pulverization, and ready fusion with the other
materials. That there is iron in the silica (and the litharge
when used) is objectionable; and the trials for its removal have
only been delayed that the investigation of a more important
point, namely, a successful process, might proceed. From
some brief experiments, I am led to believe that an unexceptionable source of silica will be obtained by acting upon this
silicate, in a state of fine division, by nitric acid and water, or
else by the use of rock crystal.

22. On some occasions I used pulverized flint glass as the
source of silica, conceiving that being already in a fusible state,
it must possess an advantage over other silica, in allowing rapid
mixture with the other materials. Allowance was made for the
oxide of lead present, and the alkali was permitted to pass, as
a substance that would probably do no harm. But a striking
effect took place, which at once showed the necessity of per-

fectly pure materials. The glass when finished and cold was
of a deep purple colour : this was immediately referred to the
manganese in the flint glass ; a supposition proved by repeating the experiment with other flint glass, and then with flint
glass of our own manufacture in which no manganese was used :
the latter glass gave no purple colour ; the former, a colour as
deep as that produced by the first flint glass.

23. Thus it appears that this very heavy glass, the silicated
borate of lead (and I find it to be the case with other heavy
glasses), has the power of developing the colour of mineral substances far beyond what flint glass possesses ; just as flint glass
surpasses in the same property plate and crown glass.   In the
case in question, the manganese, which did not give a sensible
tint to the flint glass, produced a strong colour when diluted
eight or nine times by the heavy glass, for the proportion of
flint glass used was only $\frac{10}{83}$ths of the whole.   On making a few
experiments with iron, I find that the same strong development
of colour is produced with it in these heavy glasses ; so that
the utmost care is necessary to preserve all the materials during
their preparation, and the glass in every part of the process,
from metallic contamination.

24. The use of flint glass even without manganese was also
objectionable, because of the alkali in it, which, as before stated,
was found to produce bad effects, and rendered the glass containing it very liable to tarnish.

25. Such are the materials from which the heavy optical glass
has been latterly manufactured.   When the composition had
been determined upon, the proper proportions and quantities
of each have been weighed out in a clean balance and vessels.
Thus, for the silicated borate of lead glass, consisting of single
proportionals of each substance, 24 parts of the silicate were
taken, for they contained a proportional of silica equal to 16 parts,
and in addition 8 parts of protoxide of lead : the proportional
of oxide of lead has been taken as 112 parts ; but there being
8 in the silicate, the quantity of nitrate of lead equivalent to 104
parts only was required, and this is 154·14 parts : the equivalent of dry boracic acid is 24, which being contained in 42 parts
of the crystals, that quantity was the one required.   These proportions when heated and submitted to mutual action leave only
152 parts of glass, or thereabout ; for

154·14 nitrate of lead contain   .   .   .   104 protoxide of lead.

24·00 silicate of lead contain   .   .   . $\left\{\begin{array}{l} 8 \text{ ditto.} \\ 16 \text{ silica.} \end{array}\right.$

42·00 crystallized boracic acid contain   24 dry boracic acid.

$\overline{}$

152 glass.

Hence the materials for any quantity of glass can be easily calculated; and if the above parts be ounces, about 9lbs of glass will result. The nitrate of lead is to be broken small in a clean mortar, and then the other ingredients well mixed with it in basins; the use of metal or dirty implements being carefully avoided.

26. The mixture is next melted, and made into rough glass. This preparatory operation is necessary, because from the quantity of vapourable matter which is disengaged in this part of the process, the materials, if put at once into the finishing vessel and furnace, might boil over and do injury; and the acid nature of the vapours themselves, if it did not occasion harm by acting on neighbouring iron and other parts of the furnace, would at least cause inconvenience. It is effected in a furnace which will be particularly described in the Appendix to this paper. It will be sufficient here to state, that being a close furnace, the part immediately beyond the fire-place forms a horizontal chamber, covered above by an iron plate having large circular holes; these allow crucibles to pass through them, and to stand supported on the bottom of the chamber, whilst their edges rise above the upper iron plate. In this way the fire is applied very generally to the crucibles, whilst their mouths are altogether exterior to the furnace, so that the introduction of any reducing or colouring impurity from the fire is prevented, and the greatest facility in introducing the mixture, of watching its fusion, of stirring the glass, and finally of ladling it out, is obtained. The holes through which these crucibles are inserted are five or six in number; they are never all in use at once, and those out of use are covered by crucible covers. The heat is not given altogether by flame; but, whilst coal is used in the fire-place, coke is applied between the crucibles, being introduced for that purpose, and arranged, through the unoccupied holes. The iron top of the furnace is covered by a second iron plate, or, what is better, by earthenware plates, to retain the

heat. The crucibles are of pure porcelain ware, and as thin as they can be obtained. The covers for them are evaporating dishes, considerably larger than the mouths of the crucibles: being turned upside down, they rest, when in their places, upon the neighbouring earthenware plate; not touching the crucibles, but preventing anything from falling into them, and preventing the vapours from passing into the room. The latter are, by the draught of the chimney, drawn through by the sides of the crucible into the furnace, and carried away up the flue, so as to occasion no annoyance to the operator. The covers are slung by a piece of platinum wire, which, being passed across the middle on the outside, is bent at each end round the edges, so that a rod of iron slightly curved at the extremity easily suffices to remove them when the crucible is to be opened. Great care is always taken to put them in clean situations, and that in their removal nothing shall fall from them into the glass.

27. This furnace is found to be very effectual in its action; being connected with a high flue governed by a damper, great command of the temperature is obtained. The crucibles before being used are examined as to soundness; their temperature is raised gradually, and should not be above a dull red heat when the operation commences. The mixture already described (25) is then introduced, and the crucible covered; decomposition of the nitrate of lead instantly commences; the boracic acid loses its water, all the fixed elements unite; and it is remarkable that though a considerable quantity of boracic acid usually sublimes with the water when the latter is driven off from its crystals unmixed with other substances, yet scarcely a trace seems to evaporate in the present instance, in consequence of the presence of the oxide of lead.

28. The heat should not be raised too high or the operation hastened, and then the ebullition will proceed very gradually and favourably, the rough materials being by degrees converted into glass. Before the first charge is entirely melted a second is put in, and when that is fused down, sometimes a third, according to the quantity of glass present and the soundness of the crucible. When all is fused, the temperature is allowed to rise, but not too much, lest action upon the crucible to a serious extent should occur; the glass is then well agitated and mixed by a platinum rake or stirrer, to be described hereafter. Finally,

the glass is either transferred by a platinum ladle into trays roughly turned up out of old platina foil, or into a clean deep white earthenware vessel containing much distilled water. In the latter case it is obtained in a divided state, and when drained, is dried on the sand-bath, and put up in clean bottles.

29. When a crucible has been emptied of its first portion of glass, it will serve, if carefully used, for a second, third, fourth, or for many operations; but it should be watched for cracks and casualties, that the running of the glass into the furnace may be prevented, and, if necessary, another vessel taken.

30. The rough glass thus prepared is in the next operation to be converted into an annealed and finished plate. The size must therefore be determined upon, and we will assume it as 7 inches square, and 8 tenths of an inch thick, that being the dimension of the largest plate as yet made. For the purpose of making a competent platinum vessel, a plate of that metal will be required at least 10 inches square; but if larger, it should not be cut, but either made into a tray with higher sides than is absolutely needful, or else used first in the manufacture of a larger plate of glass than the one to be described. It should be of such thickness as to weigh at least 17·5 grains to the square inch; and it is important that in its preparation a good ingot, or the good part of an ingot, of platinum has been selected, and that it has been rolled very gradually and carefully without the formation of any holes by the adhesion of dirt or hard particles, or by the dragging of the metal in the mills. The desired perfection is, I understand, best obtained by rolling the platinum between two clean plates of good copper.

31. The plate, being laid upon clean paper or a cloth on a smooth table, is to be cleansed with a cloth and a little water or alcohol, and then to be ignited at every part by a large spirit-lamp. It must next be carefully examined as to its state, and the occurrence of places upon its surface where holes are likely to exist. If the metal seems dragged in any place, an effect indicated by a roughness upon the surface, or by short lines parallel to each other but perpendicular to the course of rolling, such place should be noted or marked, for which purpose a dot of ink will be convenient. If a scale appears, or a small portion is apparently folded over, it should also be marked; and if a black spot is visible (and they are sometimes

formed by the adhesion of a particle of dirt or grit), it should
be examined, and removed by the point of a knife, if necessary,
and its place also marked. All these places and the whole
surface of the plate should then be examined for holes by a
still stronger test, namely, by holding the sheet of metal before
and close to a bright light, as a candle or lamp, in a dark
room, and every hole observed, marked. In making this examination, it must be done carefully and minutely, holding the
plate in different directions to the light (for sometimes the holes
are oblique), and being careful that no reflexion from illumined
objects, as the hands, on that side towards the face shall
give deceptive indications. In the marking, too, the indicating
spot should always be made at a certain distance from the hole,
as the fourth or the third of an inch, and on the same plate
constantly in the same direction or towards the same edge; the
holes are then easily found again, and the mark remains during
the soldering to guide the operator.

32. The holes discovered by these examinations are to be
closed by little patches of platinum soldered with gold; for gold,
like platinum, may be safely used in these experiments, when
reducing matter is absent. The gold has been used in the
finely divided state in which it is obtained by precipitation from
its solutions by means of sulphate of iron, but it must be washed
perfectly pure; the patches are formed by cutting a piece of
clean new platinum foil into small square or rectangular plates:
a sufficient heat can usually be obtained by the use of the spirit-
lamp and mouth blowpipe. In the process of soldering, a little
of the powdered gold is heaped upon the hole and slightly
flattened by some clean instrument, the spirit-lamp is applied
underneath for a moment, which causes the gold to adhere
slightly, a selected patch of platinum is laid delicately upon the
gold, and then the heat of the spirit-lamp, urged by the blowpipe, is directed beneath against the place. Usually the gold
will melt and run instantly, the platinum patch will come into
close contact with the plate, and the operation will be completed.
If well done, the fused gold will appear all the way round in
the minute angle formed by the edge of the patch, and also
faintly at the hole on the opposite side of the plate.

33. Sometimes, when the patch is large, or in the middle of
a plate, the heat obtained as above is hardly sufficient to melt

the gold freely and cause perfect adhesion. In such cases, a single or double piece of platinum foil loosely laid over the part, prevents loss of heat from the upper surface, and frequently causes such increased elevation of temperature as to render the soldering perfect and effectual. In the few cases where this expedient has not succeeded, I have resorted to the oxyalcohol blowpipe, using a small bladder of oxygen with a little attached jet for the purpose. This has never failed to produce an effectual heat, and 15 or 20 cubical inches of oxygen are sufficient for many operations.

34. This application of patches and soldering is only secure for small holes, *i. e.* such as a pin might pass through, and smaller. The patches are always to be applied on that surface of the plate which is to constitute the outside of the tray; and therefore, before the soldering begins, the two surfaces should be examined, and the most polished and perfect selected as that intended for the inside. The patches are valuable in their use far beyond what the mere application of gold to the hole would be; for the heat afterwards applied to the tray, when charged with glass, is abundantly sufficient to melt gold; in which case, if unsupported by the platinum patch, the weight of glass and the action of stirring would probably force the gold out of the hole and cause the tray to run; whereas the patch of platinum, although the gold holding it to the plate is liquid, still adheres by so strong a capillary action as to be sufficient to retain its place, and being outside is not disturbed by the motion of the stirrer. Besides, after a long application of heat, the gold and platinum combine so perfectly as to become one piece of white alloy, infusible at the heat applied.

35. The plate is now to be folded into a tray, preparatory to which, a piece of thin board is to be provided as a gauge, which in the present instance must be 7 inches square. This laid upon the plate and held tightly down, directs the foldings of the sides, and would, if placed in the middle, leave sufficient for edges one inch and a half high all round; but as the plate should serve for use several times, it is advantageous to apply the gauge a little eccentric; for then, when used for a second and third operation, its place may be shifted, and the folds not occurring where they did before, there is less chance of holes being broken through the platinum. The folds

necessary at the corners of the tray are especially likely to render the same parts unable to bear a second and third bending; but the necessity of having them in the same place may be usefully obviated by placing the gauge oblique to the sides in one direction and in another, on different occasions, and moreover gives other advantages in finishing the folding of the corners (36). These attentions, tending to the preservation of the platinum for repeated service, are very needful, in consequence of the great expense of the material: the value of the plate in question is about 6*l*. 10*s*., and when worn out, it may be sold for about half that sum. Whether it be used therefore once, twice, thrice, or four times, makes considerable difference in the expense of the resulting plates of glass.

36. When the gauge is properly placed on the platinum, the sides are raised perpendicularly: this produces four projecting folded triangular corners, which being pressed close, are then turned against the sides, and a square tray is finished, which has no aperture or orifice below its upper edge. The folding of these corners is a matter of much more consequence than might be anticipated. The plate is seldom so regular that the parts of two neighbouring sides which come together at a corner are exactly of equal height; neither is it desirable that it should be so, and the unsymmetrical position of the gauge to the plate, already recommended (35), is almost sure to prevent it. In that case, of the two sides of the folded corner, one will be higher than the other, and if the corner be so folded that its lower side is towards the tray and beneath its edge, a kind of siphon is formed which becomes charged with fluid by capillary action, and continues to discharge glass from the tray during the whole time of heating, notwithstanding that all the edges are much above the level of the fluid within. This in a long experiment is competent to occasion serious injury.

37. I have found, even when the edges of a corner have been of equal height, but below the edge of the side against which they are disposed, that still this capillary and siphon action has gone on, and the reason is not difficult to comprehend; the corners therefore have always been folded in such a manner, that their highest edge has been inwards, and both their edges above the level of the corresponding edge of the tray. To effect this, the line of their lateral flexure is not

perpendicular to the bottom of the tray, but a little outwards above, and the proper degree of inclination is easily given by using a mould upon which to bend the corners. This should be a thick square piece of wood having the four corners cut with different degrees of obliquity: when the corners of the tray are first imperfectly formed, it will be easy to ascertain by trial, which corner of this mould  will give the obliquity and position already described as necessary, after which the folding may be easily finished upon it. The accompanying sketch represents first a good and then a bad folding.

38. All occasion for changes in the folds, especially at the corners, should be avoided. The folds should be decided upon as the work proceeds, so advantageously as to make alterations unnecessary. The closer the corners are pressed, the smaller is the quantity of glass contained in them, and the less risk is there of the platinum being broken when the finished glass is taken out; but it is proper to avoid general contact between the corners and the sides against which they are disposed, otherwise welding is likely to occur during the stirring, and the platinum is injured for future experiments.

39. The tray being formed is again to be examined for holes, first by a light as before (31), and then in the following manner:—Being laid upon a sheet of bibulous paper, alcohol is to be carefully poured in until the fluid is within the fourth or the sixth of an inch of the lowest edge of the tray, so as to occasion no running over at the sides or corners. If a large hole exist, it will be rendered visible immediately; but if none such appear, a large basin or some other cover is to be placed over the tray to prevent evaporation, but without touching the vessel or its contents; and the whole is to remain undisturbed for some hours. Being then examined, the wetting of the paper will indicate a hole or a badly-folded corner, and will point out the faulty place: the tray may easily be shifted from one part of the paper to another for the discovery of any moistened places beneath. Sometimes holes occur so small that alcohol will not run in a sensible quantity through them. Suspected places of this kind and suspicious corners also should

be examined by a clean dry point of bibulous paper, which soon shows, by its change of appearance, the transmission of any fluid: but attention is required that no false indication be produced by carelessly bringing the paper near the upper edges of the platinum, especially in the folded places. These minute holes do not occasion much harm in the furnace, but no fault should be allowed to pass which care can correct.

40. When the tray is faulty, the alcohol must be removed by a small siphon, the holes soldered in the manner before described (32), and the tray again tried. When it proves good, it is, after the removal of the alcohol, to be heated red-hot in every part by the flame of a large spirit-lamp, and then reserved with care in a clean place until required.

41. If the platinum has been used before, it should first be ascertained that none of the glass from the former experiment remain on it. If there be any portion, the plate must be returned to the weak acid or pickle out of which it has been taken. If free from glass, it should then be examined as to any chemical injury it may have suffered. Any part which is altered in appearance, or has been attacked by the acid, or which tarnishes when heated to redness by the spirit-lamp, has been thus affected; and it will depend upon the extent of the action whether the plate is unfit for further use. No chemical injury is occasioned by the proper and successful performance of an experiment.

42. An examination for holes by the candle or lamp must next be made, especially in the folds at the corners and where adhesion of the platinum from welding may have occurred, and any that are discovered are to be marked as before (31). The plate should then be flattened by being put between two sheets of writing-paper upon a smooth table, and the edge of a folding knife or some other smooth substance drawn over it; but if this be done whilst old glass adheres to the plate, it is almost certain to produce injury. The holes are then to be soldered and mended, the patches being applied upon the same side as before. The gauge for the new tray is to be applied to the plate, shifted, if there be occasion, from its old position, as before intimated (35), and the folding of the tray, its completion and examination, to take place as before.

43. It is desirable never to cut the platinum smaller than can

be helped, but always to make the largest plate upon it for which it is competent. Then, when operated with a second or third time, smaller gauges may be used, and the folds will not be repeated in the same place; and if injury occurs to the metal, being generally at the sides of the tray, the middle part will still be left for the preparation of smaller plates of glass.

If such large plates of platinum are required for trays as can hardly be rolled at once, there is no difficulty in making a folded joint and rendering it tight by soldering with gold.

44. A kind of furnace, unlike the former, is now required for the completion of the glass, and its delivery in the state of an annealed plate. This furnace shall be described accurately in the Appendix. It may here be sufficient to state that it consists of a fire-place in which coals are burnt; of a part beyond, acting both as furnace and flue, in which coke is used; and of a chamber above, to be heated by the fire, though out of the course of both flame and smoke. It is in this chamber that the glass is made; so that, by the arrangement adopted, at the same time the substances are fused and access for stirring allowed, the essential condition of excluding impurity or reducing matter is also fulfilled.

45. The fire-place itself is of the ordinary construction, and fed with fuel by an aperture in front in the usual way. I have found abundant reason to be satisfied that the passage of steam beneath the bars of the grate is of considerable use; for which reason an iron trough charged with water occupies the lower part of the ash-pit. The bars are by this arrangement preserved very cool and do not burn away; they are easily kept open and clear of clinkers; the free passage of air to the fire is permitted; and the action of the furnace retained at a high point for any number of hours together.

46. That part of the furnace beneath the chamber requires peculiar and careful arrangement; for at the same time that such a heat as will soften the neighbouring materials is produced there, the bottom of the chamber in its softened state and charged with several pounds of materials, has to be firmly supported for many hours together without change of position.

47. The coke necessary in this part is introduced by two or more holes in the side of the furnace, which, when necessary, are stopped by bricks. The bottom of the chamber is sup-

ported on ledges at the sides, and upon the ends of fire-bricks in the middle, firmly placed at intervals so as neither to stop the passage for smoke and flame, nor the cross passages for the introduction of coke.

48. The value of the coke arrangement in this as in the other furnace is very great. The heat obtained by the united action of the coke and the flame from the fire-place, is abundantly sufficient, and, whilst obtained at the necessary point, does not involve that degree of mechanical action required for stoking and stirring, which is necessary with coals, and would risk the destruction of the soft thin bottom of the glass chamber. It further occasioned the perfect combustion of the smoke produced in the coal fire, which at first was so considerable in quantity that, had it continued unaltered, the experiments must have been removed from the Royal Institution; in which case they would probably have been discontinued altogether.

The flue is the same as that connected with the former furnace, and has a damper for regulating the heat, especially useful during the annealing operation.

49. The chamber was at first of cast iron, that material being selected as one which would bear a sufficient temperature without melting, would conduct and transmit the heat freely to the substances within, and could be easily obtained of the requisite form. The upper aperture was closed by plate-iron covers, and in the first trials all appeared to answer well; but when large experiments were made, and the heat was continued for a long time, the bottom gave way and became irregular: and upon endeavouring to rectify this, and place the tray of glass level by means of sand, the transmission of heat to the glass was prevented, the temperature of the iron rose, and the bottom melted. Besides these injurious liabilities, if the smallest portion of glass passed out of the tray, the moment it touched the iron it was reduced, the lead immediately caused fusion of the platinum, and in an instant the tray was destroyed, the experiment stopped, the glass rendered black and useless, and the bottom of the chamber covered with lead and rendered unfit for another operation.

50. Finally, one very curious action of the iron was discovered, which immediately caused its rejection. Plates of

glass, which seemed very good in other respects, were frequently so discoloured by dark smoky clouds as to be useless. These could not be referred to any impurity which had been left in the materials or had entered accidentally, and, as the platinum was in all such cases altered and injured, was at first supposed to be occasioned by some particular action exerted between it and glass at high temperatures. But upon every fair trial to verify such chemical action, the proofs failed, however high the temperature used, or however minutely the metal was divided. At last the cause was discovered. To understand it, it must be known that the platinum tray, with the glass in it, was either placed directly upon the bottom of the iron pan, or, for greater security, with only a plate of platinum intervening; and that the whole was covered by an evaporating basin turned upside down, forming a sort of inner chamber within the large one. In this confined state the oxygen of the portion of air present was soon abstracted by the heated metal, an oxide of iron being formed in consequence, and at the same time also a portion of carbonic oxide from the carbon in the cast iron. At the high temperature to which the experiment was raised, this carbonic oxide was competent to reduce a portion of the oxide of lead in the glass to the metallic state, itself becoming carbonic acid; but as soon as the carbonic acid so produced came in contact with the heated iron, it was again converted, according to the well-known condition of the chemical affinities at these temperatures, into carbonic oxide, and went back to the glass to repeat its evil operation and produce more metallic lead. In this way it was that the glass became sullied by smoky clouds consisting of metallic lead. It was the lead thus evolved, also, that, by alloying with the platinum, had produced the appearance of chemical action always visible in these cases; and now I knew how to account for the failure of many experiments in consequence of the formation of holes in the trays in a manner before quite inexplicable: for in the experiments purposely made to investigate this point, sometimes the glass was darkened only at the surface, the lower part being quite clear and good; and then, though the platinum tray was frequently cut through as with a knife all round level with the surface of the glass, it was quite unaltered below. At other times the superficial stain was in a greater quantity, and had collected together into little drops like fat upon hot water, and upon examination each little globule was found to be soft brilliant metallic lead. At other times a much larger globule hung from the middle of the surface into the glass, barely sustained there, and ready to sink by the least agitation when in a heated state, and in some instances the bottom of the tray was alloyed and perforated by globules of lead which had thus been formed and deposited, and the glass just running out, whilst another globule was in progress of formation at the surface exactly over the place of the hole.

51. When iron was dismissed as the material of the chamber, earthenware was resorted to. The sides were built up of brick, and the bottom formed of tiles, which resting at the sides upon ledges, and at the middle upon the fire-brick supports (47), could be replaced at pleasure. The same iron covers were used for the upper aperture of the chamber as before.

52. The use of earthenware as the material, made it far more difficult to apply a sufficient heat to the contents of the chamber than before, because of its inferiority to the iron as a conductor of heat; and a series of investigations were required to discover that substance which, at the same time that it had sufficient strength and exerted no injurious influence, was also a sufficiently good conductor. Reigate fire-stone, recommended by the builders, did not answer the purpose, and moreover in thin plates was liable to fuse and slag. Slate, however carefully heated, shivered and split not only across but parallel to its structure; and then, as soon as air intervened, it transmitted too little heat. It also softened, became curved, and let in air and smoke, and at last gradually fused, becoming unable to bear the weight of a large experiment. Yorkshire stone, rubbed down into plates $\frac{5}{8}$ths of an inch thick, answered moderately well, if the application of heat was carefully made and gradually raised. It cracked in a few places, but did not fall to pieces; and it was more difficult of fusion than the former substances. Fire-tiles of various kinds were tried; those made of Stourbridge clay answered the best, and, when about $\frac{3}{4}$ths of an inch thick and carefully heated, might be successfully used; but that which we finally arrived at was the use of plates made of the materials from which Cornish

s

crucibles are manufactured. These we obtained through the intervention of our President; they were purposely manufactured for us by Mr. Michell of Caleneck in Cornwall, a gentleman who has been ever willing and anxious to assist us in our inquiries, by supplying us with vessels of any size or form, or any other article which it was in his power to produce.

53. The Cornish plates have not much cohesion, and feel tender in the hand. They may be rubbed down to a flat surface, and resist any heat which can be applied to them in these or in much more powerful furnaces. They are therefore readily brought to any thickness, and when of about $\frac{5}{8}$ths of an inch, and supported in the furnace as before described (47), have strength to bear any weight required to be placed upon them. They do not crack, nor do they force themselves to pieces by expansion; but they are porous, as indeed are in a greater or smaller degree all the materials of which the chamber and its sides are now composed.

54. The porosity of these materials was of great importance; for it allowed of the passage of gaseous matter, and that even of a reducing nature, from the fire into the chamber. I have frequently had evidence that the sides and bottom might be considered as a very sieve-like partition between the fire, the flue, and the space called the chamber; for when the upper aperture has been closed, there has been a current through the chamber in the direction of the flame, the gaseous matter entering at the extremity nearest the fire, and passing out at the end towards the flue. In one or two cases, oxide of lead was actually reduced, and the glass thus rendered cloudy.

55. Hence it became necessary to use some certain means of maintaining an oxygenating atmosphere about the glass; to obtain which, and also to prevent any other injurious vapours from the fire entering the space beneath and within the earthenware covers (51), the expedient was adopted of allowing a current of fresh air to pass continually into that space and circulate about the glass. To effect this, a clean earthenware tube, glazed within, was inserted horizontally into the side of the furnace, in such a manner that one extremity was flush with the inside of the chamber, and of such height, that its lower edge corresponded with the level of the bottom upon which the glass in

its tray was to be placed, whilst the other end of the tube
reached to and was flush with the outside of the furnace.   A
loose piece of tube, similar in kind but smaller in diameter,
being laid upon the bottom of the chamber, and applied at its
end to the orifice of the larger one, served as a continuation
of it until the inner extremity reached to and was under the
cover of the glass experiment.   When the furnace was hot,
there was always a draft inwards through this tube; but
the quantity of air admitted was regulated by a valve (70).
The air, by first passing through the hot sides of the
furnace, and then through the shorter ignited tube serving for
connexion, was transmitted in a thoroughly heated state to
the place where its presence was required, without producing
any serious cooling effect; it there maintained a continually oxygenating atmosphere, and, judging from the effects, prevented
the draught inwards of any vapours from the fire to the space
beneath the glass covers.

56. The next point of importance, in the preparation of the
glass, is the arrangement of the tray in the furnace just
described.   To understand this, it will be necessary to say
that the glass-chamber is 25 inches long, 13 inches wide, and
8 inches deep, and that the fire being at one end, the flue is
at the other.   Plates of glass 7 inches square have been made
in it; but it would probably require a larger furnace to make
much larger pieces.

57. The bottom of the chamber being perfectly level and

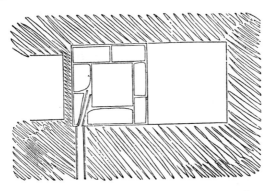

clean, the gauge-board, on which the tray was formed (35),

should be placed on the middle of the half next the fire, and then a piece of connecting air-tube taken, which being laid on the bottom of the chamber, may extend from the fixed air-tube by the side of the gauge as far as the middle, or even towards the further side of the chamber. After this, pieces of Cornish tile (53), or other clean earthenware which will not fly in the fire, contain but little iron, and are free from glaze, are to be prepared, of such size that they will fit in loosely round the gauge, covering the rest of that half of the chamber bottom, and serving to support the sides of the tray when in its place. This support to the tray is highly needful; for, otherwise, the weight of the glass, and the action of stirring, would be more than the thin and heated platinum could support. The thickness of the pieces should be, for the plate in question, about 1 inch, and they should be all uniform in that respect. They should never rise so high as the edge of the platinum, lest glass should accidentally pass from the tray to them, or impurities from them to the glass. An excellent guide to their thickness is, to make it similar to that of the intended plate. When they have been roughly arranged around the gauge, the latter should be withdrawn, and the tray itself introduced, the pieces being now finally adjusted about it. They should not be so arranged as to press against its sides; but the latter should be at liberty, though only so much, that upon the least tendency of the sides outwards, they should be supported by the pieces. The assistance thus given should be directed rather to the sides than the corners, and it is better that the latter should not be in contact with these adjuncts, but be allowed to sustain themselves, for they are strong enough for the purpose, and the corners are always those places at which, from one circumstance or another, the glass is most likely to pass outwards.

58. The piece of earthenware which is fitted nearest the mouth of the air-tube should have its angle taken off, or some other provision made, as by making the orifice of the tube oblique, that the passage of air may be uninterrupted; and on that side the tube itself may frequently form the support to the tray. If it does, and is glazed on the exterior, a piece of loose platina foil should be wrapped round it at the part where

it touches the tray, to prevent adhesion by the glaze when cold
The general disposition of the tray, the tube, and the packings,
may be seen in the sketch (57).

59. When the first set of packing pieces is properly adjusted, a second series is to be arranged over them; but these
are to be removed backward from the tray about the third or
half of an inch all round, that accidental contact with its edges
may be avoided. Their thickness should be sufficient to raise
them level with, or rather above, the edges of the tray. All
these adjusting pieces are to be rendered perfectly clean and
free from dust before they are applied. Their use is not only
to afford support and assistance to the platinum tray, but also
to sustain the glass covers, and likewise, by retaining the heat
upon the bottom of the chamber, prevent much of the inconvenience that would otherwise occur at the times of stirring
the glass.

60. The tray-covers have, up to this period, consisted of
inverted evaporating basins, suspended at pleasure, in the
manner before described, by platina wires (26). When the
platinum trays used have been sufficiently small to admit of the
arrangement in our present furnace, two, and even three covers
have been used simultaneously, each prepared with its own
platinum suspension; but of such size, that the larger could be
placed over and enclose the smaller, without touching it. In
such cases the temperature of the glass, after being lowered
by stirring, or in any other way, rose very rapidly; but with the
large plate of 7 inches square, the furnace would admit of but
one glass cover of sufficient size, and the only additional
assistance which could be obtained was that which was given
by putting a similar but smaller cover on the outside and
above the principal one.

61. The first and important cover is to be selected of such
dimensions, that when in its place and resting by its edges
upon the packing pieces, it shall fully enclose the platinum tray
and its charge, not only for the purpose of accumulating heat
and confining an oxygenating atmosphere within, but also
sheltering the glass, and preventing any oxide of iron from the
chamber covers, or dirt from other sources, falling into it.
These covers, when hot, are raised and removed by means of
clean iron rods, which being sufficiently thick to have abundant

strength, and no injurious degree of elasticity, are made taper at one extremity, and slightly curved there. This end is easily introduced beneath the platinum suspension wire, and as easily withdrawn when the cover is removed.

62. All these matters being preliminarily arranged, the final disposition of the tray and its charge is made. The air-tube is carefully wiped, and its external aperture closed by a clean loose plug of dry sponge. The tray is for the last time freed from dust by inversion and blowing upon it, and is put into its place. The quantity of rough glass necessary for the required plate, about 8lbs in the present instance (30), is carefully weighed out, and then introduced by an evaporating basin, or some other means which shall not allow of the admission of any reducing or colouring matter, or permit any portion of glass to pass beyond the edges of the tray. The tray-covers are then to be arranged in their places; the iron covers of the chamber likewise adjusted, and over all are to be placed a set of thick earthenware tiles, which have been fitted together so as to constitute a general covering to the whole, well calculated to retain heat.

63. The ensuing part of the process is one in which the precise order and most advantageous proceedings have not yet been ascertained. Variations have been made up to the very last experiment, and it is only by still more extensive experience that the arrangement will ultimately be settled.

64. A fire being lighted in the furnace, and some coke put beneath the glass chamber, the temperature is gradually raised. In about an hour the bottom of the chamber begins to appear ignited, and in four hours the top iron covers are usually dull red-hot. These appearances are useful as indications of the progress of the operation. When the furnace has been heated for the first half hour, then every care is taken that the temperature may be fully sustained to the end of the experiment; and besides the ordinary kind of attention to the fire, particular care is taken that coke be supplied, by the lateral holes, to the part beneath the chamber; for, if the fuel there be allowed to burn out, the heat soon falls, notwithstanding the flame from the coals. Although the fire may seem quickly to have attained its best condition, yet the temperature continues to rise in the chamber long afterwards; for, from the

quantity of lateral brick-work to be heated, it is usually many hours before the sides of the chamber are so hot, that the tray and its contents have attained their highest temperature.   At the same time it must be understood that the heat of the glass is very much governed, especially at the early part of an experiment, by the number of tray-covers over it, and rises far more rapidly, and much higher, with two or three covers than with one.

65. Perhaps the glass may with propriety be examined once, early in the experiment, for the purpose of ascertaining that the tray and its contents are safe; but usually it is left for six or eight, or a greater number of hours, that the whole may fuse, the temperature rise, and the bubbles escape. When the glass is to be examined, the tile and iron covers are to be removed from over that half of the chamber containing it, by which, consequently, the tray-covers are exposed; these are next to be carefully raised, one by one, using the iron instrument before described, for the purpose (61), and, as they are removed, are to be carefully put into the further part of the chamber, which still remains covered, where they will be retained in a heated state. This prevents their cracking and falling to pieces, as they would do if brought into the open air. If the experiment, and consequently the covers, are upon so large a scale that the latter cannot all be placed in this situation, then the exterior ones may be placed upon the top of the heated covers and tiles; but the particular cover, which immediately encloses the glass, being of great importance, must be put into the further safe part of the furnace, that it may be carefully preserved from injury, and ready to be replaced over the glass with the least possible disturbance.

66. The moment the last cover is removed, the glass is exposed to any falling substance from the iron plates, or tiles, or other sources, so that extreme attention is required at such times to keep the place free from dust, and to perform every requisite operation as quietly as possible. The current of hot air which rises from the chamber, ascending and striking against the ceiling, frequently causes, by change of temperature and mechanical agitation, the separation of small particles of matter, which, descending, cause risk of injury to the glass; for which reason it may sometimes be needful to have a temporary

shelter fixed over the furnace, either of tin plate, clean boards, or some other material which shall not throw off scales or impurities of any kind.

67. If, by any unfortunate accident, a fragment of matter does fall into the glass, it should be instantly removed. It certainly will not sink, because of the great density of the glass, and may be taken out, usually with facility, by touching it, and the glass in its neighbourhood, with the platinum stirrer (28), or the bottom of the platinum ladle (28). In carrying it and the adhering glass away, great attention should be given, that none of the latter fall over the sides of the tray; since such portion might be a means of introducing impurity hereafter, or of cementing the tray and the earthenware together in a very inconvenient and injurious manner.

68. If, also, it should be observed at this time, that there is a superabundance of glass in the tray, and not sufficient distance between its surface and the edges of the platinum, the excess should be ladled out (28), an operation easily performed, but which must be done with care.

69. When the glass is ascertained to be in a proper condition, and that there is no appearance of any portion of it outside the tray, the covers are to be replaced, the chamber closed, and the heat continued. If the tray-covers be glazed, some precaution is required in their arrangement; for on putting the second cover over the first, if they are left in contact by a portion of glazed surface, they will be found, upon their next removal, to adhere at that place. They should never be put in contact therefore with each other, or, if that cannot be avoided, a piece of old platina foil should be laid upon the place where the contact is necessary (58).

70. Whilst the glass is covered and subjected to a high temperature, there is, as before stated, an inward current of fresh air passing continually to and about it through the air-tube, during the whole time of the experiment (55).

It was necessary to apply a valve to the external orifice of this tube to regulate the supply; for the draught was so considerable, that the glass was cooled by it, and much dust carried in. Finding reason to believe that even when very much diminished, the quantity of soots and dust in a London atmosphere, and especially in that portion of it taken from an expe-

rimental room in which a powerful furnace was at work, were
competent to do much harm in eighteen or twenty-four hours,
by giving colour and forming striæ, experiments were made on
the means of cleansing the entering air. It was found easy to
effect this, by the assistance of two or three Woulfe's bottles,
or two or three jars, inverted one within another, using at the
same time portions of diluted sulphuric acid, or such solutions
of salts in the vessels as would not supply any moisture to the
air, but rather take water with the dust from it. In these cases
the air did not bubble through the liquid, but only passed close
to its surface, and had time to deposit its dust during its passage through the enclosed spaces above the fluid; but finally
a still simpler arrangement was used, consisting merely of a
plug of clean dry sponge fitted into the end of the tube, which,
at the same time that it allowed sufficient air to pass, seemed,
from the appearance of the tube afterwards, to have excluded
every impurity.

71. There are two conditions of the finished glass, each of
great importance, which it is the object of the process to secure
in this state of the substance. One, and the most essential, is
the absence of all striæ and irregularities of composition; the
other, the absence of even the most minute bubbles. The first
is obtained by agitation and perfect mixture of the whole; the
latter, principally by a state of repose: so that the means required to be successful on both points are directly opposed to
each other. Were the glass absolutely incapable of change by
the long-continued action of heat, it would be easy first to render it uniform by stirring, and then to leave it in a quiescent
state, until the bubbles had disappeared; but I am not yet fully
assured of the fact which is necessary to this order of proceedings. That the glass, as far as proportions are concerned, if
changed at all, is altered only in an extremely minute and inappreciable degree, is shown by some experiments, in which, after
a portion had been prepared and heated for many hours, and
also stirred well, the resulting piece was divided into smaller
portions, and these heated at different temperatures, in platinum
trays, for sixteen hours. Three portions were heated as powerfully as the furnace would admit of; three only to redness, which
may be considered as a very low heat; and three to an intermediate degree: all were cooled slowly and annealed for an

equal time. The specific gravities of each after the experiments were as follows:—

Highest heat   . 5·4206 5·4211 5·4203 Mean sp. gr. 5·42066
Intermediate heat 5·4253 5·4242 5·4255   „   „   5·42500
Least heat . . 5·4258 5·4262 5·4235   „   „   5·42516
Original glass . 5·4247 5·4261 . . .   „   „   5·42540

72. Here, notwithstanding the irregularities between the similar experiments, there seems, from the comparison of the mean specific gravities, to be a gradual though minute diminution of density, as the glasses have been more powerfully heated; and I found also, that when glass was so well stirred as to leave no doubt that it was thoroughly well mixed, yet being left in the furnace at a high temperature for eight or nine hours, it contained striæ.

73. On the other hand, first to render the glass perfectly free from bubbles and clear, and then to stir out the irregularities of composition, I have not found to be a practicable process; because the stirring, in the manner in which I have yet performed it, tends to introduce bubbles into the glass; and though these are small, still they are objectionable. Hence a mixed process has been adopted, which, as I have before stated, is subject to correction from future experiments. To render the process as far as it has been carried sufficiently intelligible to others, I will first describe the circumstances connected with stirring, and their influence upon striæ; and afterwards, the plans adopted for the dispersion of bubbles.

74. It is not a small degree of stirring and agitation which is sufficient to make a fluid of mixed materials homogeneous; especially when the mixture is not exceedingly fluid, but has, like tar or syrup, a considerable degree of tenacity. An idea of the extent to which it must be carried, and of the general nature of striæ in fluids, may be gained by taking a glass full of clear saturated syrup, made from white sugar, putting a few drops of water into it, and stirring the whole together. It may then be remarked how slow the striæ are in disappearing; and when they are apparently destroyed, if the whole be left for some hours, it will frequently happen that a separation will take place into a lower heavy, and a superincumbent light portion, which, when stirred together again, produce striæ. In the glass, the stirring must be in the utmost degree

perfect; for if there be the least difference in different parts, it is liable to form striæ: nor are the different portions allowed by the process to arrange themselves according to their specific gravities, in which case one part might perhaps be removed from another, after the glass was finished and cold; but the ascending and descending currents which inevitably take place in the fluid matter, are certain to arrange the irregularities in such a manner as to produce the strongest possible bad effect.

75. The instrument used for stirring has hitherto consisted of a piece of plate platinum, which for the seven-inch glass (taken as illustrating the process) is $6\frac{1}{4}$ inches in length and $\frac{3}{4}$ths of an inch in breadth. It is perforated with various irregular holes, that, when drawn through the glass like a rake, it may effectually mix the parts. A piece of thick platinum wire, about 13 inches long, is riveted to it, and the extremity of this screwed into the end of a clean iron rod which answers the purpose of a handle. No small or cellular apertures should be allowed in this stirrer; for they will frequently retain air or moisture, which may cause bubbles in the heated matter and do much harm. A little gold, therefore, should be applied to the part where the stem is attached, and fused, so that all hollows may be filled up. Stirrers of different dimensions are to be provided for different-sized plates of glass. Before being used, they should be steeped in dilute nitric acid, and also heated to redness in the spirit-lamp, just previous to their immersion in the glass for the first time in each experiment.

76. When a stirring is to be performed, the tiles and iron covers are removed from the first part of the chamber (44. 49. 65), the tray-covers also taken off and put into the back part of the chamber (61. 65), the glass quickly examined, to give assurance that all is in good condition, and then the stirring commenced. The stirrer should be put in gently, that no air may be carried down with it, and then drawn through the glass quickly but steadily, so as to mingle effectually, but not to endanger forcing the substance over the edges of the tray or to run the risk of involving air-bubbles. The chamber and its contents are cooled by the necessary exposure to the atmosphere, and therefore, when the agitation has been continued until the glass is so much lowered in temperature as to become thick, it should be discontinued, the stirrer carefully removed,

the tray-covers replaced, the chamber covers restored to their situation, and the temperature allowed to rise for fifteen or twenty minutes, when the operation may be renewed.

77. All the precautions against loose particles, dust, and soot, that were before spoken of (66), should be adopted in this operation. In the act of stirring, the instrument should not be struck carelessly against the bottom or sides of the tray; for the platinum in this highly heated state is very soft, and a hole would readily be forced through it; nor should it be brought forcibly against the corners, for the metal is in such a favourable condition for welding, that the least blow upon a double part causes adhesion. By merely allowing the stirrer, when ignited, to sink upon the bottom of the tray rather more hastily than usual, it has adhered to the place; and when, for safety, an underlying plate of platinum was used (50), it was always found welded to the tray at the places which the stirrer had touched a little more forcibly than the adjacent parts, and could not afterwards be separated without leaving holes in the metal. This circumstance was the principal occasion of the advantages afforded by the use of the underlying plate being given up.

78. The heat which has to be borne during the operation of stirring is very considerable, especially upon the hands; but at such a moment no retreat from the work, because of mere personal inconvenience, can be allowed. But the circumstance renders the use of a cover for the stirring hand very advantageous. I have found a loose linen bag, into which the hand could go freely, more convenient for this purpose than a glove; for being in contact with the skin at distant parts only, the hand is preserved at a much lower temperature. Two small holes in it, one at the front and the other at the top, allow the handle of the stirrer to pass obliquely through, by which arrangement it is easily held with firmness, and the bag itself prevented from slipping towards the glass. It should not be larger than to cover the wrist, or it will embarrass the movements; and it should be very stiffly starched and ironed, that no fibrous particles may fly from it to the glass during the stirring.

79. The glass which, adhering to, is brought away with the stirrer, indicates, by its appearance, the general character

and state of that in the tray; but during its examination, the experimenter must carefully refrain from touching it; for if the finger, or any other organic substance, come into contact with it, the next time the instrument is immersed in the ignited glass, the part touched will produce bubbles. It is therefore of importance that the stirrer be preserved perfectly clean from one stirring to another, for which purpose it may be deposited so that the platinum shall be received in an evaporating basin, the mouth of which is afterwards covered over.

80. In entering upon the considerations relative to the bubbles, it will be evident, from the nature of the materials and the quantity of elastic matter originally present, that these air cavities are at first very numerous. The larger ones soon ascend to the surface, and, breaking, are dissipated without inconvenience; but the smaller ones rise with far less readiness, and the smallest have so little power of elevation, that the general currents in the liquid appear sufficient to carry them downwards, or in any other direction, and thus retain them for any period within the mass. A useful idea of the length of time required for very minute bubbles to ascend through a fluid having some tenacity, may be gained by the person who will take a glassful of clear concentrated white sugar syrup, and beat it up with a little air, until a portion of the latter is in extremely minute bubbles. If these are allowed to remain undisturbed, it will be observed, that though the larger bubbles rise quickly, and the smaller soon after, the smallest will continue for many hours under the surface, destroying the pellucidness of the fluid; and this will be the case although there are none of those descending currents, resulting from difference of temperature, which in the glass assist in retaining the bubbles beneath the surface.

81. From the great length of time which it required to liberate the bubbles even from small pieces of glass, and when no stirring was practised, I was induced to conclude that the evolution of gaseous or vaporous matter had not ceased upon the first fusion of the materials, but that the glass itself when highly heated continued to evolve small portions for some time. It occurred to me also, that in that case its formation might be hastened and the final separation advanced by mixing some extraneous and insoluble substance with the glass, to act as a nucleus, just as pieces of wood, or paper, or grains of sand

operate when introduced into soda water or sparkling cham-
pagne; in which cases they cause the gas, which has a tendency
to separate from the fluid, to leave it far more quickly and perfectly than if they had not been present.

82. The substance I resorted to for this purpose was platinum
in the spongy state. It was chosen as being a body solid at
high temperatures, uninfluenced by the glass, easily reduced
to powder, and likely to retain its finely divided condition during
the operation: its preparation is described in the Appendix.
In experiments made expressly to ascertain its action, it was
found to assist powerfully in the evolution and separation of
the bubbles, and afterwards to sink so completely to the bottom, that not a particle remained suspended in the mass. Even
stirring does not render it injurious; for the particles, by that
action, are welded to the bottom, and the glass ultimately as free
from mixture with them as if they had never been present.

83. The spongy metal should be perfectly pure. It is easily
reduced to powder by rubbing it with a clean finger on clean
paper. No attrition with a hard substance should be allowed,
as that burnishes the metal, and takes away the roughness,
which is highly advantageous in assisting the evolution of the
bubbles. When reduced to powder, it should be again heated
upon a piece of platinum foil in the flame of a spirit-lamp.

84. The quantity of powdered platinum which I have usually
employed has been about 7 or 8 grains for every pound weight
of glass. But in order to effect its more general and perfect
diffusion, I have usually mixed it with ten or twelve times its
bulk of pulverized glass. For this purpose, some of the rough
glass, the same in composition with that to be perfected, has
been crushed small in a clean agate mortar, and the finer parts
separated from the coarser on an inclined and shaken sheet of
paper. The former have been then mixed little by little with
the platinum, and rubbed slightly with the finger, to effect perfect separation of the metal; and then the coarser parts have
been added, to increase the bulk. In this state it was ready
for use.

85. The time of introducing this prepared platinum is, like
the times of stirring, as yet under investigation. It has usually
been sprinkled from the platinum ladle (28) over the surface of
the well-fused and highly-heated glass, at the period of the first

stirring. This method has the advantage of bringing the assisting substance into contact with the glass when the latter is highly disposed to throw off its adhering gaseous matter, and also allows of thorough mixture; but it also causes the addition of fresh glass after the concoction of the materials has been proceeding for many hours; and it likewise occasions the introduction of many bubbles formed by the air in the interstices of the powder.

86. On other occasions the prepared mixture of platinum and glass has been introduced into the tray at the period when it was charged with the due quantity of rough glass, and before the application of fire. Particular attention was then paid to its general diffusion throughout the charge, and on these occasions its action commenced the moment the glass in contact with it was fluid. I am inclined to believe the latter will ultimately prove the better method of proceeding, both for the greater length of time during which the platinum can act, and for the facility and convenience of its introduction.

87. In either mode of appliance the platinum has been found highly serviceable; and in every case since its use, where stirring has not been necessary, the resulting glass has proved to be perfectly free from bubbles.

88. As already mentioned, the best periods for stirring and repose have not been finally determined. Stirring introduces bubbles, and therefore should, if possible, be avoided towards the conclusion of the experiment. Rest, or at least that condition in which there is no other motion than what is due to the currents produced by slight differences of temperature, causes striæ even after very careful mixture (71. 72), and is therefore equally to be feared; and whatever other variations may have been adopted, I have always found it important to apply a careful concluding stirring. The following may be considered as the order of an experiment. If the spongy platinum has not been introduced into the tray with the rough glass, then about the sixth hour after lighting the fire it is added in the manner already directed (85), and the glass well stirred (76). At about the twelfth hour the stirrings are recommenced, for the purpose of making the mixture perfect, and are repeated every 20 or 30 minutes, according to the fusibility of the glass and the state of the heat (60), for eight or nine times. The glass is

then allowed to remain at rest for six or eight hours, that bubbles may ascend and be dissipated, after which it is well stirred twice or thrice more with particular attention, that, if possible, no air may be introduced, being thus finally mixed for the last time.

89. The concluding mixture is peculiar, in that it has to be continued until the glass is so cold and thick that no ascending and descending currents can be formed in it; after which the temperature is not again to be allowed to rise; hence the operation requires certain preliminary arrangements. The first point necessary is to clear out a considerable quantity of slag from the flue furnace, or that part beneath the chamber (47). This slag results from the fused ashes of all the coke which has been consumed there, with other portions that have passed on from the coal fire. It is to be drawn on to the bars of the furnace by a fire-rake which will pass into the passages beneath the chamber. If not taken out in its fused state, it would be impossible afterwards to remove it without risk of great injury to the furnace. At the same time that the slag is removed, all the coke is likewise to be withdrawn. All the fuel in the fire-bars is also to be brought out of the furnace; and if the bars are embarrassed with clinkers, they are to be loosened. These things being done quickly and quietly, and the furnace apertures closed, a few moments are to be allowed for the little dust that may have been agitated to settle, and then the chamber is to be opened and the glass stirred. The heat will have fallen very little during the preceding operations, and the glass may be well mixed; but with this precaution, that when once the stirrer is beneath the surface, it should not again be taken out until the conclusion. By opening the feed-hole or the ash-pit, air may now be allowed freely to enter the furnace, and will rapidly lower its temperature, especially at such parts as the bottom of the pan, which are thin and at this moment exposed to the atmosphere on both surfaces. The temperature of the glass will fall in a corresponding degree, and the stirring being all this while continued, though more slowly if convenient, the substance will gradually thicken, until at last motion will endanger its being pushed out of the tray, and then the stirrer is to be carefully withdrawn. No currents in the glass need be feared, for the temperature cannot now rise higher. But a

single cover being put over the tray, and the outer orifice of the air-tube closed by a good cork, the whole may be left a few minutes to cool still further for perfect security, until, the glass being supposed to have arrived at the state of a thick paste, the annealing should commence. Then the ash-pit, the fire-place, and all the other apertures to the furnace are to be closed; the second glass cover put into its place; the chamber shut up by its iron and tile covers; a layer of bricks arranged close together over the whole upper surface of the chamber and furnace; the damper of the flue closed to prevent air passing through the fire-place, and the whole left to cool gradually for several days.

90. The interval between the common temperature and that at which the glass begins to lose solidity and acquire softness, is so much less with this variety than with flint glass, that it is probable a much shorter period of time is required for its perfect annealing than for the latter. That no failure might occur in this point, however, four days and nights have been allowed for the annealing of the large plates. If everything were left as just described, the contents of the chamber would be warm on the sixth or even the seventh day, so gradually do the arrangements allow it to cool; but on the morning or the evening of the third day, according to circumstances, the damper in the flue is withdrawn a very little to allow the passage of a small quantity of air, and by this means the cooling facilitated and regulated.

91. When the furnace and its contents are cold, the chamber is opened: if the experiment has been well conducted, everything will be found loose, and unaltered in disposition from what they were when first arranged. The earthenware supports are to be removed, and the tray taken out. After examining the glass itself, the exterior of the tray should be carefully observed, whether there be any appearance of leakages either through imperceptible holes or at the corners; and such places as can be rectified by a patch should be noted in reference to the future use of the platinum.

92. An operation which, to be successful, requires much care, is then to be performed; namely, the separation of the platinum from the glass. The tray should be placed on clean smooth paper upon a cloth. The corners are one by one to be opened by a blunt smooth knife, or some softer instrument, from the side towards which they were folded; and being then carefully

T

pulled outwards by their extremities, will usually open, so that the platinum becomes single again. Then proceeding from corner to corner, the platinum will peel or strip easily from the sides of the glass, and will remain adhering by the bottom only. From time to time, as fragments of glass are formed, they should be blown away or otherwise removed, that they may not cut the metal. If now the glass be placed a little over the edge of the table and firmly held, the platinum may gradually be separated from the bottom in the same manner as from the sides, and the glass and the metal finally divided from each other without any injury to the former, and very little to the latter.

93. Immediately upon the separation of the platinum, and before it can receive any mechanical injury beyond what it was impossible to avoid, it is to be put into a pickle consisting of nitric acid and water, and left there for several days. The dilute acid acts upon the adhering glass, dissolving and loosening it, and the plate is thus rendered fit for future operations (41). The stirrers also, when no longer required in an experiment, should be taken from their iron handles and put into the same pickling liquor. In this way the platinum is perfectly cleaned, and being afterwards washed carefully in pure water and ignited, is again ready for use.

94. Such is the nature of the process as practised at present, by which plates of heavy optical glass seven inches square and eight pounds in weight have been prepared. I am encouraged to believe that it will admit of improvement, perhaps even to the full extent of our desires; but it will require time and patience to effect it. As I have before said, we are in the course of our experiments only; and up to the last have seen reason to vary the arrangements, and still intend to make alterations. Everything agrees to convince me that the size of the plate is not a circumstance involving any additional difficulty; but that, on the contrary, it will probably be safer to make a large than a small experiment. We can at pleasure obtain a glass perfectly free from striæ, unexceptionable in hardness, and with less colour than crown glass; but it is the simultaneous absence of all striæ and bubbles, with at the same time that degree of hardness and colour which will render the glass fit for optical purposes, that I am aiming at, and that I trust shortly to obtain.

95. As soon as the plates of glass are removed from the

platinum and briefly examined, they are sent to Mr. Dollond, who then enters upon the discharge of his particular duties in the Committee, by cutting, examining, and even working them into telescopes. It is not, however, my place to detail this gentleman's exertions (as a member of the Glass Sub-committee) in the cause of science. They will, I trust, appear in due season; and I hope that the want of perfect success on my part will not long be a cause of delay.

§ 2. *General qualities of the heavy Optical Glasses.*

96. A great variety of glasses have been formed by the use of different proportions of ingredients. They vary importantly from each other, though by no means to the extent of the difference existing between any of them and flint glass. The *specific gravity* rises very high in borate of lead, consisting of single proportions, *i. e.* nearly 24 by weight of boracic acid and 112 of oxide of lead; it is often as high as 6·39 or 6·4, being double that of some specimens of flint glass. In silicated borate of lead, which, in addition to the former quantities, contains 16 parts, or a proportional of silica, it is about 5·44. As the proportion of oxide of lead diminishes, so also does the specific gravity lessen, and it is in some of the specimens as low as 4·2; still permitting by the proportions present such fusibility and other qualities as consist with the process described. The specific gravity of Guinand's heavy flint glass is about 3·616; that of a specimen of ordinary flint glass 3·290; that of plate glass 2·5257; and that of crown glass 2·5448.

97. The *refractive* and *dispersive* powers of the glasses increase with their specific gravity, as was to be expected. The powers of two of them, namely, borate of lead and silicated borate of lead, consisting always, if not otherwise expressed, of single proportionals, have been ascertained by Mr. Herschel, and are as follows:—

	Bor. Lead.	Sil. Bor. Lead.
Angle of glass prism	29° 6′	. . 30° 26′
Refractive index for extreme red rays	$\mu = 2{\cdot}0430$	. . 1·8521
Refractive index for maximum yellow	. $\mu = 2{\cdot}0652$	. . 1·8735
Refractive index for extreme violet ::	. $\mu = 2{\cdot}1223$	. . 1·9135
Dispersive index $= \dfrac{\delta\mu}{\mu - 1} =$	0·0740	. . 0·0703

These intense powers upon light are not accompanied by any circumstance rendering the glass optically unfit for the compensation of the dispersive powers of crown or plate glass.

Three object-glasses have been constructed for the express purpose of ascertaining this point; and all of them tend to demonstrate that the compensation or correction may be effected with equal if not greater facility than with flint glass.

98. One important circumstance connected with the application of these glasses to the purposes for which they are designed, is their *colour*. The great power they have of developing strong tints from metallic impurities, has been already described and illustrated (22, 23), and creates a difficulty in the way of obtaining them unobjectionably free from colour. The usual colour is more or less of yellow, and is perhaps almost altogether, if not quite, dependent upon the presence of a little iron. Like many of those dependent upon mineral substances, it is very much heightened by elevation, and lessened by diminution of temperature. It is rapidly and permanently diminished by increasing the proportions either of the silica or the boracic acid. The silicated borate of lead has latterly been obtained of such faint tint by the precautions, relative to impurities, already described, that when 9 inches in thickness, white paper looked at through it in open daylight resembled in appearance and depth of tint the surface of a lemon. Glass consisting of 1 proportional $=112$ oxide of lead, 1 proportional $=16$ silica, and $1\frac{1}{2}$ proportional $=36$ boracic acid, when 7 inches in thickness and examined in the same manner, did not give a colour surpassing that of pale roll sulphur. The triborate-of-lead glass is almost as colourless as good flint glass, but might perhaps be found objectionable on other accounts.

99. As there is a certain quantity of light intercepted by glass which is altogether dependent upon and in proportion to its colour, it is evident that this property of the heavy glasses must be considered in relation to their use in telescopes; but there appears no reason for supposing they will ultimately prove inapplicable on this account. The colour of the glass already obtained is far less in depth than that of the crown glass constantly used in the construction of telescopes, which yet intercepts by its colour no important quantity of

light; and if two plates 8 or 10 inches long, one of the yellow heavy glass and the other of crown glass, be looked through edgeways, it will be seen in a moment that the crown glass intercepts by far the most light. The colour of the glass is of no consequence, otherwise than as causing a loss of light from interception; for the tinge which is cast over objects looked at through a telescope constructed with it is scarcely perceptible to the most acute eye, and quite unimportant. When to these circumstances is added the reasonable expectation entertained of removing a large proportion of the little remaining colour by the use of purified silica (21), it need not be anticipated that experience will prove the glass faulty in this respect.

100. There is one very important action of the glass upon light, however, which may perhaps interfere more with its application, in telescopes at least, than any other, *i. e.* its *reflective power.* This is very strong in all the heavy glasses, far stronger than in flint, and exceedingly surpassing the similar power of crown glass. It is in proportion, as might have been expected, to the refractive power and the density of the glasses, all these properties increasing with the oxide of lead. The loss of light occasioned by the reflexion from the two surfaces of a plate through which a ray is passed, appears to me to be greater than from the united action of both colour and bubbles in a piece of glass 7 inches thick.

I endeavoured to ascertain the comparative quantities of light reflected by these heavy and other glasses, in some photometrical experiments made upon the principle of similar shadows, measuring only the reflexion from the first surface of the different glasses, that from the second surface being destroyed. The ray was made incident in all the cases at an angle of 45°. It was obtained from a small single-wicked lamp, *a*; and when reflected, its intensity was measured by the distance of a similar lamp, *b*, whose direct light cast the comparative shadow. The uniformity of the two lights, or at least of their relation to each other, was established by trials before and after the experiments with the reflecting surfaces, and each surface was tried two or three times, at intervals, and in a mixed manner; so that no anticipation of the result could in any case bias the mind. The following Table shows the results, small decimals being neglected:—

Inches.

Light *a* direct	10·70 . . .	1	. . . 1
Light *a* reflected by glass 5 . . .	36·75 . . .	11·80 . . .	$\frac{1}{11·8}$
1 . . .	40·69 . . .	14·46 . . .	$\frac{1}{14·4}$
4 . . .	43·46 . . .	16·50 . . .	$\frac{1}{16·5}$
9 . . .	47·31 . . .	19·56 . . .	$\frac{1}{19·5}$
6 . . .	50·31 . . .	22·12 . . .	$\frac{1}{22·1}$
7 . . .	51·63 . . .	23·29 . . .	$\frac{1}{23·3}$
3 . . .	52·69 . . .	24·26 . . .	$\frac{1}{24·2}$
8 . . .	54·33 . . .	25·80 . . .	$\frac{1}{25·8}$
2 . . .	54·56 . . .	26·02 . . .	$\frac{1}{26·}$

The first column refers to the glasses below; the second gives the distance of the measuring flame *b*; the third, the preceding numbers squared and reduced to the direct light as unity; and the fourth, consequently, the proportion of the light *a* reflected by the first surface of each glass. No. 5 was glass consisting of 1 proportional of oxide of lead, $\frac{1}{2}$ a proportional of silica, and $1\frac{1}{2}$ proportional boracic acid. No. 1 was composed of 1 oxide of lead, 1 silica, and $1\frac{1}{2}$ boracic acid. No. 4, of 1 oxide of lead, $1\frac{1}{2}$ silica, and $1\frac{1}{2}$ boracic acid. No. 9 was flint glass; No. 6, 7 and 3, different pieces of crown glass; and No. 8 and 2, different pieces of plate glass. 1, 3, 5, 6 and 7, were natural surfaces; 2, 4, 8 and 9, polished surfaces.

The deficiency of light resulting from the increased reflecting power, though considerable, may easily be compensated for by slightly increasing the area of the plate; and the power of obtaining plates of any size is professed to be given by the general process: but whether that expedient involves any other objections, it will be for the optician to determine.

101. In *hardness*, these glasses differ from each other as much as in any other quality, and indeed more. The borate of lead is very soft; the biborate of lead is harder, and the triborate equal to flint glass in hardness. The silicated borate of lead is softer than flint glass; but the glass consisting of 1 proportional oxide of lead, 1 of silica, and $1\frac{1}{2}$ proportional of boracic acid, is as hard as ordinary flint glass, at the same

time that it has that degree of fusibility, colour, and other properties, which makes it a very promising variety.

102. The hardness increases with the diminution of the oxide of lead; but the *fusibility* diminishes in the same proportion; and this is a property which it is essential to preserve to a certain degree for the removal of striæ and bubbles. The borate of lead is so fusible as to soften and lose its form under the surface of boiling oil. The silicated borate, and the glass consisting of the proportions above mentioned, are quite fusible enough to allow of the processes necessary for the removal of striæ and bubbles.

103. The fusibility of these glasses, and of glass generally, must not be confounded with their relative tendency to soften by elevation of temperature. It is not that glass which softens first, that becomes most fluid at a certain given high temperature; for glasses, like other substances, vary in their readiness to pass into the fluid state. Hence it has often occurred amongst the variety of compositions tried for glasses, that when the resulting substances have been placed side by side on platinum foil, and heated, that which first softened did not when heated highly become so fluid as some other specimens that longer resisted the first impression of heat. It has, however, always been found that those glasses which when subjected to a rising temperature, most slowly passed from the solid to the fluid state, were also those which when subjected to long annealing processes, were least liable to assume a crystalline structure; and thus very useful indications of the probable qualities of compounds under investigation were often obtained.

104. A most important consideration relative to the application of these glasses to the construction of telescopes, is their liability to change and injury by the action of substances usually occurring in an ordinary atmosphere. When the value of a good object-glass is considered, frequently amounting to many hundred pounds, this point will be thought of no little consequence; and when it is known that even flint and plate glass are frequently injured in this way, a little anxiety for the capability of resistance in the heavy glasses may readily be allowed, since they contain so much less of the substance (silica) which confers the power of resistance, and so much

more of that (oxide of lead) which is considered as the vulnerable part, than does either of the former kinds.

105. The superficial changes of glass which interfere with its optical uses are of two kinds. The one is shown by a tarnish upon the surface, which when strong is iridescent. It is quickly produced by the intentional presence of sulphuretted hydrogen, which acting upon the oxide of lead present, reduces it, and forms a sulphuret of lead. It takes place only with flint glass, and is in every case produced either by sulphuretted hydrogen or other sulphuretted vapours. In plate glass the change is of another kind, and is shown by the appearance of minute vegetations or crystallizations, which spread, obstructing the light wherever they occur. Mr. Dollond, who has shown me cases of both kinds of injury in flint and plate glass, is inclined to believe that the latter has, during his long experience, proved most injurious.

106. From the commencement of the experiments it was expected that these heavy optical glasses would tarnish more than flint glass; but as specimens of borate of lead and other dense compounds of that metal had been retained in an ordinary atmosphere, without any particular precautions, for long periods of time, yet without tarnishing, there was encouragement to continue the investigations; and though when specimens were put into atmospheres purposely contaminated with sulphuretted hydrogen, they tarnished quickly, and much more than any flint glass, yet it did not follow that they would of necessity tarnish in the telescope; especially as, being (from the construction of the achromatic object-glass) enclosed by the tube and the crown or plate glass lens, they would be considerably protected, and at the same time would admit of the intentional application of extraneous chemical protectors.

107. The kind of protection which occurs to the mind is the application of such substances to the interior of the tube as, having a strong attraction for sulphuretted vapours, should continually retain the atmosphere within free from their presence. Carbonate of lead, precipitated borate of lead or finely-ground litharge, mixed with the pigment which is usually applied to blacken the inside of the telescope that all extraneous light may be absorbed, will probably effect this purpose completely.

108. A very curious and important influence of alkali in facili-

tating the tarnish of glasses containing oxide of lead, was discovered during the course of these investigations; and when the quantity of lead in flint glass is increased but a little beyond the ordinary proportions, its effect is powerfully manifested. Ordinary flint glass consists of 33·28 oxide of lead, 51·93 silica, and 13·77 potassa; the rest of the substances present, being in very small quantity, may be disregarded. Here the oxide of lead is 33·28 hundredths of the whole; and if it be only a little increased, for the purpose of giving greater dispersive power, the glass is liable to tarnish in an ordinary town atmosphere. Such is the case with a˙ specimen of Guinand's glass, which I have analyzed, and which contains 43·05 oxide of lead, 44·3 silica, and 11·75 potassa. But provided the alkali be away, the quantity of oxide of lead may be enormously increased; and a glass containing 64 per cent. of oxide of lead, in combination with 36 per cent. of silica, has not tarnished by an exposure for eighteen months on the same shelves with flint glasses that have tarnished. The following case will point out the effect still more strongly :—A combination of equal weights of silica and oxide of lead was formed, and the compound has shown no tendency to tarnish in an ordinary atmosphere since February 1828. Eight parts of this was fused with as much pearlash as was equivalent to 1 part of potassa, and a glass was formed which has since become much tarnished. But other 8 parts being fused with 3 parts more of oxide of lead, so as almost to double the proportion of the latter, gave a glass without alkali, which does not yet exhibit the slightest trace of tarnish.

109. Hence the reason why the absence of alkali has been earnestly insisted upon in the preparation of the ingredients for the heavy optical glasses (18. 24). Hence the reason also why heavy flint glass, as already mentioned, has tarnished equally with some of the heavier glasses, though containing so much less lead, and of such inferior specific gravity. This influence of alkali is associated with, and perhaps directly referable to, another circumstance affecting the liability of change in the glass; I mean the action of water or of aërial moisture, which is frequently considerable, and appears to be dependent upon the alkali present.

110. If a small quantity of flint glass be very finely pulverized

in an agate mortar, then placed upon a piece of turmeric paper, and moistened with a drop of pure water, strong indications of free alkali will be obtained. The same effect is produced by using plate glass; and if the pulverization be very perfect, the alkali can be detected in glasses containing far smaller quantities of that substance than either of those mentioned. This experiment, due to Mr. Griffiths, shows that in whatever state of combination the alkali may be, it can still act upon, and is subject to, the action of moisture; and that flint glass is by no means a compound resulting from very strong chemical affinities, is also shown by an experiment which I made many years ago; namely, that if flint glass be pulverized exceedingly fine, the powder will indicate the presence of sulphuretted hydrogen in the air by becoming blackened, almost as readily as carbonate of lead. Glass may be considered rather as a solution of different substances one in another, than as a strong chemical compound; and it owes its power of resisting agents generally to its perfectly compact state, and the existence of an insoluble and unchangeable film of silica, or highly silicated matter upon its surface.

111. The half-combined and hygrometric state of the alkali appears to be the cause of the deposited film of moisture which is well known to adhere to ordinary glass when exposed to the atmosphere at common temperatures. This film is highly calculated to condense any portion of sulphuretted vapours which may be floating in the atmosphere, and thus bring them into contact with the oxide of lead under the most favourable conditions for the production of that action which is the direct cause of tarnish. Now from this cause of action the heavy glass is free; and hence a satisfactory reason to me why the heavy glasses have suffered so little when left with common care in an ordinary atmosphere.

112. An extraordinary difference exists between the electrical relations of this glass and other glasses, due principally to the same absence of alkali. Ordinary glasses, either flint, plate or crown, will, from the hygrometric film of moisture upon the surface, freely conduct electricity under common circumstances. Thus if a gold-leaf electrometer be diverged, and then touched with them in their ordinary state, the electricity is instantly discharged, even though the hand be two or three feet from the part touching the instrument. If a similar experiment be

made with these heavy glasses, they have no sensible power of discharging the electricity, but insulate as perfectly as sealing-wax or gum-lac. If one of these plates of glass, without any previous warming and drying, be lightly brushed or wiped with flannel or silk, it instantly becomes strongly electrical, and retains its electricity for a long time; but it would be almost impossible to develope electricity by such slight means with flint or plate or even crown glass in a similar state. Hence the glass makes as good an electrophorus as lac or resin, and may probably be found hereafter to answer many useful electrical purposes. But the great point at present in view, is the proof which such electrical properties give of the absence of that film of moisture which is so constant upon other glasses.

113. All these circumstances are favourable to the opinion that the heavy glass will not be found objectionable in the construction of telescopes, because of any undue tendency to tarnish, and especially when precautions are taken to protect it from sulphuretted vapours in the manner before described (107). No difficulty can be anticipated in preserving the air within a limited and enclosed space free from such contamination: to preserve it dry, if that had been necessary, under the different circumstances of varying temperature and the inevitable change of the air more or less frequently, would have been a far more difficult task.

114. The other kind of superficial change, *i. e.* the corrosion or crystallization which takes place principally on plate glass, is doubtless also due to the alkali present. Sometimes, indeed, specimens of glass may be found where the alkali being too abundant, a similar but more extensive action has taken place over the whole of the surface, and the glass falls off in scales. Whether the alteration be due to the action of the alkali on the water only, or on the carbonic acid and other substances it finds in the air, or to its united action on all together, is of little consequence at present, as the substance on which it depends is altogether absent from the glass under consideration.

115. Among the great number of glasses made, there are several of different composition, which have been selected, because of their general characters and properties, for more extensive trial and investigation when time will permit. Of these it would be useless to speak at present, as what might be

stated of them now would probably require correction from future experiments. Up to this period the attention has been devoted, as it still must be for a while, to the establishment of a process which, competent to produce with certainty a glass fitted for optical purposes, may have the philosophy and practice of every part so fully ascertained, as to be capable of description in a manner sufficiently clear to enable any other person, with moderate care, to obtain the same results without the labour of long and tedious investigation.

APPENDIX.

Rough-glass furnace.—The only furnace for making rough glass which has been constructed, answers its purpose exceedingly well; and though if a second were to be made, it should be upon a larger scale, yet I think it better to describe the tried one accurately, than to direct alterations which have not been experimentally approved of; especially as there seems to be nothing which, in principle, need differ in a larger furnace. An iron box (Plate II.) 30 inches long, 14 inches wide, and $8\frac{1}{2}$ inches deep, forms the principal part of the exterior: it is open entirely at the top, and at the bottom also, in the fore part, where a fire-grate is to be placed. It has a common iron furnace door in front, the aperture of which is 8 inches wide by 6 inches high; and at the opposite end, or back of the furnace, a flanched aperture, $6\frac{1}{2}$ inches by $4\frac{1}{2}$ for a piece of funnel pipe to connect the furnace with a powerful flue. The sides of this box, and such part of the bottom as is not appropriated for the fire-grate, are lined with fire-stone $1\frac{1}{2}$ inch in thickness, except in the fire-place, where it is $2\frac{5}{8}$. The grate is 12 inches long by 8 wide; and the part above it is closed by a fire-tile 2 inches thick and 12 inches square, which resting on the edges of the lining, finishes the portion intended for the coal fire, leaving it $5\frac{1}{2}$ inches in depth from the covering tile to the grate. The other part is covered by an iron plate, $17\frac{1}{2}$ inches long, 13 inches wide, and $\frac{5}{8}$ths of an inch thick, which, resting upon the edges of the lining, encloses a space of 16 inches long, 10 inches wide, and 5 inches deep, for the reception of crucibles. This plate is formed with circular holes, about 3 inches, or rather more, in diameter, arranged as in fig. 1, that the crucibles inserted through them may leave plenty of room for the intervention of coke and flame. As many round crucible-covers belong to the plate as there are holes, serving to close such of them as are not occupied by crucibles.

As the plate becomes very hot when in use, it is necessary to have a second above it, which may be formed of sheet-iron with corresponding holes, and when put into its place, separated from the first, a little space, by pieces of tobacco pipe, or other convenient substance, to include a layer of air. But it is much better for the retention of heat, and also for its superior cleanliness, that this second plate should consist of pieces of earthenware fitted to each other, so as to cover the surface of the iron plate, from which it should also be separated by a short interval.

The crucibles used are 5 inches high outside, $3\frac{1}{4}$ inches diameter at the top, and 2 inches diameter at the bottom. They are of pure porcelain biscuit, perfectly white and clean. They should be made as thin as possible, of the finest and most refractory kind of ware, and baked at a high temperature. We have some crucibles made about thirty years ago for Mr. Hatchett, which, though not of the size required, are precisely the right kind of ware. They have been used many times in succession without cracking or being importantly acted upon by the glass, and no sensible degree of impurity was given to it from them.

When these crucibles are arranged in the furnace, they should be supported by little stands of earthenware, formed out of brick or Cornish tile, so that their edges shall rise about $\frac{1}{2}$ or $\frac{1}{3}$rd of an inch above the surface of the upper covering plate, that no impurity may enter them. The holes in the plate should be of such dimensions that, when hot, the crucibles may fit loosely, that they may be uninjured, and also that there may be room between for the vapours that are evolved from the mixture to pass away.

The covers to the crucibles are evaporating basins about $4\frac{1}{2}$ inches in diameter. They are slung with their edges downwards by pieces of platinum wire sufficiently strong for the purpose, which being first bent at the middle into an angle, are then stretched across the outside of the basins, and have their ends bent round the opposite edges of the latter. The bent extremity of an iron rod passed under the loop thus formed over the middle of the bottom, serves to raise and remove any cover from place to place. When a crucible is in use, the cover should be arranged over it in such a manner as not to touch the vessel, but rest by its edges on the earthenware plate around.

The platinum stirrers in use with this furnace have been before described (28. 75), fig. 3. The platinum ladle consists of a small crucible of that metal riveted to a platinum wire, and that made fast by a screw to an iron rod (fig. 4).

The use and manner of working this furnace will be well understood from the above description, and what has before been said (26, &c.). The crucible should never be suddenly heated or cooled. The coke may be fed and arranged at such of the crucible holes as are out of use at the time. Because of the very valuable effects of a trough of water under the fire-bars (45) experienced in the larger furnace, one is constantly used with that just described.

Finishing furnace.—This furnace on the outside is a parallelopiped, principally of brickwork, built against a wall; it is 64 inches in length, from the fire front to the beginning of the flue, against which it is built, 45 inches wide, and 28 inches high (figs. 5, 6 & 7). It is the only one that has yet been built, and, for the reasons before given, shall be described exactly as it is. The fire-place is at one end, and the course of the flame and smoke is directly from that to the other end, and then immediately into the upright flue. The fire-place is 15 inches from back to front, 13 inches wide, and $11\frac{1}{2}$ inches from the arched roof to the bars. Its outward side, or that from the wall, is $18\frac{1}{2}$ inches in thickness of brickwork, which is intended to give stability to the structure. The mouth of the fire-place is an aperture 8 inches by 6 inches, made in a piece of fire-stone 7 inches inwards from the front of the brickwork: its lower edge is level with a fire-stone sill, which, extending forwards from the fire-place to the outer surface of the brickwork, forms a shelf, on which two bricks stand, that serve in place of a door to close the mouth of the furnace. The ash-pit is 25 inches long, 12 inches wide under the fire, and 10 inches high to the bars. A trough made of rolled iron, riveted together, and $5\frac{1}{4}$ inches high on the sides, occupies its lower part. This being preserved full of water, is sustained at the boiling temperature by the radiation of heat and the hot ashes which fall into it.

From the back part of the fire-place, and 2 inches above the level of the fire-bars, the brickwork is carried on horizontally until close to the stack. The sides of this part are perpendicular, and 12 inches apart: they are continued upwards to the top of the brickwork 14 inches unbroken, except that at 5 inches from the bottom they are thrown back $\frac{1}{3}$rd of an inch so as to form a ledge there. This ledge is for the purpose of receiving the edges of certain fire-tiles, which, when put in, form the top of the flue and at the same time the bottom of the glass chamber; but the whole is so constructed, that the tiles can be put in and taken out at pleasure without disturbing the rest of the work. The side, or rather end of the chamber nearest the fire, is constructed of a fire-tile, which terminates and faces the brick arch over the fire-place, and extends from the surface of the brickwork downwards 9 inches to the side ledges before described: the further end of the chamber is finished in a similar way, and beyond that the flue is carried in the most convenient and direct manner, but without any unnecessary contraction, into the stack or chimney. The length of this upper aperture, afterwards constituting the chamber, is 25 inches, its breadth $12\frac{3}{4}$ inches. When the bottom tiles are in their places, they leave a depth of 5 inches for that part of the furnace or flue beneath the chamber, which is also 38 inches from the fire to the end, and, with the exception of certain supports in it, is 12 inches wide.

These supports are built in with the bottom of the flue. They are essential to the permanency and regularity of the bottom of the glass chamber, and require considerable nicety in their arrangement. They consist of fire-bricks placed up on end, so that their narrowest surfaces are towards the ends of the furnace, their sides or broadest exposed surfaces parallel with the sides of the furnace itself. They rise to the same height above the bottom of the flue as the ledges on the sides of the brickwork, or 5 inches; and with them form the support for the bottom tiles. There are three of them in the furnace, placed in a line equidistant from the two sides of the flue; and being $2\frac{1}{2}$ inches thick, they leave spaces for the passage of flame and the reception of coke, which are $4\frac{3}{4}$ inches in width. The first of these is 2 inches from the back edge of the fire, and in that direction extends 4 inches; the second is 4 inches from the first; and 6 inches beyond that one is the third.

During the action of the furnace, coke is supplied to this part, and arranged through two holes level with this space, and wrought in the side of the furnace by leaving out a brick. They are made to occur nearly opposite the spaces between the supports seen when looking across the course of the flame, and are stopped by the insertion of loose bricks, and a piece of paper put before the place, which adheres from the pressure inwards of the atmosphere. These holes, being in the thickness of the walls of the furnace, are 17 inches long.

The tiles which form the bottom of the chamber and top of the flue, are of Cornish ware (52. 53), or at least the one which constitutes the half nearest the fire is of that material; but the other, which is not so highly heated, and never has to be removed, may be some other ware, and $2\frac{1}{4}$ inches in thickness. The tile nearest the fire has to transmit heat to the glass, and if of Cornish ware, and being supported as described, is abundantly strong when $\frac{3}{4}$ths of an inch in thickness. It should be nicely adjusted by grinding (53), and when fitted in, the edges should be made close by a little fire-lute.

There is a part of the furnace not yet mentioned, which must be arranged as the structure is raised. This is the air-tube (55). It is of glazed porcelain, and passes horizontally through the side of the furnace, so that its inner aperture is 2 inches from the end of the glass chamber, and its lower edge level with the upper surface of the Cornish tile constituting the bottom, whilst the outer end is flush with the outside surface of the brick-work. Its length is 17 inches, its internal diameter $\frac{7}{8}$ths of an inch. The short pieces of adjusting tube (55) are 6, 7, and 8 inches in length, and $\frac{7}{10}$ths of an inch internal diameter: their ends are usually finished obliquely.

All those parts of the furnace which are in contact with or near the fire, are of the best fire-bricks laid in loam; but the sides of that part of the cavity already described, which form the glass chamber, are fire-tiles, and they rise about an inch above the neighbouring brickwork, forming a raised edge all round, which, at the same time that it better excludes dirt than if level with the rest of the work, also allows the covers of the glass chamber to apply more closely. These covers are three wrought-iron plates, each $\frac{1}{4}$th of an inch in thickness and 16 inches long; but their widths vary, and are 7, 10, and 12 inches. These put side by side cover the mouth of the chamber, but, varied in juxtaposition, allow of more or less of the chamber being opened at once, according to whatever the experiment may require; each has a short solid handle fixed to the middle of the upper surface.

Besides the iron covers, there are a set of earthenware covers, consisting of six square tiles each $1\frac{3}{8}$ inch thick (62). These are notched to receive the handles of the iron covers, and being put together over them, constitute a covering of earthenware, which very importantly assists in retaining the heat.

The tiles and brick used in the annealing process (89) are the ordinary dry varieties, with some pieces of various sizes, to allow of the close adjustment of the whole.

The earthenware supporting blocks (57) required for the arrangement and support of the platinum tray, should be formed out of some kind of flat unglazed ware containing as little iron as possible, and should be of various thicknesses, sizes, and forms, although parallelopipeds are the most usual. They should not be of such substance as is liable to fly or send off anything when heated; and when any portion of glass adheres to them, it should either be cleared off or the piece thrown away. The Cornish tile before described (52. 53) is excellent for this use, and may be sawn, rasped or ground into any shape required.

The glass-covers (60, &c.) that have yet been used were merely inverted evaporating basins. They answer the required purpose exceedingly well, except that, when large, they are too strong, too heavy, and too deep. Some covers for the purpose are therefore in progress, and as they only have to support their own weight and hold together, they are to be thin. The covers should be of very refractory and highly baked ware; it may be desirable to have them very slightly glazed, to keep them clean, and prevent the absorption of any substance which might send off vapours injurious to the glass.

The fire tools required for this furnace will suggest themselves. Amongst the rest should be a pair of tongs which will readily lay hold either of the earthenware tile or the iron covers;

U

a slag and coke rake (89); and a stoking iron, with its extremity bent, for the purpose of breaking the clinkers off the bars from beneath upwards.

Preparation of spongy platinum.—The platinum used for this preparation should be pure, and may be the refuse pieces resulting from such plate and foil as has been in use for trays in former experiments. This, after being taken out of the pickle (93), and condemned as useless for other purposes in the glass house, should be trimmed from all alloyed parts, if any such are adhering to it, and then digested in a Florence flask, with a mixture of five measures of strong hydrochloric acid, one measure of strong nitric acid, and three measures of water. But little heat should be applied at first until the action diminishes. According to Dr. Wollaston, one ounce of platinum will be dissolved by about four ounce measures of such acid, and it is advantageous to have a considerable excess of platinum present. The solution obtained is to be precipitated by a strong solution of muriate of ammonia; a bright yellow pulverulent substance will fall, and a mother-liquid having more or less colour remain. The precipitate being allowed to subside, the liquor is to be poured off, and the former then washed with two or three portions of water. The washing liquors and the mother-water may afterwards be concentrated together; but it is better not to prepare spongy platinum for this particular use from these fluids, but only from the precipitate which falls on adding the muriate of ammonia.

The yellow precipitate, when washed, is to be dried on a filter, or in a basin, and then decomposed by the application of a dull red heat. This may be done in a clean white earthenware crucible. The heat should be continued until vapours cease to arise; but this will be found a long operation, in consequence of the low temperature which is to be applied, and the exceedingly bad conducting power of platinum for heat when in this spongy state. The reduction may also be performed by putting the precipitate upon a piece of platinum foil in a layer about $\frac{1}{6}$th of an inch in thickness, and covering it with another piece of foil; a spirit lamp will then suffice to reduce the metal, but the foil and powder must be turned occasionally, that both sides may be exposed to the flame. The platinum will appear as a dull grey spongy metallic mass. It should be broken up, mingled, and then again heated to ensure the dissipation of all volatile matter.

After this is done, the platinum should be rubbed to powder by the clean finger on clean paper (83), heated slightly a third time, and then preserved in a clean and well-stopped bottle.

On a Peculiar Class of Optical Deceptions *.

THE pre-eminent importance of the eye as an organ of perception confers an interest upon the various modes in which it performs its office, the circumstances which modify its indications, and the deceptions to which it is liable, far beyond what they otherwise would possess. The following account of a peculiar ocular deception, which, in a greater or smaller degree, is not uncommon, and which, if looked for, may be observed with the utmost facility, may therefore prove worthy of attention; and I am the more inclined to hope so, because in some points it associates with an account and explanation of an ocular deception given by Dr. Roget in the 'Philosophical Transactions' for 1825, page 121.

The following are some cases of the appearance in question. Being at the magnificent lead mills of Messrs. Maltby, two cog-wheels were shown me moving with such velocity, that if the eye were retained immoveable, no distinct appearance of the cogs in either could be observed; but, upon standing in such a position that one wheel appeared behind the other, there was immediately the distinct though shadowy resemblance of cogs moving slowly in one direction.

Mr. Brunel, jun. described to me two small similar wheels at the Thames Tunnel: an endless rope, which passed over and was carried by one of them, immediately returned and passed in the opposite direction over the other, and consequently moved the two wheels in opposite directions with great

---

* Quarterly Journal of Science, 1831, vol. i. p. 205.

I take the opportunity here of pointing out that, three years prior to my paper, Professor Plateau had published an account of the chief fact, in the fourth volume of the 'Correspondence Mathématique et Physique' of M. Quetelet, p. 393. I was of course unaware of the circumstance. Further observations by M. Plateau will be found in the 'Annales de Chimie,' 1831, xlviii. p. 281.

but equal velocities.   When looked at from a particular position, they presented the appearance of a wheel with immoveable radii.

When the two wheels of a gig or carriage in motion are
looked at from an oblique position, so that the line of sight
crosses the axle, the space through which the wheels overlap
appears to be divided into a number of fixed curved lines,
passing from the axle of one wheel to the axle of the other, in
general form and arrangement resembling the lines described by
iron filings between the opposite poles of a magnet.   The effect
may be obtained at pleasure by cutting two equal wheels out
of white cardboard (Plate III. fig. 1), each having from twelve
to twenty or thirty radii, sticking them on a large needle two
or three inches apart, revolving them between the fingers, and
looking at them in the right direction against a dark or black
ground ; the greater the velocity of the wheels the more perfect
will be the appearance (fig. 2).

When the dark-coloured wheel of a carriage is moving on a
good light-coloured road, so that the sun shines almost directly
on its broadside, and the wheel and its shadow are looked at
obliquely, so that the one overlaps the other in part, then, in
the overlapping part, luminous or light lines will be perceived
curved more or less, and conjoining the axle and its shadow,
if the wheel and shadow are superposed sufficiently; or, tending
to do so, if they are superposed only in part: the more rapid
the motion the more perfect is the appearance.   The effect
may be easily observed by making a pasteboard wheel like one
of those just described, blackening it, sticking it on a pin, and
revolving it in the sunshine, or in candlelight, before a sheet
of white paper (fig. 3).   If the wheel be converted into a
tetotum or top, by having a pin thrust through its centre, and
spun upon a sheet of white paper, the effect produced by the
wheel and its shadow will be obtained with facility, and in form
will resemble fig. 2.   In all these cases no rims are required ;
the spokes or radii produce the effect.

If a carriage wheel running rapidly before upright bars, as
a palisade or railing, be observed, the attention being fixed
upon the wheel, peculiar stationary lines will appear: those
perpendicular to the nave or axis will be straight, but the
others curved ; and the curve will be greatest in those which

are furthest from the upper straight line. These curves are the same in form as those already described and explained by Dr. Roget*, and the appearance itself is produced in a similar manner; but the phenomena are distinct and the causes different. The effect at present referred to is best observed when the velocities are great, whereas that explained by Dr. Roget takes place only when the velocities are moderate. It is probable that some of the appearances briefly mentioned by an anonymous writer in the ' Quarterly Journal of Science,' First Series, vol. x. p. 282, and already referred to by Dr. Roget, were of the kind now to be explained; for though the description is not accurate either for the effects which form the object of this paper, or that explained by Dr. Roget,— and is, indeed, inconsistent with the observation or explanation of any of the phenomena,—it probably had its origin in the occurrence of some of both kinds under the eyes of the writer. The effect is easily obtained by revolving a white pasteboard wheel before a black or dark ground, and then, whilst regarding the wheel fixedly, traversing the space before it with a grate also cut out of white pasteboard. By altering the position of the grate and direction of its motion, it will be seen that the straight lines in the wheel are always parallel to the bars of the grate, and that the convexity of the curved lines is always towards that side of the grate where its motion coincides in direction with the motion of the radii of the wheel. By varying the velocity of the wheel and grate, the curves change in their appearance, and the whole or any part of the system, as described and figured by Dr. Roget, may be obtained at pleasure.

I have had a very simple apparatus constructed, by which these and many other analogous appearances can be shown in great perfection and variety. One board was fixed upright upon the middle of another, serving as a base; the upright board was cut into the shape represented in fig. 4; the middle, and the two extreme projections, forming points of support, were supplied with little caps cut out of copper-plate and bent into shape (fig. 5), so that, when in their places, they offer four bearings for the support of two axes, one on each side the middle. The axes are small pieces of steel wire tapered at the

* Philosophieal Transactions, 1825, p. 131.

extremities; each has upon it a little roller or disc of soft wood, which, though it can be moved by force from one part of the axis to another, still has friction sufficient to carry the latter with it when turned round. These axes are made to revolve in the following manner:—A circular copper plate about 4 inches in diameter has three pulleys of different diameter fixed upon its upper surface, whilst its lower surface is covered with a piece of fine sand-paper attached by cement. A hole is made through the centre of the plate and pulleys, and guarded by brass tube, so fitted as to move steadily but freely upon an upright steel pin fixed in the middle of the centre wooden support; hence when the plate is in its place, it rests upon the two rollers belonging to the horizontal axes, whilst it is rendered steady by the upright pin. It can easily be turned round in a horizontal plane, and it then causes the two axes with their rollers to revolve in opposite directions, and the velocities of these can be made either equal to each other, or to differ in almost any ratio by shifting the rollers upon the horizontal axes nearer to, or further from the centre of the stand.

To produce motions of the axes in the same direction, an aperture was cut in the lower part of the upright board; a roller, turned for it, which loosely fitted within the aperture; and a steel pin or rod passed as an axis through the roller. The roller hangs in its place by endless lines made of thread, passing under it, and over little pulleys fixed on the horizontal axes; when, therefore, it is turned by the projecting pin, it causes the revolution of the axes. The variation in velocities is obtained by having the roller of different diameters in different parts, and by having pulleys of different dimensions. This description will be easily understood by reference to the figures.

This apparatus had to carry wheels either with cogs or spokes; which was contrived in the following manner:—The wheels were cut out of cardboard, were about 7 inches in diameter, and were formed with cogs or spokes at pleasure. A piece of cork, being the end of a phial cork, about the tenth of an inch in thickness, was then fastened by a little soft cement to the middle of the wheel, and a needle run through both, and then withdrawn. These wheels could at any time be put upon the axes, and, being held sufficiently firm by the friction of the cork, turned with them. By these arrangements the axes could be changed, or the wheels shifted, or the velocities altered without the least delay.

The beauty of many of the effects obtained by this apparatus has induced me to describe it more particularly than I otherwise should have done. The appearance which I first had shown to me by Mr. Maltby was exhibited very perfectly; two equal cog-wheels (fig. 6) were mounted, so as to have equal opposite velocities; when put into motion, which was easily done by the thumb and finger applied to the upper pulley of the horizontal copper plate, they presented each the appearance of a uniform tint at the part corresponding to the series of cogs or teeth, provided that the eye was so placed as to see the whole of both wheels; but when a position was taken up so that the wheels were visually superposed, then, in place of a uniform tint, the appearance of teeth or cogs was seen—misty but perfectly stationary, whatever the degree of velocity given to the wheel. By cutting the cogs or teeth in the wheel nearest to the eye, deeper (fig. 7), the eye could be brought into the prolongation of the axes of the wheels, and then the spectral cog-wheel appeared perfect (fig. 8). The number of intervals thus occurring was exactly double the number of teeth in either wheel: thus a wheel with twelve teeth produced twenty-four black, and twenty-four white alternations. When one wheel was made to move a little faster than the other, by shifting the wooden roller on its axis, then the spectrum travelled in the direction of that wheel having the greatest velocity; and with more rapidity the greater the difference between the velocities of the two wheels. When the wheels were looked at so that they visually superposed each other in part, the effect took place only in those parts: and it was striking and extraordinary to observe two uniform tints mingling, and instantly breaking out into the alternations of light and shade which I have described. There are many variations in the curvature and other appearances obtained by altering the position of the eye, which will be at once understood when observed, and which for brevity's sake I refrain from describing.

Wheels were then fixed on the machine, consisting of radii

or spokes, each having twelve, equal in length and width (fig. 1). When revolved alone, each wheel gave, with a certain velocity, a perfectly regular tint; but when visually superposed, there appeared a fixed wheel, having twenty-four spokes, equal in dimensions to the original spokes. Variations of the position of the eye, or of the relative velocity of the two wheels, caused alterations similar to those I have referred to with the cog-wheels.

In observing these effects, either the wheels should be black or in shade, whilst the part beyond is illuminated; or else the wheels should be white and enlightened, whilst the part beyond is in deep shade. The cog-wheels present nearly a similar appearance in both cases, though in reality the parts of the spectrum which appear darkest by the one method are lightest by the other. The spoke-wheels give a spectrum having white radii in the first method and dark radii in the second. Placing the wheels between the eye and the clouds, or a white wall, or a lunar lamp, answers well for the first method; and for the second, merely reversing the position and allowing the light to shine on the parts of the wheel towards the eye, whilst the background is black, or in obscurity, is all that is required. Strictly, the phenomena should be viewed with one eye only, but it is not often that vision with two eyes disturbs the effects to any extent.

The cause of these appearances, when pointed out, is sufficiently obvious, and immediately indicates many other effects of a similar kind, and equally striking, which are dependent upon it. The eye has the power, as is well known, of retaining visual impressions for a sensible period of time; and in this way, recurring actions, made sufficiently near to each other, are perceptibly connected, and made to appear as a continued impression. The luminous circle visible when a lighted coal or taper is whirled round—the beautiful appearances of the kaleidophone—the uniform tint spread by the revolution of one of the spoke- or cog-wheels already described —are a few of the many effects of this kind which are well known.

But during such impressions, the eye, although to the mind occupied by an object, is still open, for a large proportion of time, to receive impressions from other sources; for the original object looked at is not in the way to act as a screen, and shut out all else from sight; the result is, that two or more objects may seem to exist before the eye at once, being visually superposed. The schoolboy experiment of seeing both sides of a whirling halfpenny at the same moment,—the appearances produced by the thaumatrope,—and the transparency of the revolving cog- or spoke-wheels referred to, in consequence of which other objects are seen through the shaded parts,—are all effects of this kind ; two or more distinct impressions, or sets of impressions, being made upon the eye, but appearing to the perception as one.

So it is in the appearances particularly referred to in this paper : they are the natural result of two or more impressions upon the eye, really, but not sensibly, distinct from each other. If, whilst the eye is stationary, a series of cogs like those represented by the continuous outline (fig. 9) pass rapidly before it, they produce a uniform tint to the eye : and for the purpose of following out the description, let it be supposed the cogs are in shade between the eye and a white background; the tint is then a hazy, semitransparent grey. If another series of cogs, represented by the dotted outline, and close to the first, so as to give no sensible angular difference in the dimensions of the cogs, pass with equal velocity in the same direction, it will produce its corresponding tint. If the two sets of cogs be visually superposed in part, as in the figure, there will be no alteration in the uniformity of the tint. If the cogs of one set be more or less to the right or left of the other, then the superposed part will approach more or less to the tint of the shaded and uncut part of the cardboard wheel, and be less transparent. But if, instead of the motion being equal, the velocities are unequal, then total changes of the appearance supervene ; the spectrum (if I may so call it) of the superposed parts becomes alternately light and dark, and the alternations take place more or less rapidly as the velocities of the two sets of cogs differ more or less from each other.

When the cogs move in opposite directions, the uniform tint which each alone can produce is soon broken up in the superposed parts into lighter and darker portions, and when the velocities of both are equal, the spectrum is resolved into a certain number of light and dark alternations, which are perfectly fixed (fig. 10), and which, to the mind, offer a singular contrast to the rapidly moving state of the wheels, and to the variations which their velocity may undergo without altering the visible result.

This effect, strange as it at first appears, will be easily understood by reference to fig. 9. Suppose the eye directed to the part *l* beyond the cogs, and the sets of cogs to be moving with equal velocities in the opposite directions, indicated by the arrow heads: the part *l* will be eclipsed by the cogs *a* and *b* simultaneously, and for exactly the same time, for they begin to cover it and they leave it together; *l* therefore is alternately open to and shut from the eye for equal times; for what these cogs have done, will be performed by all the other cogs in turn, and the cogs are equal in area to the spaces between: half the light, therefore, from that part of the background comes to the eye, and produces a corresponding impression. But with respect to the point *d*, although the cog *b* is just leaving it exposed, the cog *a* is just beginning to eclipse it; and by the time the latter has passed over, the edge of the cog *e* will be upon the spot, and that cog will therefore hide it until *f* comes up; so that in fact the point *d* is always hidden, no light comes from that part of the background, and it consequently appears dark —*l'* is circumstanced just as *l* was, for the cogs *a* and *e* cover it simultaneously, and so do all the other cogs in pairs; it is therefore a light space in the spectrum: *d'* is a repetition in everything of *d*, and is a dark space. The parts intermediate between the maxima of light and darkness will, by examination, be found to be eclipsed for intermediate periods, and to appear more or less dark in consequence, so that the appearance of the spectrum belonging to the visually superposed parts of the two sets of cogs is as in fig. 10.

In the case of equal wheels with radii, the fixed spectrum produced when the wheels superpose each other has twice the number of radii of either wheel, that being of course the number of times which the radii coincide with each other in one revolution. Fig. 11 represents the fixed spectrum produced by two equal wheels of eight radii each. When the radii or spokes are narrow, the difference in the intensity of tint between the middle and the edges of each image of a spoke is so slight as to be scarcely perceptible. But as this circumstance and many others will explain themselves immediately they are experimentally observed, it is unnecessary to dwell minutely upon them here.

A very simple experiment will render the whole of these effects perfectly intelligible. If a little rod of white cardboard 5 or 6 inches long, and one-thirtieth of an inch wide, be moved to and fro from right to left before the eye, an obscure or black background being beyond, it will spread a tint, as it were, over the space through which it moves (fig. 12). A similar rod held and moved in the other hand will produce the same effect; but if these be visually superposed, *i. e.* if one be moved to and fro behind the other, also moving, then in the quadrangular space included within the intersection of the two tints will be seen a black line sometimes straight, and connecting the opposite angles of the quadrangle; at other times oval or round, or even square, according to the motions given to the two cardboard rods (fig. 13).

This appearance is visible even when the rods are several inches or a foot apart from each other, provided they are visually superposed. It is produced exactly as in the former case, and the black line is in fact the path of the intersecting point of the moving rods. As their motions vary, so does the course of this point change, and wherever it occurs, there is less eclipse of the black ground beyond than in the other parts, and consequently less light from that spot to the eye than from the other portions of the compound spectrum produced by the moving rods.

In this experiment the eye should be fixed, and the part looked at should be between the planes in which the rods are moved. The variation produced by using black rods, and looking at a white ground, will suggest itself. Those who find it difficult to observe the effect at first, will instantly be able to do so if the rod nearest the eye is black, or held so as to throw a deep shade: the line is then much more distinct; but the explanation is not quite the same, though nearly so—it will suggest itself. Two bright pins or needles produce the effect very well in diffuse daylight; and the line produced by the shadow of one on the other, and that belonging to the intersection, are easily distinguished and separated.

If, whilst a single bar is moved in one hand, several bars or a grate is moved in the other, then spectral lines, equal to the number of bars in the grate, are produced. If one grate is moved before another, then the lines are proportionably numerous; or if the distances are equal, and the velocity the same, so that many spectral lines may coincide in one, that one is so much the more strongly marked. If the bars used be serpentine or curved, the lines produced may be either straight or curved at pleasure, according as the positions and motions are arranged, so as to make the intersecting point travel in a straight, or a curved, or in any other line.

The cause of the curious appearance produced, when spoke- or cog-wheels revolve before each other, already described, will now be easily understood; the spokes and cogs of the wheels produce precisely the same effect as the bars held in the hand, and the fixedness of the position of the spectrum depends upon the recurrence of the intersecting or hiding positions, exactly in the same place with equal wheels, provided the opposite motions of each be of equal velocity, and the eye be fixed.

When wheels were used in the little machine described (fig. 4), having equal but oblique teeth, and the obliquity in the same direction, the spectrum was also marked obliquely; but when the obliquity was in opposite directions, the spectrum was marked as with straight teeth.

When equal wheels were revolved with opposite motions, one rather faster than the other, the spectrum travelled slowly in the direction of the fastest wheel; when the difference in velocity between the two wheels was made greater, the spectrum travelled faster. These effects are the necessary consequence of the transference of the intersecting points already described, in the direction of the motion of the fastest wheel.

When one wheel contained more cogs than the other, as, for instance, twenty-four and twenty-two, then with equal motions the spectrum was clear and distinct, but travelled in the direction of the wheel having the greatest number of teeth. When the other wheel was made to move so much faster as to bring an equal number of cogs before the eye, or rather any one part of the eye, in the same time as the other, the spectrum became stationary again. The explanations of these variations will suggest themselves immediately the effects are witnessed.

When the motion of the wheels upon the machine is in the same direction, the velocities equal, and the eye placed in the prolongation of the axis of the wheels, no particular effect takes place. If it so happens that the cogs of one coincide with those of the other, the uniform tint belonging to one wheel only is produced. If they project by the side of each other, it is as if the cogs were larger, and the tint is therefore stronger. But when the velocities vary, the appearances are very curious; the spectrum then becomes altogether alternately light and dark, and the alternations succeed each other more rapidly as the velocities differ more from each other.

When wheels with radii are put upon the machine, it is easy to observe, in perfection, the optical appearance already referred to, as exhibited by carriage wheels, &c. (fig. 2). They should be looked at obliquely, so as to be visually superposed only in part; and provided the wheels are alike, and both revolving in the same direction with equal velocity, they immediately assume the form described, passing in curves from the axis of one wheel to the axis of the other, and much resembling in disposition the curves formed by iron filings between two opposite poles of a magnet.

If the wheels revolve in opposite directions, then the spectral lines, originating at each axis as a pole, have another disposition, and instead of running the one set into the other, are disposed generally like the filings about two similar magnetic poles, as if a repulsion existed : not that the curves or the cause are the same, but the appearances are similar. A very little attention will show that all these lines are the necessary consequence of the travelling of successive intersecting points ; and any one of them may be followed out by experimenting with the two pasteboard rods already described, these being moved in the hand as if each were the spoke of a wheel.

All these effects may be simply exhibited by cutting out two equal pasteboard wheels without rims, passing a pin as an axis through each, spinning one upon a mahogany or dark table, and then spinning the other between the fingers over it, so that the two may be visually superposed. If the appearances are observed by a lamp or candle, the wheels should be so held to the light that the shadow of the upper may not fall upon the lower, otherwise the effects are complicated by similar sets of lines which appear upon the lower wheel, and are produced by the shadow of the upper. These are the same in form and disposition as the former, and are even more distinct; they should be viewed, not through the upper wheel, but directly upon the lower; their explanation has in part been given, and will be sufficiently evident.

The form which the appearance occasionally assumes when a carriage wheel is revolving before upright bars, is exceedingly well shown by the little machine described (fig. 4), when mounted with a single wheel carrying several equal radii at equal distances. The bars of the grate should be equidistant, the intervals between them being about that between the extremities of two contiguous spokes of the wheel. The varied appearances produced by varying the motion of the wheel and grate, both in direction and velocity, will be better understood from a few easy experiments than from any description.

The lines which thus occur may any one of them be imitated by the two cardboard bars held and moved in the hand; the whole system may then be obtained at once if one of the independent wheels (fig. 1) be revolved by the pin between the fingers, and a single pasteboard bar (of equal width with the radii) passed once, not too rapidly, before it; by returning the bar the lines are seen a second time. Should the eye not readily catch the appearance, a black instead of a white single bar may be used, or a shadow be thrown by an opake bar from a candle, or the sun, upon the revolving wheel; and then, to extend and follow out the forms, the bar should be moved to and fro slowly before the revolving wheel, to the extent of one half or the whole length of a radius, when it will immediately be seen that all the lines produced, even when a grate is used, are merely the courses of so many points of intersection between the radii of the wheel and the bars passing before or behind it.

A variation in the mode of observing many of these curious spectra, but which still further supports the explication given, is to cast the shadows of the revolving wheels, either by sun or candlelight, upon a screen, and observe their appearance. The way in which the cogs or radii of the wheels shut out more or less of a background from the eye, as already described, will enable them, to an equal degree, to intercept light, which would otherwise fall upon a screen. When the two equal cog-wheels are revolved so as to have the shadows cast upon a white screen, that shadow exhibits all the appearances and variations observed when the eye is looking by the wheels in shade at a white background. The shadow is light where the wheels appear dark, for there the light has passed by the cogs; and dark where the wheels appear light, for there the cogs have intercepted most of the rays. The screen should be near to the wheels, that the shadow may be sharp; and it is convenient to have one wheel of rather smaller radius than the other, or else to place them obliquely to the sun for the purpose of distinguishing the shadow of each wheel, and showing how beautifully the spectrum breaks out where they superpose. When the spoke-wheels are revolved they also cast a shadow, presenting either the appearance of fixed or moving radii according to the circumstances already described. When the two small spoke-wheels upon one pin are revolved in an oblique direction, their shadow exhibits very beautifully the lines often seen in the wheels of carriages.

During these experiments the attention cannot but be drawn to the observation of the figures produced by the shadow of one wheel upon the face of the other. These are frequently very beautiful, and combining as they often do with the designs produced, as already described, are occasionally more striking than any of the appearances yet spoken of. Mr. Wheatstone is, however, engaged in an inquiry of a much more general and important kind, which includes these effects, and which, I trust, he will soon give to the public.

Several of the effects with wheels already described, and some new ones, may be obtained with great simplicity, by means of reflexion, in a very striking manner. If a white cardboard wheel, with equal radii, be fixed upon a pin, and rotated between the fingers before a glass, so that the wheel and its reflected image may visually superpose in part, the fixed lines will be seen, like those of fig. 2, passing in curves between the axis of the wheel and the reflected image. If the person gradually recede from the glass, but still look through the wheel in his hand at the reflected image, *i. e.* still retain them superposed, which is best done by bringing the revolving wheel close to the eye, he will see the lines or radii of the reflected image gradually become straight, and when from three feet to any greater distance from the glass, will see the spectrum of the reflected image, having as many dark radii upon it as there are radii in the wheel he is revolving. Whatever the velocity, or however irregular the motion of the wheel, these lines are perfectly stationary. The explanation of the change of form and ultimate appearance of the whole, and of the number and fixed position of the lines, will be so evident when the experiment is made, in conjunction with what has been said, as to require no further statement here.

A very striking deception may be obtained in this way, by revolving a single cog-wheel (fig. 6) between the fingers before the glass, when from twelve to fifteen or eighteen feet from it. It is easy to revolve the wheel before the face so that the eyes may see the glass through or between the cogs, and then the reflected image appears as if it were the image of a cog-wheel, having the same number of cogs, but perfectly still, and every cog distinct; instead of being the image of one in such rapid motion, that by direct vision the cogs cannot be distinguished from each other, or their existence ascertained. The effect is very striking at night if a candle be placed just before the face, and near to it, but shaded by the wheel; in the reflexion the wheel is then well illuminated, and the reflected face or shadow forms a good background against which to observe the effect.

I have, perhaps, already rendered this paper longer than necessary; but the singularity of the appearances, and the facility with which they may be observed, have induced me to suppose that many persons would like to repeat the experiments, and must be my excuse for some further variations in the mode of experimenting.

A disc of cardboard, about two inches and a half in diameter, was cut into a wheel like fig. 16; another disc, rather larger, was cut into a similar wheel, and then the radii of one were twisted obliquely like the wings of a ventilator, and the radii of the other similarly set, but in the opposite direction: a small hole being made in the centre of each, a large pin was passed through that of the smaller wheel, and then a small piece of cork passed on to the pin to hold the wheel near the head, but free to turn; two or three beads were then added, the second wheel put on, and then a second piece of cork; the end of the pin was then stuck into a quill or a pencil, and thus was formed an apparatus very like a child's windmill, except that it had two sets of vanes, each revolving in opposite directions. On walking across a room towards a window, or a candle, with this little toy in the hand, or blowing at it slightly from the mouth, the lines were beautifully seen, being either stationary or moving, according to the relative velocity of the two wheels. This could be altered at pleasure by inclining the vanes more or less, or by blowing towards the centre of the wheels, or towards the edges when the larger hind wheel received more propulsive force.

Spinners or whirligigs formed of discs of cardboard stuck upon pins, and upon which radii, either straight or curved, or other forms, had been drawn in bold lines with black ink, when spun upon a sheet of paper, and then looked at through the moving fingers or through equidistant bars of pasteboard moved before them, show a great many of the effects.

Finally, a couple of open radial wheels (fig. 1) upon pins or wires, if revolved between the fingers in different positions and directions, show a great many of these effects extremely well. Their shadows may be thrown upon each other, or upon the wall; one may be held near the eye, when it acts like a grate with parallel bars; and if one side of each wheel is black whilst the other is white, still greater variety may be obtained. They will be quite sufficient, when employed in a few experiments, to make anything in this description clear, which I may have left obscure.

The curious appearance exhibited by the wheel animalcule has such a resemblance to some of those described in this paper, that they inevitably associate in the mind of a person who has witnessed both effects. This little insect has been well described by Mr. Baker* and others, and can only be viewed distinctly under a high magnifying power; it then presents an elongated sack-like form (fig. 17), either attached by the posterior part to the side of the vessel containing the water in which it exists, or else floating in the fluid. When the effect in question is observable, there is seen the appearance of two wheels, one on

* Baker on the Microscope, vol. ii. p. 266; see slso Leeuwenhoek, Phil. Trans., Nos. 283, 295, 337 ; and Adams on the Microscope, p. 548.

each side of the head; they seem formed of deep teeth or short radii, perhaps fourteen or fifteen in number; the form of these teeth is not sharp or well defined, but hazy at the edges; the interval between them is perhaps rather more than the width of the teeth; the teeth are not distinctly set on to a nave or axis, but appear sometimes even to melt away or attenuate at the part towards the centre, and sometimes appear as independent portions, *i. e.* as much separated from the centre part or supposed place of attachment as from the neighbouring teeth.

These parts are never seen as wheels, except in motion; the animal is sometimes seen without them, the parts which produce the appearance being then either retracted and drawn inwards, or disposed in other forms, for the animal is of a very changeable nature. The motion of the wheels is continuous, as if they were spinning constantly in one direction upon their axis; the velocity is such as to carry the teeth rapidly before the eye, but is not enough to confound the impression of one tooth with that of its neighbours, and therefore they may be distinctly seen. Both wheels move usually in the *same direction*; and when the head of the animal is towards the observer, the direction is generally the same as that of the hands of a clock. Baker states, however, that he has seen them move in opposite directions, and also has seen the motion first discontinued, and then reversed, in the same wheel. The velocity is not always the same, but varies with the efforts of the animal to catch its food. Whatever the mechanism of the parts, the result is, that currents are established in the water towards the head of the animal, which currents pass off outward from the edges of the apparent wheels; and little particles floating in the water may be seen to pass towards the head, and be suddenly thrown off at the edges of the wheels with considerable force.

So striking are the appearances of these animalcula, that men of much practice in microscopical observation are at this day convinced they do possess wheels, which actually revolve continuously in one direction. The struggle in Mr. Baker's mind between the evidence of his senses and his judgment, illustrates this point in so lively a manner, that I may be excused quoting his account of it:—" As I call these parts *wheels*, I also term the motion of them a rotation, because it has exactly the

appearance of being such. But some gentlemen have imagined
there may be a deception in the case, and that they do not
really turn round, though indeed they seem to do so. The
doubt of these gentlemen arises from the difficulty they find in
conceiving how or in what manner a wheel or any other form,
as part of a living animal, can possibly turn upon an axis supposed to be another part of the same living animal, since the
wheel must be a part absolutely distinct and separate from the
axis whereon it turns; and then say they, how can this living
wheel be nourished, as there cannot be any vessels of communication between that and the part it goes round upon, and which
it must be separate and distinct from? To this I can only
answer, that place the object in whatever light or manner you
please, when the wheels are fully protruded they never fail to
show all the visible marks imaginable of a regular turning
round; which I think no less difficult to account for, if they do
not really do so. Nay, in some positions you may, with your
eye, follow the same cogs or teeth whilst they seem to make a
complete revolution; for the other parts of the insect being very
transparent, they are easily distinguished through it. As for
the machinery, I shall only say, that no true judgment can be
formed of the structure and parts of minute insects by imaginary
comparisons between them and larger animals, to which they
bear not the least similitude. However, as a man can move his
arms or his legs circularly as long and as often as he pleases by
the articulation of a ball and socket, may there not possibly be
some sort of articulation in this creature whereby its wheels or
funnels are enabled to turn themselves quite round?

   " It is certain all appearances are so much on this side of the
question, that I never met with any who did not, on seeing it,
call it a *rotation*; though, from a difficulty concerning how it
can be effected, some have imagined they might be deceived.
M. Leeuwenhoek also declared them to be *wheels* that *turn
round* (*vide* Phil. Trans., No. 295). But I shall contend with
nobody about this matter: it is very easy for me, I know, to be
mistaken, and so far possible for others to be so too, that I am
persuaded some have mistaken the *animal itself*, which perhaps
they never saw; whilst, instead thereof, they have been examining one or other of the several *water-animalcules* that are
furnished with an apparatus commonly called *wheels*, though

they turn not round, but excite a current by the *mere vibration of fibrillæ* about their edges."

Notwithstanding the evidence adduced by Mr. Baker, which, as I have said, is admitted by some at the present day, it must be evident, from a consideration of the nature of muscular force, and the condition of continuity under which all animals exist, that the rotation cannot really occur. The appearances are altogether so like some of those exhibited in the experiments already described, that I feel no doubt the wheels must be considered not as having any real existence, but merely as spectra, produced by parts too minute, or else having too great a velocity when in use by the animal to be themselves recognized. It is not meant that they are produced by toothed or radiated wheels; for that supposition would take for granted what has already been considered as impossible—continual revolution of one part of an animal whilst another part is fixed; but arrangements may be conceived, which are perfectly consistent with the usual animal organization, and yet competent to produce all the effects and appearances observed. Thus, if that part of the head of the animal were surrounded by fibrillæ, endowed each with muscular power, and projecting on all sides, so as to form a kind of wheel; and if these fibrils were successively moved in a tangental direction rapidly the one way, and more slowly back again, it is evident that currents would be formed in the fluid, of the kind apparently required to bring food to the mouth of the animal; and it is also evident, that if the fibrils, either alone or grouped many together, had any power of affecting the sight, so as to be visible, they would be less visible at the part through which they were rapidly moving, than that through which they were slowly returning; and at that place, therefore, an interval would appear, which would seem to travel round the wheel, in consequence of the successive action of the fibrils. But if, instead of the whole group of fibrils acting in succession as one series, they were to be divided by the will or powers of the animal into fifteen or sixteen groups, the action being in every other respect the same, then there would be the appearance of fifteen or sixteen dark spaces, and as many light ones disposed as a wheel; and these would continue to travel round in one direction, so long as the animal continued the alternate action of the fibrils. This may be illustrated by supposing fig. 14 to represent a fixed circular brush, with long hairs, and the little dots to be the sections of so many wires, forming the arms of a frame which, when turned round, shall carry the hairs of the brush forward a little, and then, letting them go, allow them to return quickly to their first position. If this frame be turned continually round, it would cause the brush, when looked at from a distance, to appear as a revolving toothed wheel, although in reality it had no circular motion. Now, what is performed here by the wire-arms at the outer extremity of the hairs, and the natural elasticity of the latter, may, in the wheel animalcula, be effected at the roots of the fibrillæ by muscular power; and in this or some similar way the animal may have the power of urging the current necessary to supply food, and, at the same time, producing the spectrum of a continually revolving wheel, or even the more complicated forms discovered by Leeuwenhoek (fig. 15), without requiring any powers beyond those which are within the understood laws of Nature, and known to exist in the animal structure *.

Royal Institution, Dec. 10, 1830.

[In Mr. Whitock's carpet and fringe-manufactory at Edinburgh, they were covering a cord with silk. The cord was of two strands, differently coloured, slightly twisted, and was turning rapidly round on its axis. In many places it looked like a party-coloured cord *perfectly still.* This was from the continual recurrence of portions different from their neighbours, in the same place; they were not visible all the way round, but only above or below, or in some particular part of their revolution.—July 1833. M. F.]

ADDITIONAL NOTE†.

In consequence of the necessity I was under of sending the paper (page 291) referred to in the above 'Proceedings' to press by a certain time, I was unable to pursue many of the beautiful combinations of form, colour, and appearance to which the experiments led, especially as they promised only amusement, and little more of instruction than the paper itself contained;

* See in relation to this subject, Horner, on the Dædaleum, Phil. Mag. 1843, iv. 36.

† Quarterly Journal of Science, 1831, vol. i. 334.

but one or two varieties in the appearances, which have occurred to me since, are so striking, that I am glad of the opportunity of noticing them briefly in the same Number with the paper. At page 304 I have described the singular appearance produced when the reflected image of a revolving cog-wheel, held before a glass, is observed through the cog-wheel itself. If, in such a wheel, a little nearer the centre, a series of regular apertures be cut, so as to represent cogs and their intervals, but the number different by 1.2.3, or any small quantity, from the number of the cogs, then, upon making the experiment as before, that series of cogs in the revolving wheel through which the eye looks will appear to stand still, but the other series will travel in the spectrum: upon changing the eye to the other series of apertures, then the quiescent part of the spectrum will move, and the moving part become quiescent. If two or three series more of such apertures be cut in the wheel, concentric one to another, but the number of intervals varying in each, then a great variety of changes are produced, as the eye looks through one part or another of the wheel. The series of cogs in the spectrum move with different velocities, or in opposite directions, changing with the slightest motion of the eye. Two or three persons looking through different parts of the wheel see appearances entirely different; yet all these deceptive appearances result from a single reflexion of a single wheel, moving in a constant direction and with uniform velocity.

By the application of colours and coloured foils, very curious effects occur, which are endless in their variety. As an illustration, let a wheel with a single series of cogs at the edge, and with intervals equal to the cogs, have a circle of colour applied between the cogs and the centre of the wheel; let the part below the cogs be green, and the part below the spaces red; the coloured circle will consist of green and red alternately. If this wheel be revolved before the glass, the green and red mingle, and the reflexion observed in the ordinary way will exhibit one uniform colour; but if the reflexion be observed from between or behind the cogs, the green and red immediately separate, and besides having the appearance of fixed cogs, there is also the appearance of fixed unmingled colours. If the interval be equal to only half a cog, and three colours be applied, the three colours may, after being mingled by rotation, be again developed, and it is easy in this way to separate many colours from each other. The experiment in illustration of Newton's theory of colour, by painting the head of a top and spinning it, is well known; by the means just described the experiment can be still further extended, and the colours separated one from another, even while the whole system remains in motion.

The combination of other forms than wheels by the apparatus described, page 294, produces very beautiful effects. The application of colours here also is so evident as to need no illustration. The variation of the proportion of the interval to the remaining pasteboard causes many curious appearances, especially when the shadows produced in sunlight are observed.

Since the printing of the paper, a friend has referred me to the article 'Animalcula' in Brewster's Encyclopædia, where an opinion on the appearance of these creatures is given, nearly the same as that I have ventured. Speaking of the opinions of those who suppose them to be true revolutions, it is said, "Yet notwithstanding our respect for the skill and talents of such renowned naturalists, we cannot deny that we think the production of the vortex is more probably effected by the simple motion of the fibrilla—that it may ensue from their rapidly bending in regular or alternate succession, or by some analogous means."

Trevelyan's *Experiments on the Production of Sound during the Conduction of Heat**.*

[Read Friday evening, April 29, 1831.]

Mr. Trevelyan had remarked that when a heated poker was laid down upon a table, so that the knob rested upon it, whilst the hot part was supported by an interposed block of cold lead, regular musical notes were frequently produced. By extending the experiments, he found that a better form than that of a poker might be used for the hot metal: a piece of brass about four inches long, one inch and a quarter broad, and half an inch thick, should have a groove of one-eighth of an inch in width, formed down the middle of one of the broad faces, and then that face bevelled from the edges of the groove on each

* Quarterly Journal of Science, 1831, ii. 119.

side. Being now placed with the groove downwards upon a table, and shaken, it rocks to and fro, and is in right condition for the experiment. It is convenient to fasten a brass wire, terminated by a knob, to one end of this rocker, so as to act as a prolongation of an axis: it renders the whole arrangement steady and regular in action. When this piece of metal is used instead of the poker, musical sounds are almost always produced. The surface of the lead upon which it rests should be clean.

The peculiar effects exhibited in these experiments depend upon the occurrence of isochronous vibrations performed by the rocker. When by loading the rocker these are rendered slow, they become visible; but when they occur with sufficient rapidity, they produce the necessary result, a musical note, of higher or lower pitch, as the vibrations or tappings are more or less numerous. It often happens that other and extraneous sounds, as those due to the ringing of the metal, the vibration of the table, or subdivisions of the whole vibrating system, mingle with the true sound produced by the blows of the rocker; these were referred to and illustrated, and a method shown of easily distinguishing the latter from the former. It consisted in pressing perpendicularly with a small stick or pointed metal rod on the back of the rocker, exactly over the groove, so as to make the vibrations quicker, but not to disturb their regularity; the true sound of the beats of the rocker immediately rises in pitch, and may be sometimes made to pass through an octave or more at pleasure, falling again as the pressure is removed.

As the sound is evidently due to the rapid blows of the rocker, the only difficulty was to discover the true cause of the sustaining power by which the rocker was continued in motion, whilst any considerable difference of temperature existed between it and the block of lead beneath; this Mr. Faraday referred to the ultimate expansion and contraction, as Professor Leslie and Mr. Trevelyan have done generally; but he gave a minute account of the manner in which, according to his views, such expansion and contraction could produce the effect. When the heated rocker is reposing upon a horizontal ridge of lead, it touches at two points, which are heated and expanded, and form, as it were, two hills; when one side of the rocker is raised, the point relieved from its contact is instantly cooled by the neighbouring portions of lead, the expansion ceases, and the hill falls. When the rocker, therefore, is left free, the raised side descends through a greater space than that through which it was lifted, and also to a lower level than the other side; in consequence of which a momentum is given to it, which carries its centre of gravity beyond the point to which it would pass if there had been no alteration in the heights of the sustaining points. It is this additional force which acts as a maintaining power; and recurs twice in each vibration, *i. e.* once on each side. The force is gained by the whole rocker being lifted bodily by the point on which it is for the time supported, and comes into play by the side of the rocker which is descending, having a greater space to fall through than that which is passed over by the mere force of its momentum during its previous rise. A curious consequence of this action is, that the force which really lifts the rocker is on one side of the centre of gravity, whilst the rising side of the rocker itself is on the other.

This, however, is not the only maintaining cause or mechanical force generated by the alternate expansion and contraction of the lead. If the vertical direction of the forces be put out of consideration for a time, and the two points of support be examined, it will be found that whilst the rocker is quiescent, both points (with their neighbouring parts) being heated, will expand and compress the lateral portions of the lead, until the tension of the latter is equal to their own. When one side of the rocker is raised, the point that it rested upon is instantly cooled, and therefore contracts; but as the neighbouring parts retain their tension, they move towards the contracting part, the other point of support moving with the rest. When the rocker returns in its oscillation, it reheats and re-expands the first point of support, whilst the second, now out of contact, is cooled and contracted, and the first point, therefore, moves towards the second. A necessary consequence of this mutual relation of the points is, that the one under process of heating is always moving towards the other which is under process of cooling, and, consequently, towards a perpendicular from the centre of gravity; but as it is at the same time the supporting point to the rocker, that supporting point is, by irresistible impulse, carried in a direction under and towards the line passing from the centre of gravity towards the earth, at the same instant that the centre of gravity of the rocker is, by the momentum of the latter, moving in the opposite direction: hence a very simple maintaining power, sufficient, whenever the rocker continues to vibrate, to compensate for the loss of force in each half of the vibration which would occur if the rocker and lead were of the same temperature. Mr. Faraday illustrated the sustaining force of the lateral motion of the points of support, by placing a rocker on a piece of lead, and the latter on a board. A pair of sugar-tongs was held tightly by the bend against the edge of the board, so that the line from the tongs towards the rocker was perpendicular to the axis of the latter. On making the limbs of the sugar-tongs vibrate in the manner of a tuning-fork, they communicated longitudinal vibrations of equal duration and number to the board, and through it to the lead and points supporting the rocker; which latter itself immediately acquired vibratory motion isochronous with the vibrations of the tongs, and by successive blows upon the lead produced sound; upon removing the rocker, and repeating the other parts of the experiment, no sound was produced.

Experiments with other metals were then made. A piece of curved silver plate being heated and placed on an iron triblet, rocked and sang in the manner of the others; this is an effect which working silversmiths have long known. The superiority of lead, as the cold metal, was referred to its great expansive force by heat, combined with its deficient conducting power, which is not a fifth of that of copper, silver, or gold; so that the heat accumulates much more at the point of contact in it, than it could do in the latter metals.

Mr. Trevelyan's paper had been read to the Royal Society of Edinburgh, but is not yet published. Mr. Faraday stated that Mr. Trevelyan had very liberally allowed him the use of a written copy.

On a Peculiar Class of Acoustical Figures; and on certain Forms assumed by groups of particles upon vibrating elastic Surfaces*.

[Read May 12, 1831.]

1. THE beautiful series of forms assumed by sand, filings, or other grains, when lying upon vibrating plates, discovered and

* Philosophical Transactions, 1831, p. 299.

developed by Chladni, are so striking as to be recalled to the minds of those who have seen them by the slightest reference. They indicate the quiescent parts of the plates, and visibly figure out what are called the nodal lines.

2. Afterwards M. Chladni observed that shavings from the hairs of the exciting violin bow did not proceed to the nodal lines, but were gathered together on those parts of the plate the most violently agitated, *i. e.* at the centres of oscillation. Thus when a square plate of glass held horizontally was nipped above and below at the centre, and made to vibrate by the application of a violin bow to the middle of one edge, so as to produce the lowest possible sound, sand sprinkled on the plate assumed the form of a diagonal cross; but the light shavings were gathered together at those parts towards the middle of the four portions where the vibrations were most powerful and the excursions of the plate greatest.

3. Many other substances exhibited the same appearance. Lycopodium, which was used as a light powder by Oersted, produced the effect very well. These motions of lycopodium are entirely distinct from those of the same substance upon plates or rods in which longitudinal vibrations are excited.

4. In August 1827, M. Savart read a paper to the Royal Academy of Sciences*, in which he deduced certain important conclusions respecting the subdivision of vibrating sonorous bodies from the forms thus assumed by light powders. The arrangement of the sand into lines in Chladni's experiments shows a division of the sounding plate into parts, all of which vibrate isochronously, and produce the same tone. This is the principal mode of division. The fine powder which can rest at the places where the sand rests, and also accumulate at other places, traces a more complicated figure than the sand alone, but which is so connected with the first, that, as M. Savart states, "the first being given, the other may be anticipated with certainty; from which it results that every time a body emits sounds, not only is it the seat of many modes of division which are superposed, but amongst all these modes there are always two which are more distinctly established than all the rest. My object in this memoir is to put this fact beyond a doubt, and to study the laws to which they appear subject."

* Annales de Chimie, xxxvi. p. 187.

5. M. Savart then proceeds to establish a secondary mode of division in circular, rectangular, triangular and other plates; and in rods, rings, and membranes. This secondary mode is pointed out by the figures delineated by the lycopodium or other light powder; and as far as I can perceive, its existence is assumed, or rather proved, exclusively from these forms. Hence much of the importance which I attach to the present paper. A secondary mode of division, so subordinate to the principal as to be always superposed by it, might have great influence in reasonings upon other points in the philosophy of vibrating plates; to prove its existence therefore is an important matter. But its existence being assumed and supported by such high authority as the name of Savart, to prove its non-existence, supposing it without foundation, is of equal consequence.

6. The essential appearances, as far as I have observed them, are as follows:—Let the plate before mentioned (2), which may be three or four inches square, be nipped and held in a horizontal position by a pair of pincers of the proper form, and terminated, at the part touching the glass, by two pieces of cork; let lycopodium powder be sprinkled over the plate, and a violin bow be drawn downwards against the middle of one edge so as to produce a clear full tone. Immediately the powder on those four parts of the plate towards the four edges will be agitated, whilst that towards the two diagonal cross lines will remain nearly or quite at rest. On repeating the application of the bow several times, a little of the loose powder, especially that in small masses, will collect upon the diagonal lines, and thus, showing one of the figures which Chladni discovered, will also show the principal mode of division of the plate. Most of the powder which remains upon the plate will, however, be collected in four parcels; one placed near to each edge of the plate, and evidently towards the place of greatest agitation. Whilst the plate is vibrating (and consequently sounding) strongly, these parcels will each form a rather diffuse cloud, moving rapidly within itself; but as the vibration diminishes, these clouds will first contract considerably in bulk, and then settle down into four groups, each consisting of one, two, or more hemispherical parcels (53), which are in an extraordinary

condition; for the powder of each parcel continues to rise up at the centre and flow down on every side to the bottom, where it enters the mass to ascend at the centre again, until the plate has nearly ceased to vibrate. If the plate be made to vibrate strongly, these parcels are immediately broken up, being thrown into the air, and form clouds, which settle down as before; but if the plate be made to vibrate in a smaller degree, by a more moderate application of the bow, the little hemispherical parcels are thrown into commotion without being sensibly separated from the plate and often slowly travel towards the quiescent lines. When one or more of them have thus receded from the place over which the clouds are always formed, and a powerful application of the bow is made, sufficient to raise the clouds, it will be seen that these heaps rapidly diminish, the particles of which they are composed being swept away from them, and passing back in a current over the glass to the clouds under formation, which ultimately settles as before into the same four groups of heaps. These effects may be repeated any number of times, and it is evident that the four parts into which the plate may be considered as divided by the diagonal lines are repetitions of one effect.

7. The form of the little heaps, and the involved motion they acquire, are no part of the phenomena under consideration at present. They depend upon the adhesion of the particles to each other and to the plate, combined with the action of the air or surrounding medium, and will be resumed hereafter (53). The point in question is the manner in which fine particles do not merely remain at the centres of oscillation, or places of greatest agitation, but are actually driven towards them, and that with so much the more force as the vibrations are more powerful.

8. That the agitated substance should be in very fine powder, or very light, appears to be the only condition necessary for success; fine scrapings from a common quill, even when the eighth of an inch in length or more, will show the effect. Chemically pure and finely divided silica rivals lycopodium in the beauty of its arrangement at the vibrating parts of the plate, although the same substance in sand or heavy particles proceeds to the lines of rest. Peroxide of tin, red lead, vermilion, sulphate of baryta, and other heavy powders

when highly attenuated, collect also at the vibrating parts. Hence it is evident that the nature of the powder has nothing to do with its collection at the centres of agitation, provided it be dry and fine.

9. The cause of these effects appeared to me, from the first, to exist in the medium within which the vibrating plate and powder were placed, and every experiment which I have made, together with all those in M. Savart's paper, either strongly confirm, or agree with this view. When a plate is made to vibrate (2), currents (24) are established in the air lying upon the surface of the plate, which pass from the quiescent lines towards the centres or lines of vibration, that is, towards those parts of the plates where the excursions are greatest, and then proceeding outwards from the plate to a greater or smaller distance, return towards the quiescent lines. The rapidity of these currents, the distance to which they rise from the plate at the centre of oscillation, or any other part, the blending of the progressing and returning air, their power of carrying light or heavy particles, and with more or less rapidity or force, are dependent upon the intensity or force of the vibrations, the medium in which the vibrating plate is placed, the vicinity of the centre of vibration to the limit or edge of the plate, and other circumstances, which a simple experiment or two will immediately show, must exert much influence on the phenomena.

10. So strong and powerful are these currents, that when the vibrations were energetic, the plate might be inclined 5°, 6°, or 8° to the horizon, and yet the gathering clouds retain their places. As the vibrations diminished in force, the little heaps formed from the cloud descended the hill; but on strengthening the vibrations they melted away, the particles ascending the inclined plane on those sides proceeding upwards, and passing again to the cloud. This took place when neither sand nor filings could rest on the quiescent or nodal lines. Nothing could remain upon the plate except those particles which were so fine as to be governed by the currents, which (if they exist at all) it is evident would exist in whatever situation the plate was placed.

11. M. Savart seems to consider that the reason why the powder gathers together at the centres of oscillation is, " that the amplitude of the oscillations being very great, the middle

of each of those centres (of vibration) is the only place where the plate remains nearly plane and horizontal, and where, consequently, the powder may reunite ; whilst the surface being inclined to the right or left of this point, the parcels of powder cannot stop there." But the inclination thus purposely given to the plate, was very many times that which any part acquires by vibration in a horizontal position, and consequently proves that the horizontality of any part of the plate is not the cause of the powder collecting there, although it may be favourable to its remaining there when collected.

12. Guided by the idea of what ought to happen, supposing the cause now assigned were the true one, the following amongst many other experiments were made. A piece of card about an inch long and a quarter of an inch wide was fixed by a little soft cement on the face of the plate near one edge, the plate held as before at the middle, lycopodium or fine silica strewed upon it, and the bow applied at the middle of another edge ;

Fig. 1.

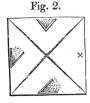

the powder immediately advanced close to the card, and the place of the cloud was much nearer to the edge than before. Fig. 1 represents the arrangement; the diagonal lines being those which sand would have formed, the line at the top *a* representing the place of the card, and the × to the right place where the bow was applied. On applying a second piece of card, as at *b*, the powder seemed indifferent to it or nearly so, and ultimately collected as in the first figure: *c* represents the place of the cloud when no card is present.

13. Pieces of card were then fixed on the glass in the three angular forms represented in fig. 2; upon vibrating the plate, the fine powder always went into the angle, notwithstanding its difference of position in the three experiments, but perfectly in accordance with the idea of currents intercepted more or less by the card. When two pieces of card were fixed on the plate, as in fig. 3 *a*, the powder proceeded into the angle, but not to the edge of the glass, remaining about $\frac{1}{8}$th of an inch from it ; but on closing up that opening, as at *b*, the powder went quite up into the corner.

14. Upon fixing two pieces of card on the plate as at *c*, fig. 3, the powder between them collected in the middle very nearly as if no card had been present; but that on the outside of the cards gathered close up against them, being able to proceed so far in its way to the middle, but no further.

Fig. 3.

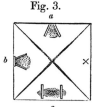

15. In all these experiments the sound was very little lowered, the form of the cross was not changed, and the light powders col- lected on the other three portions of the plate, exactly as if no card walls had been applied on the fourth; so that no reason appears for supposing that the mode in which the plate vibrated was altered, but the powders seem to have been carried forward by currents which could be opposed or di- rected at pleasure by the card stops.

16. A piece of gold-leaf being laid upon the plate, so that it did not overlap the edge, fig. 4, the current of air towards the centre of vibration was beautifully shown; for, by its force, the air crept in under the gold-leaf on all sides, and raised it up into the form of a blister; that part of the gold-leaf corresponding to the centre of the locality of the cloud, when light powder was used, being frequently a sixteenth or twelfth of an inch from the glass. Lycopodium or other fine powder, sprinkled round the edge of the gold-leaf, was carried in by the entering air, and accumulated underneath.

Fig. 4.

17. When silica was placed on the edge of another glass plate, or upon a book, or block of wood, and the edge of the vibrating plate brought as nearly as possible to the edge of the former, fig. 5, part of the silica was al- ways driven on to the vibrating plate, and collected in the usual place; as if in the midst of all the agitation of the air in the neighbourhood of the two edges, there was still a current towards the centre of vibration, even from bodies not themselves vibrating.

Fig 5.

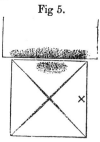

18. When a long glass plate is supported by bridges or strings at the two nodal lines represented in fig. 6, and made

to vibrate, the lycopodium collects in three divisions; that between the nodal lines does not proceed at once into a line equidistant from the nodal lines and parallel to them, but advances from the edges of the plate towards the middle by paths, which are a little curved and oblique to the edges where they occur near the nodal lines, but are almost perpendicular to it elsewhere, and the

Fig. 6.

powder gradually forms a line along the middle of the plate; it is only by continuing the experiment for some time that it gathers up into a heap or cloud equidistant from the nodal lines. But upon fixing card walls upon this plate, as in fig. 7, the course of the powder within the cards was directly parallel to them and to the edge, instead

Fig. 7.

of being perpendicular, and also directly towards the centre of oscillation. To prove that it was not as a weight that the card acted, but as an obstacle to the currents of air formed, it was not moved from its place, but bent flat down outwards, and then the fine powder resumed the courses it took upon the plate when without the cards. Upon raising the cards the first effect was reproduced.

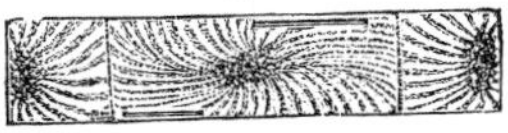

19. The lycopodium sprinkled over the extremities of such a plate proceeds towards places equidistant from the sides and near the ends, as at *a*, fig. 8; but on cementing a piece of paper to the edge, so as to form a wall about one quarter or one third of an inch high, *b*, the powder immediately moved up to it, and retained this new place. In a longer narrow plate, similarly arranged, the powder could be made to pass to either edge, or to the middle, according as paper interceptors to the currents of air were applied.

Fig. 8.

20. Plates of tin, four or five inches long, and from an inch to two inches wide, fixed firmly at one end in a horizontal position, and vibrated by applying the fingers, show the progress of the air and the light powders well. The vibrations are of comparatively enormous extent, and the appearances are consequently more instructive.

21. If a tuning-fork be vibrated, then held horizontally

Y

with the broad surface of one leg uppermost, and a little lycopodium be sprinkled upon it, the collection of the powder in a
cloud along the middle, and the formation of the involving
heaps also in a line along the middle of the vibrating steel bar,
may be beautifully observed.   But if a piece of paper be
attached by wax to the side of the limb, so as to form a fence
projecting above it, as in the former experiments (19), then
the powder will take up its place close to the paper; and if
pieces of paper be attached on different parts of the same leg,
the powder will go to the different sides, in the different parts,
at the same time.

22. The effects under consideration are exceedingly well
shown and illustrated by membranes.   A piece of parchment
was stretched and tightly tied, whilst moist, over the aperture
of a funnel five or six inches in diameter; a small hole was
made in the middle, and a horse-hair passed through it, but with
a knot at the extremity that it might thereby be retained.   Upon
fixing the funnel in an upright position, and after applying a
little powdered resin to the thumbs and fore-fingers, drawing
them upward over the horse-hair, the membrane was thrown
into vibration with more or less force at pleasure.   By supporting the funnel on a ring, passing the horse-hair in the opposite direction through the hole in the membrane, and
drawing the fingers over it downwards, the direction in which
the force was applied could be varied according to circumstances.

23. When lycopodium or light powders were sprinkled upon
this surface, the rapidity with which they ran to the centre,
the cloud formed there, the involving heaps, and many other
circumstances, could be observed very advantageously.

24. The currents which I have considered as existing upon
the surface of the plate, membranes, &c. from the quiescent
parts towards the centres or lines of vibration (9), arise necessarily from the mechanical action of that surface upon the air.
As any particular part of the surface moves upwards in the
course of its vibration, it propels the air and communicates a
certain degree of force to it, perpendicular or nearly so to the
vibrating surface; as it returns, in the course of its vibration,
it recedes from the air so projected, and the latter consequently
tends to return into the partial vacuum thus formed.   But as

of two neighbouring portions of air, that over the part of the plate nearest to the centre of oscillation has had more projectile force communicated to it than the other, because the part of the plate urging it was moving with greater velocity, and through a greater space, so it is in a more unfavourable condition for its immediate return, and the other, *i. e.* the portion next to it towards the quiescent line, presses into its place. This effect is still further favoured, because the portion of air thus displaced is urged from similar causes at the same moment into the place left vacant by the air still nearer the centre of oscillation; so that each time the plate recedes from the air, an advance of the air immediately above it is made from the quiescent towards the vibrating parts of the plates.

25. It will be evident that this current is highly favourable for the transference of light powders towards the centre of vibration. Whilst the air is forced forward, the advance of the plate against the particles holds them tight; but when the plate recedes, and the current exists, the particles are at that moment left unsupported except by the air, and are free to move with it.

26. The air which is thus thrown forward at and towards the centre of oscillation, must tend by the forces concerned to return towards the quiescent lines, forming a current in the opposite direction to the first and blending more or less with it. I endeavoured, in various ways, to make the extent of this system of currents visible. In the experiment already referred to, where gold-leaf was placed over the centre of oscillation (16), the upward current at the most powerful part was able to raise the leaf about one tenth of an inch from the plate. The higher the sounds with the same plate or membrane, *i. e.* the greater the number of vibrations, the less extensive must be the series of currents; the slower the vibrations, or the more extensive the excursion of the parts from increased force applied, the greater the extent of disturbance. With glass plates (2. 12) the cloud is higher and larger as the vibrations are stronger, but still not so extensive as they are upon the stretched membrane (22), where the cloud may frequently be seen rising up in the middle and flowing over towards the sides.

27. When the membrane stretched upon the funnel (22) was

made to vibrate by the horse-hair proceeding downwards, and
a large glass tube, as a cylindrical lamp-glass, was brought
near to the centre of vibration, no evidence of a current entirely through the lamp-glass could be perceived ; but still the
most striking proofs were obtained of the existence of carrying
currents by the effects upon the light powder, for it flew more
rapidly under the edge, and tended to collect towards the axis
of the tube ; it could even be diverted somewhat from its course
towards the centre of oscillation.   A piece of upright paper,
held with its edge equally near, did not produce the same
effect ; but immediately that it was rolled into a tube, it did.
When the glass chimney was suspended very carefully, and at
but a small distance from the membrane, the powder often
collected at the edge, and revolved there ; a complicated action
between the currents and the space under the thickness of the
glass taking place, but still tending to show the influence of
the air in arranging and disposing the powders.

28. A sheet of drawing-paper was stretched tightly over a
frame so as to form a tense elastic surface nearly three feet by
two feet in extent.   Upon placing this in a horizontal position,
throwing a spoonful of lycopodium upon it, and striking it
smartly below with the fingers, the phenomena of collection at
the centre of vibration, and of moving heaps, could be obtained
upon a magnificent scale.   When the lycopodium was uniformly
spread over the surface, and any part of the paper slightly
tapped by the hand, the lycopodium at any place chosen could
be drawn together merely by holding the lamp-glass over it.
It will be unnecessary to enter into the detail of the various
actions combining to produce these effects ; it is sufficiently
evident, from the mode in which they may be varied, that they
depend upon currents of air.

29. A very interesting set of effects occurred when the
stretched parchment upon the funnel (22) was vibrated under
plates ; the horse-hair was directed downwards, and the membrane, after being sprinkled over with light powder, was
covered by a plate of glass resting upon the edge of the
funnel ; upon throwing the membrane into a vibratory state,
the powder collected with much greater rapidity than without
the plate ; and instead of forming the semi-globular moving
heaps, it formed linear arrangements, all concentric to the

centre of vibration.   When the vibrations were strong, these assumed a revolving motion, rolling towards the centre at the part in contact with the membrane, and from it at the part nearest the glass; thus illustrating in the clearest manner the double currents caged up between the glass and the membrane.   The effect was well shown by carbonate of magnesia.

30. Sometimes, when the plate was held down very close and tight, and the vibrations were few and large, the powder was all blown out at the edge; for then the whole arrangement acted as a bellows; and as the entering air travelled with much less velocity than the expelled air, and as the forces of the currents are as the squares of the velocity, the issuing air carried the powder more forcibly than the air which passed in, and finally threw it out.

31. A thin plate of mica laid loosely upon the vibrating membrane showed the rotating concentric lines exceedingly well.

32. From these experiments on plates and surfaces vibrating in air, it appears that the forms assumed by the determination of light powders towards the places of most intense vibration, depend, not upon any secondary mode of division, or upon any immediate and peculiar action of the plate, but upon the currents of air necessarily formed over its surface, in consequence of the extra-mechanical action of one part beyond another.   In this point of view the nature of the medium in which those currents were formed ought to have great influence over the phenomena; for the only reason why silica as sand should pass towards the quiescent lines, whilst the same silica as fine powder went from them, is, that in its first form the particles are thrown up so high by the vibrations as to be above the currents, and that if they were not thus thrown out of their reach they would be too heavy to be governed by them; whilst in the second form they are not thrown out of the lower current, except near the principal place of oscillation, and are so light as to be carried by it in whatever direction it may proceed.

33. In the exhausted receiver of the air-pump, therefore, the phenomena ought not to occur as in air; for as the force of the currents would be there excessively weakened, the light powders ought to assume the part of heavier grains in the air.

Again, in denser media than air, as in water for instance, there was every reason to expect that the heavier powders, as sand and filings, would perform the part of light powders in air, and be carried from the quiescent to the vibrating parts.

34. The experiments in the air-pump receiver were made in two ways. A round plate of glass was supported on four narrow cork legs upon a table, and then a thin glass rod with a rounded end held perpendicularly upon the middle of the glass. By passing the moistened fingers longitudinally along this rod the plate was thrown into a vibratory state ; the cork legs were then adjusted in the circular nodal line occurring with this mode of vibration ; and when their places were thus found they were permanently fixed. The plate was then transferred into the receiver of an air-pump, and the glass rod by which it was to be thrown into vibration passed through collars in the upper part of the receiver, the entrance of air there being prevented by abundance of pomatum. When fine silica was sprinkled upon the plate, and the plate vibrated by the wet fingers applied to the rod, the receiver not being exhausted, the fine powder travelled from the nodal line, part collecting at the centre, and another part in a circle, between the nodal line and the edge. Both these situations were places of vibration, and exhibited themselves as such by the agitation of the powder. Upon again sprinkling fine silica uniformly over the plate, exhausting the receiver to twenty-eight inches, and vibrating the plate, the silica went from the middle towards the nodal line or place of rest, performing exactly the part of sand in air. It did not move at the edges of the plate, and as the apparatus was inconvenient and broke during the experiment, the following arrangement was adopted in its place.

35. The mouth of a funnel was covered (22) with a well-stretched piece of fine parchment, and then fixed on a stand with the membrane horizontal ; the horse-hair was passed loosely through a hole in a cork, fixed in a metallic tube on the top of the air-pump receiver ; the tube above the cork was filled to the depth of half an inch with pomatum, and another perforated cork put over that ; a cup was formed on the top of the second cork, which was filled with water. In this way the horse-hair passed first through pomatum and then water, and by giving a little pressure and rotary motion to the upper

cork during the time that the horse-hair was used to throw the membrane into vibration, it was easy to keep the pomatum below perfectly in contact with the hair, and even to make it exude upwards into the water above. Thus no possibility of the entrance of air by and along the horse-hair could exist, and the tightness of all the other and fixed parts of the apparatus was ascertained by the ordinary mode of examination. A little paper shelf was placed in the receiver under the cork to catch any portion of pomatum that might be forced through by the pressure, and prevent its falling on to the membrane.

36. This arrangement succeeded : when the receiver was full of air, the lycopodium gathered at the centre of the membrane with great facility and readiness, exhibiting the cloud, the currents, and the involving heaps. Upon exhausting the receiver, until the barometrical gauge was at twenty-eight inches, the lycopodium, instead of collecting at the centre, passed across the membrane, towards one side which was a little lower than the other. It passed by the middle just as it did over any other part; and when the force of the vibrations was much increased, although the powder was more agitated at the middle than elsewhere, it did not collect there, but went towards the edges or quiescent parts. Upon allowing air to enter until the barometer stood at twenty-six inches, and repeating the experiments, the effect was nearly the same. When the vibrations were very strong, there were faint appearances of a cloud, consisting of the very finest particles, collecting at the centre of vibration; but no sensible accumulation of the powder took place. At twenty-four inches of the barometer the accumulation at the centre began to appear, and there was a sensible, though very slight effect visible of the return of the powder from the edges. At twenty-two inches these effects were stronger; and when the barometer was at twenty inches, the currents of air within the receiver had force enough to cause the collection of the principal part of the lycopodium at the centre of vibration. Upon again, however, restoring the exhaustion to twenty-eight inches, all the effects were reproduced as at first, and the lycopodium again proceeded to the lower or the quiescent parts of the membrane. These alternate effects were obtained several times in succession before the apparatus was dismounted.

37. In this form of experiment there were striking proofs of the existence of a current upwards from the middle of the membrane when vibrating in air (24), and the extent of the system of currents (26) was partly indicated. The powder purposely collected at the middle by vibrations, when the receiver was full of air, was observed as to the height to which it was forced upwards by the vibrations; and then the receiver being exhausted, the height to which the powder was thrown by similar vibrations was again observed. In the latter cases it was nothing like so great as in the former, the height not being two-thirds, and barely one-half, the first height. Had the powder been thrown up by mere propulsion, it should have risen far higher *in vacuo* than in air: but the reverse took place; and the cause appears to be, that in air the current had force enough to carry the fine particles up to a height far beyond what the mere blow which they received from the vibrating membrane could effect.

38. For the experiments in a denser medium than air, water was chosen. A circular plate of glass was supported upon four feet in a horizontal position, surrounded by two or three inches of water, and thrown into vibration by applying a glass rod perpendicular to the middle, as in the first experiment *in vacuo* (34); the feet were shifted until the arrangement gave a clear sound, and the moistened brass filings sprinkled upon the plate formed regular lines or figures. These lines were not however lines of rest, as they would have been in the air, but were the places of greatest vibration; as was abundantly evident from their being distant from that nodal line determined and indicated by the contact of the feet, and also from the violent agitation of the filings. In fact, the filings proceeded from the quiescent to the moving parts, and there were gathered together; not only forming the cloud of particles over the places of intense vibration, but also settling down, when the vibrations were weaker, into the same involving groups, and in every respect imitating the action of light powders in air. Sand was affected exactly in the same manner; and even grains of platina could be in this way collected by the currents formed in so dense a medium as water.

39. The experiments were then made under water with the membranes stretched over funnels (22) and thrown into vibration

by horse-hairs drawn between the fingers. The space beneath the membrane could be retained, filled with air, whilst the upper surface was covered two or three inches deep with water; or the space below could also be filled with water, or the force applied to the membrane by the horse-hair could be upwards or downwards at pleasure. In all these experiments the sand or filings could be made to pass with the utmost facility to the most powerfully vibrating part, that being either at the centre only, or in addition, in circular lines, according to the mode in which the membrane vibrated. The edge of the funnel was always a line of rest; but circular nodal lines were also formed, which were indicated, not by the accumulation of filings upon them, but by the tranquil state of those filings which happened to be there, and also by being between those parts where the filings, by their accumulation and violent agitation, indicated the parts in the most powerful vibratory state.

40. Even when by the relaxation of the parchment from moisture, and the force upwards applied by the horse-hair, the central part of the membrane was raised the eighth of an inch or more above the edges, the circle not being four inches in diameter, still the filings would collect there.

41. When in place of parchment common linen was used, as becoming tighter rather than looser when wetted, the same effects were obtained.

42. Both the reasoning adopted and the effects described were such as to lead to the expectation, that if the plate vibrating in air was covered with a layer of liquid instead of sand or lycopodium, that liquid ought to be determined from the quiescent to the vibrating parts and be accumulated there. A square plate was therefore covered with water, and vibrated as in the former experiments (2. 6); but all endeavours to ascertain whether accumulation occurred at the centres of oscillation, either by direct observation, or the reflexion from its surface of right-lined figures, or by looking through the parts, as through a lens, at small print and other objects, failed.

43. As, however, when the plate was strongly vibrated, the well-known and peculiar crispations which form on water at the centres of vibration, occurred and prevented any possible decision as to accumulation, it was only when these were absent and the vibration weak, and the accumulation therefore small,

that any satisfactory result was to be expected; but as even then no appearance was perceived, it was concluded that the force of gravity combined with the mobility of the fluid was sufficient to restore the uniform condition of the layer of water after the bow was withdrawn, and before the eye had time to observe the convexity expected.

44. To remove in part the effect of gravity, or rather to make it coincide with, instead of oppose the convexity, the under surface of the plate was moistened instead of the upper, and by inclining the plate a little, the water made to hang in drops at *a* or *b* or *c*, fig. 9, at pleasure. On applying the bow at ×, and causing the plate to vibrate, the drops instantly disappeared, the water being gathered up and expanded laterally over the parts of the plate from which it had flowed. On stopping the vibration, it

Fig. 9.

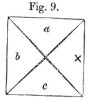

again accumulated in hanging drops, which instantly disappeared as before on causing the plate to vibrate, the force of gravity being entirely overpowered by the superior forces excited by the vibrating plate. Still, no visible evidence of convexity at the centres of vibration were obtained, and the water appeared rather to be urged from the vibrating parts than to them.

45. The tenacity of oil led to the expectation that better results would be obtained with it than with water. A round plate, held horizontally by the middle (6. 42), was covered with oil over the upper surface, so as to be flooded, except at ×, fig. 10, and the bow applied at × as before, to produce strong vibration. No crispation occurred in the oil, but it immediately accumulated at *a*, *b*, and *c*, forming fluid lenses there, rendered

Fig. 10.

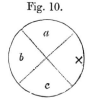

evident by their magnifying power when print was looked at through them. The accumulations were also visible on putting a sheet of white paper beneath, in consequence of the colour of the oil being deeper at the accumulations than elsewhere; and they were also rendered beautifully evident by making the experiment in sunshine, or by putting a candle beneath the plate, and placing a screen on the opposite side to receive the images formed at the focal distance.

46. When the vibration of the plate ceased, the oil gradually flowed back until of uniform depth. On renewing the vibration, the accumulations were re-formed, the phenomena of accumulation occurring with as much certainty and beauty as if lycopodium powder had been used.

47. To remove every doubt of the fluid passing from the quiescent to the agitated parts, centres of vibration were used, nearly surrounded by nodal lines. A square plate, fig. 11, being held at *c*, and the bow applied at ×, gave with sand, nodal lines, resembling those in the figure. Then clearing off the sand, putting oil in its place, and producing the same mode of vibration as before, the oil accumulated at *a* and *b*, forming two heaps or lenses as in the former experiment (45).

Fig. 11.

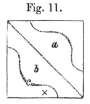

48. The experiment made with water on the under surface (44) was now repeated with oil, the round plate being used (45). The hanging drop of oil rose up as the water did before, but the lateral diffusion was soon limited; for lenses were formed at the centres of vibration just as when the oil was upon the upper surface, and, as far as could be ascertained by general examination, of the same form and power. On stopping the vibration, the oil gathered again into hanging drops; and on renewing it, it was again disposed in the lens-like accumulations.

49. With white of egg the same observable accumulation at the centres of vibration could be produced.

50. Hence it is evident that when a surface vibrating normally, is covered with a layer of liquid, that liquid is determined from the quiescent to the vibrating parts, producing accumulation at the latter places; and that this accumulation is limited, so that if purposely rendered too great by gravity or other means, it will quickly be diminished by the vibrations until the depth of fluid at any one part has a certain and constant relation to the velocity there and to the depth elsewhere.

51. From the accumulated evidence which these experiments afford, I think there can remain no doubt of the cause of the collection of fine powders at the centres or lines of vibration of plates, membranes, &c. under common circumstances; and that no secondary mode of division need be assumed to account for them. I have been the more desirous of accumulating experi-

mental evidence, because I have thought on the one hand that the authority of Savart should not be doubted on slight grounds; and on the other, that if by accident it be placed in the wrong scale, the weight of evidence against it should be such as fully to establish the truth and prevent a repetition of the error by others.

52. It must be evident that the phenomena of collection at the centres or lines of greatest vibration are exhibited in their purest form at those places which are surrounded by nodal lines; and that where the centre or place of vibration is at or near to an edge, the effects must be very much modified by the manner in which the air is there agitated. It is this influence, which, in the square plates (6. 12) and other arrangements, prevents the clouds being at the very edge of the glass. They may be well illustrated by vibrating tin plates under water over a white bottom, and sprinkling dark-coloured sand or filings upon various parts of the plates.

On the peculiar Arrangement and Motions of the Heaps formed by Particles lying on vibrating Surfaces.

53. The peculiar manner in which the fine powder upon a vibrating surface is accumulated into little heaps, either hemispherical or merely rounded, and larger or smaller in size, has already been described (6. 28), as well also as the singular motion which they possess, as long as the plate continues in vibration. These heaps form on any part of the surface which is in a vibratory state, and not merely under the clouds produced at the centres of vibration, although the particles of the clouds always settle into similar heaps. They have a tendency, as heaps, to proceed to the nodal or quiescent lines, but are often swept away in powder by the currents already described (6). When on a place of rest, they do not acquire the involving motion. When two or more are near together or touch, they will frequently coalesce and form but one heap, which quickly acquires a rounded outline. When in their most perfect and final form, they are always round.

54. The moving heaps formed by lycopodium on large stretched drawing-paper (28), are on so large a scale as to be very proper for critical examination. The phenomena can be exhibited also even by dry sand on such a membrane, the sand

being in large quantity and the vibrations slow. When the surface is thickly covered by sand from a sieve, and the paper tapped with the finger, the manner in which the sand draws up into moving heaps is very beautiful.

55. When a single heap is examined, which is conveniently done by holding a vibrating tuning-fork in a horizontal position, and dropping some lycopodium upon it, it will be seen that the particles of the heap rise up at the centre, overflow, fall down upon all sides, and disappear at the bottom, apparently proceeding inwards; and this evolving and involving motion continues until the vibrations have become very weak.

56. That the medium in which the experiment is made has an important influence, is shown by the circumstance of heavy particles, such as filings, exhibiting all these peculiarities when they are placed upon surfaces vibrating in water (39); the heaps being even higher at the centre than a heap of equal diameter formed of light powder in the air. In water, too, they are formed indifferently upon any part of the plate or membrane which is in a vibratory state. They do not tend to the quiescent lines; but that is merely from the great force of the currents formed in water as already described (38), and the power with which they urge obstacles to the place of greatest vibration.

57. If a glass plate be supported and vibrated (6), its surface having been covered with sand enough to hide the plate, and water enough to moisten and flow over the sand, the sand will draw together in heaps, and these will exhibit the peculiar and characteristic motion of the particles in a very striking manner.

58. The aggregation and motion of these heaps, either in air or other fluids, is a very simple consequence of the mechanical impulse communicated to them by the joint action of the vibrating surface and the surrounding medium. Thus in air, when in the course of a vibration, the part of a plate under a heap rises, it communicates a propelling force upwards to that heap, mingled as it is with air, greater than that communicated to the surrounding atmosphere, because of the superior specific gravity of the former; upon receding from the heap, therefore, in performing the other half of its vibration, it forms a partial vacuum, into which the air, round the heap, enters with

more readiness than the heap itself; and as it enters, carries in the powder at the bottom edge of the heap with it. This action is repeated at every vibration, and as they occur in such rapid succession that the eye cannot distinguish them, the centre part of the heap is continually progressing upwards; and as the powder thus accumulates above, whilst the base is continually lessened by what is swept in underneath, the particles necessarily fall over and roll down on every side.

59. Although this statement is made upon the relation of the heap, as a mass, to the air surrounding it, yet it will be seen at once that the same relation exists between any two parts of the heap at different distances from the centre; for the one nearest the centre will be propelled upward with the greatest force, and the other will be in the most favourable state for occupying the partial vacuum left by the receding plate.

60. This view of the effect will immediately account for all the appearances; the circular form, the fusion together of two or more heaps, their involving motion, and their existence upon any vibrating part of the plate. The manner in which the neighbouring particles would be absorbed by the heaps is also evident; and as to their first formation, the slightest irregularities in the powder or surface would determine a commencement, which would then instantly favour the increase.

61. It is quite true, that if the powder were coherent, that force alone would tend to produce the same effect, but only in a very feeble degree. This is sufficiently shown by the experiments made in the exhausted receiver (36). When the barometer of the air-pump was at twenty-eight inches, that in the air being about $29 \cdot 2$ inches, the heaps, or rather parcels, formed very beautifully over the whole surface of the membrane; but they were very flat and extensive compared with the heaps in air, and the involving motion was very weak. As the air was admitted, the vibration being continued, the heaps rose in height, contracted in diameter, and moved more rapidly. Again, in the experiments with filings and sand in water, no cohesive action could assist in producing the effect; it must have been entirely due to the manner in which the particles were mechanically urged in a medium of less density than themselves.

62. The conversion of these round heaps into linear concentric involving parcels, in the experiment already described

(29. 31), when the membrane was covered by a plate of glass, is a necessary consequence of the arrangements there made, and tends to show how influential the action of the air or other including medium is in all the phenomena considered in this paper. No incompatible principles are assumed in the explication given of the arrangement of the forces producing the two classes of effects in question; and though by variation of the force of vibration and other circumstances, the one effect can be made, within certain limits, to pass into the other, no anomaly or contradiction is thus involved, nor any result produced, which, as it appears to me, cannot be immediately accounted for by reference to the principles laid down.

Royal Institution, March 21, 1831.

Appendix.

On the Forms and States assumed by Fluids in contact with vibrating Elastic Surfaces.

63. When the upper surface of a plate vibrating so as to produce sound (2. 6) is covered with a layer of water, the water usually presents a beautifully crispated appearance in the neighbourhood of the centres of vibration. This appearance has been observed by Oersted [*], Wheatstone [†], Weber [‡], and probably others. It, like the former phenomena which I have endeavoured to explain, has led to false theory, and being either not understood or misunderstood, has proved an obstacle to the progress of acoustical philosophy.

64. On completing the preceding investigation, I was led to believe that the principles assumed would, in conjunction with the cohesion of fluids, account for these phenomena. Experimental investigation fully confirmed this expectation, but the results were obtained at too late a period to be presented to the Royal Society before the close of the Session; and it is only because the philosophy and the subject itself is a part of that received into the Philosophical Transactions in the preceding paper, that I am allowed, by the President and Council, the privilege of attaching the present paper in the form of an Appendix.

[*] Lieber's Hist. of Natural Phenomena for 1813.

[†] Annals of Philosophy, N. S. vi. p. 82.   [‡] Wellenlehre, p. 414.

65. The general phenomenon now to be considered is easily produced upon a square plate nipped in the middle, either by the fingers or the pincers (2. 6), held horizontally, covered with sufficient water on the upper surface to flow freely from side to side when inclined, and made to vibrate strongly by a bow applied to one edge, ×, fig. 12, in the usual way.

Fig. 12.

Crispations appear on the surface of the water, first at the centres of vibration, and extend more or less towards the nodal lines, as the vibrations are stronger or weaker. The crispation presents the appearance of small co-noidal elevations of equal lateral extent, usually arranged rectangularly with extreme regularity; permanent* (in appearance), so long as a certain degree of vibration is sustained; increasing and diminishing in height, with increased or diminished vibration; but not affected in their lateral extent by such variations, though the whole crispated surface is enlarged or diminished at those times. If the plate be vibrated, so as to produce a different note, the crispations still appear at the centre of vibration, but are smaller for a high note, larger for a low one. The same note produced on different sized plates, by different modes of vibration, appears to produce crispations of the same dimension, other circumstances being the same.

66. These appearances are beautifully seen when ink diluted with its bulk of water is used on the plate.

67. It was necessary, for examination, both to prolong and enlarge the effect, and the following were found advantageous modes of producing it. Plates of crown-glass, from eighteen to twenty-two inches long, and three or four inches wide, were supported each by two triangular pieces of wood acting as bridges (18), and made to vibrate by a small glass rod or tube resting perpendicularly at the middle, over which the moist fingers were passed. By sprinkling dry sand on the plates, and shifting the bridges, the nodal lines were found (usually about one-fifth of the whole length from each end), and their places marked by a file or diamond. Then clearing away the sand, putting water or ink upon the plate, and applying the rod or fingers, it was easy to produce the crispations and sustain

* Weber's Wellenlehre, p. 414.

them undisturbed, and with equal intensity for any length of time.

68. By making a broad mark, or raising a little ledge of bees' wax, or a mixture of bees' wax and turpentine, it was easy to confine the pool of water to the middle part of the plate, fig. 13, where, of course, the crispations were most powerfully produced. Such a barrier is often useful to separate the wet and dry parts of the glass, especially when a violin bow is used as the exciter.

Fig. 13.

69. In other experiments, deal laths, two, three, or four feet long, one inch and a half wide, and three-eighths or more of an inch in thickness, were used instead of the glass plates. These could be made to vibrate by the fingers and wet rod (67), and by either shifting the bridges or changing the lath, an almost unlimited change of isochronous vibrations, from that producing a high note to those in which not more than five or six occurred in a second, could be obtained. The crispations were formed upon a glass plate attached to the middle of the lath, by two or three little pellets of soft cement*.

70. Obtained in this way the appearances were very beautiful, and the facilities very great. A glass plate, from four to eight inches square, could be covered uniformly with crispations of the utmost regularity; for, by attaching the plate with a little method, and at points equidistant from the centre of the bar, it was easy to make every part travel with the same velocity, and in that respect differ from and surpass the bar which sustained it. The conoidal heaps constituting the crispation could be so enlarged by slowness of vibration, that three or four occupied a linear inch. The glass plate could be removed, and another of different form or substance, and with other fluids, as mercury, &c., substituted in an instant.

71. In using laths, it is necessary to confine the parts bearing upon the bridges, either by slight pressure of the fingers, or by loops of string, or by weights. The exciting glass rod need not necessarily rest upon the middle of the bar or plate, but may be applied with equal effect at some distance from it. Long laths may be made to subdivide in their mode of vibration, according as the rod is applied to different places, and the

* Equal parts of yellow wax and turpentine.

z

pressure given by the exciting moist fingers is varied; with each change of this kind an immediate change of the crispation is observed.

72. This form of apparatus was enlarged until a board eighteen feet long was used, the layer of water being now three-fourths of an inch in depth and twenty-eight inches by twenty inches in extent. The sides of the cistern were very much inclined, so that the water should gradually diminish in depth, and thus reflected waves be prevented. The vibrations were so slow as to be produced by the direct application of the hand, and the heaps were each from an inch to two inches in extent. Though of this magnitude, they were identical in their nature with those forming crispations on so small a scale as to appear merely like a dullness on the surface of the water.

73. In these experiments the proportion of water requires a general adjustment, the crispations being produced more readily and beautifully when there is a certain quantity than when there is less. For small crispations, the water should flow upon the surface freely. Large crispations require more water than small ones. Too much water sometimes interferes with the beauty of the appearance, but the crispation is not incompatible with much fluid, for the depth may amount to eight, ten, or twelve inches (111), and is probably unlimited.

74. These crispations are equally produced upon either the under or the upper surface of vibrating plates. When the lower surface is moistened, and the bow applied (65), the drops which hang down by the force of gravity are rippled; but being immediately gathered up as described in the former paper (44), a certain definite layer is produced, which is beautifully rippled or crispated at the centres of vibration.

75. Most fluids, if not all, may be used to produce these crispations, but some with particular advantages; alcohol, oil of turpentine, white of egg*, ink, and milk produce them. White of egg, notwithstanding its viscosity, shows them readily and beautifully. Ink has great advantages, because, from its colour and opacity, the surface form is seen undisturbed by any reflexion from the glass beneath; its appearance in sunshine is exceedingly beautiful. When diluted ink is used for large crispations, upon tin plate or over white paper, or mer-

* Wheatstone.

cury, the different degrees of colour or translucency corresponding to different depths of the fluid, give important information relative to the true nature of the phenomena (78. 85. 97).
Milk is, for its opacity, of similar advantage, especially when a
light is placed beneath; and being more viscid than water, is
better for large arrangements (72. 98), because it produces less
splashing.

76. Oil does not show small crispations readily (120), and
was supposed to be incapable of forming them, but when
warmed (by which its liquidity is increased) it produces them
freely.   Cold oil will also produce large crispations, and for
very large ones would probably be better than water, because
of its cohesion.   The difference between oil and white of egg
is remarkable; for the latter, from common observation, would
appear to be a thicker fluid than oil: but the qualities of cohesion differ in the two, the apparent thickness of white of egg
depending upon an elastic power (probably due to an approach
to structure), which tends to restore its particles to their first
position, and co-existing with great freedom to move through
small spaces, whilst that of oil is due to a real difficulty in removing the particles one by another.   It is possible that the
power of assuming, more or less readily, the crispated state,
may be a useful and even important indication of the internal
constitution of different fluids.

77. With mercury the crispations are formed with great
facility, and of extreme beauty, when a piece of amalgamated
tin or copper plate, fixed on a lath (69), is flooded with the
fluid metal, and then vibrated.   A film quickly covers the
metal, and then the appearances are not so regular as at first;
but on removing the film by a piece of paper, their regularity
and beauty are restored.   It is more convenient to cover the
mercury with a little very dilute acetic or nitric acid; for then
the crispations may be produced and maintained for any length
of time with a surface of perfect brilliancy.

78. When a layer of ink was put over the mercury, the acid
of the ink removed all film, and the summits of the metallic
heaps, by diminishing the thickness of the ink over them, became more or less visible, producing the appearance of pearls
of equal size beautifully arranged in a black medium.   When
mercury covered with a film of dilute acid was vibrated in the

sunshine, and the light reflected from its surface received on a screen, it formed a very beautiful and regular image ; but the screen required to be placed very near to the metal, because of the short focal lengths of the depressions on the mercurial surface.

79. It is sometimes difficult to arrive by inspection at a satisfactory conclusion of the forms and arrangements thus presented, because of multiplied reflexion and the particular condition of the whole, which will be described hereafter (95). When observed, well formed with vibrations so slow as to produce three or four elevations in a linear inch (70), they are seen to be conoidal heaps rounded above, and apparently passing into each other below by a curvature in the opposite direction. When arranged regularly, each is surrounded by eight others; so that, a single light being used, nine images may be sent from each elevation to the eye. These are still further complicated, when transparent fluids are used, by reflexions from the glass beneath. The use of ink (75) removes a good deal of the difficulty experienced, and the production of slow, regular, sustained vibrations, more (67. 69).

80. These elevations I will endeavour to distinguish henceforth by the term *heaps*.

81. The crispation on the long plate of glass described (67) always ultimately assumed a rectangular arrangement, *i. e.* the heaps were equidistant, and in rows parallel or at right angles to each other. The rows usually form angles of 45° to the sides of the plate at the commencement; but if the vibration be continued, the whole system usually wheels round through 45° until the rows coincide with the edges of the plate.

82. The lateral dimension of the heaps remained constant, notwithstanding considerable variations in the force of vibration. But it was soon found that variation in the depth of water affected their number; that with less water the heaps were smaller, and with more water larger, though the sound therefore and the number of vibrations in a given period remained the same. The number of heaps could be reduced to eight or increased to eleven and a half in the three inches by a change in no other condition than the depth of fluid.

83. With the above plate (67.81) the appearances were usually in the following order, the pool of water being quadrangular or

nearly so, and the exciting rod resting in the middle of it.
Ring-like linear heaps concentric to the exciting rod first
form to the number of six or seven; these may be retained by
a moderated state of vibration, and produce intervals which
measured across the diameter of the rings are to the number
of ten in three inches, with a certain constant depth of water.
By increasing the force of vibration, the altitude of these elevations increases, but not their lateral dimension; and then linear
heaps form across these circles and the plate, and parallel to
the bridges, having an evident relation to the manner in which
the whole plate vibrates.   These, which like all other of these
phenomena are strongest at the part most strongly vibrating,
soon break up the circles, and are themselves broken up, producing independent heaps, which at first are irregular and changeable, but soon become uniform and produce the quadrangular
order; first at angles of 45° to the edges of the plate, but gradually moving round until parallel to them.   So the arrangement continues, unless the force be so violent as to break it up
altogether : if the vibratory force be gradually diminished, then
the heaps as gradually fall, but without returning through the
order in which they were produced.   The following lines may
serve to indicate the course of the phenomena.

Fig. 14.

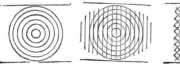 

When perfectly formed, the heaps are also to the number of
ten in three inches with the same depth of water as that which
produced the rings.   The intervals between the rings and the
heaps are the same, other influential circumstances remaining
unaltered.

84. Another form of heaps occasionally occurred, but always
passing ultimately into those described.   These heaps were
grouped in an arrangement still very nearly rectangular,
and at angles of 45° to the sides of the plate, but were contracted in one direction, and elongated in the other; these
directions being parallel to the sides and ends of the plate.

If the marks in fig. 15 be supposed to represent
the tops of the heaps, an idea of the whole will
be obtained. Three inches along these heaps
included eight, but across them it included
fifteen nearly. These numbers are therefore the
relation of length to breadth. But along the
lines of the quadrilateral arrangement three
inches included eleven heaps, which, notwithstanding the difference in form, is the same number that was produced by the
same plate, with the same depths of water, when the heaps were
round; therefore an equal number of heaps existed in the
same area in both cases; and the departure from perfect rectangular arrangement, and also the ratio of $1:2$, is probably
due to some slight influence of the sides of the plate.

Fig. 15.

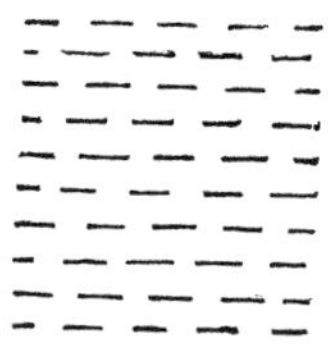

85. When mercury covered with a film of very dilute nitric
acid is vibrated (77), the rectangular arrangement is constantly
obtained. When vibrated under dilute ink (78), it is still more
beautifully seen and distinguished. The tin plate sustaining
the mercury was square, and when the whole surface was
covered with crispations, the lines of the rectangular arrangement were always at angles of $45°$ to its edges.

86. When sand is sprinkled uniformly over a plate on which
large water crispations are produced, *i. e.* four, five or six in
the inch, it gives some very important indications. It immediately becomes arranged under the water, and with a little
method may be made to yield very regular forms.
It is always removed from under the heaps,
passing to the parts between them, and frequently producing therefore the accompanying
form, fig. 16, of great regularity. As the sand
figure remains when the vibration has ceased,
it allows of the determination of position, the
measurement of intervals, &c. very conveniently.

Fig. 16.

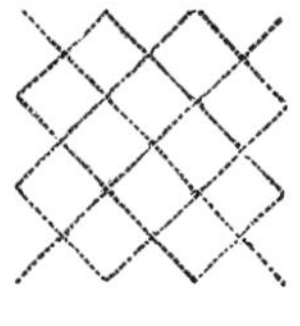

87. Very often the lines of sand are not continuous, but
separated with extreme regularity into portions, as represented
fig. 17. The portions of these lines were sometimes, with
little sand on the plate, very small, fig. 18; and when more
sand was present they were thickened occasionally, fig. 19;
then assuming the appearance of heaps arranged in straight
lines at angles of $45°$ to the lines regulating the position of the

water-heaps which formed them, and just double in number to the latter. At other times, the sand, instead of being deficient at the intersecting angle, would accumulate there only, fig. 20;

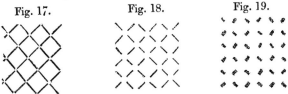

Fig. 17.   Fig. 18.   Fig. 19.

and at other times would accumulate there principally, but still show the original form by a few connecting particles, fig. 21.

88. When the heaps were of the form described (84), the sand was still washed from under them; it did not however assume lines parallel to the rectangular arrangement of the heaps, but was arranged as in fig. 22.

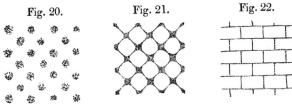

Fig. 20.   Fig. 21.   Fig. 22.

89. When only the circular linear heaps (83) were produced, the sand assumed similar circular forms, concentric and alternating with the water elevations.

90. On strewing a little lycopodium over the water for the purpose of gaining information relative to what occurred at the surface during the crispation, it moved about over the fluid in every possible direction, whilst the crispations existed of the utmost steadiness beneath. The same thing occurred with pieces of cork on very large crispations (98). But when much lycopodium was put on, so that the particles retained each other in a steady position, then it formed lines* parallel to the arrangement of the heaps, the powder being displaced from the parts over the heaps, and taking up an arrangement perpendicularly over the sand beneath. As the lycopodium forms float on the water they are easily disturbed, and in no respect approach as to beauty and utility to the forms produced by the sand; but lycopodium may be used with smaller crispations than sand.

          * Wheatstone.

91. The crispations are much influenced by various circumstances. They tend to commence at the place of greatest vibration; but if the quantity of fluid is too little there, and more abundant elsewhere, they will often commence at the latter place first. Their final arrangement is also much affected by the form of the plate, or of the pool of water on which they occur. When the plates or pools are rectangular, and all parts vibrate with equal velocity, the lines of heaps are at angles of 45° to the edges. But when semicircular and other plates were used, the arrangement, though quadrangular, was unsteady, often breaking up and starting by pieces into different and changing positions.

92. When mercury was used (77), the film formed on it after a few moments had great power, according to the manner in which it was puckered, of modifying the general arrangement of new crispations.

93. When a circular plate, supported by cork feet attached where a single nodal line would occur, was covered with water and vibrated by a rod resting upon the middle, the crispations extended from the middle towards the nodal line; these were sometimes arranged rectangularly, but had no steadiness of position, and changed continually. At other times the heaps appeared as if hexagonal, and were arranged hexagonally, but these also shifted continually. These and many other experiments (83) showed that the direction and nature of the vibration of the plate (*i. e.* of the lines of equal or varying vibrating force) had a powerful influence over the regularity and final arrangement of the crispations.

94. The beautiful appearance exhibited when the crispations are produced in sunshine, or examined by a strong concentrated artificial light, has been already referred to (78, 79). When the reflected image from any one heap is examined (for which purpose ink (75) or mercury (77) is very convenient), it will be found not to be stationary, as would happen if the heap was permanent and at rest; nor yet to form a vertical line as would occur if the heap were permanent but travelled to and fro with the vibrating plate; but it moves so as to re-enter upon its course, forming an endless figure, like those produced by Dr. Young's piano-forte wires, or Wheatstone's kaleidophone, varying with the position of the light and the observer,

but constant for any particular position and velocity of vibration. Upon placing the light and the eye in positions nearly perpendicular to the general surface of the fluid, so as to avoid the direct influence of the motion of vibration, still the luminous, linear, endless figure was produced, extending more or less in different directions, according to the relation of the light and eye to the crispated surface, and occasionally corresponding in its extent one way to the width of the heap, *i. e.* to the distance between the summit of one heap and its neighbours, but never exceeding it. The figure produced by one heap was accurately repeated by all the heaps when the vibrating force of the plate was equal (70) and the arrangement regular.

95. The view which I had been led to anticipate of the nature of the heaps, from the effects described in the former paper, were, that each heap was a permanent elevation, like the cones of lycopodium powder (53. 58), the fluid rising at the centre, but descending down the inclined sides, the whole system being influenced, regulated, and connected by the cohesive force of the fluid. But these characters of the reflected image, with others of the effects already described, led to the conclusion, that notwithstanding the apparent permanency of the crispated surface, especially when produced on a small scale, as by the usual method, the heaps were not constant, but were raised and destroyed with each vibration of the plate; and also that the heaps did not all exist at once, but (referring to locality) formed two sets of equal number and arrangement, fig. 23, never existing together, but alternating with, and being resolved into each other, and by their rapidity of recurrence giving the appearance of simultaneous and even permanent existence. Provided this view were confirmed, it seemed as if it would be easy to explain the production of the heaps, their regular arrangement, &c., and to deduce their recurrence, dimensions, and many other points relative to their condition.

Fig. 23.

96. On producing a water crispation, having four or five heaps in a linear inch, placing a candle beneath, and a screen of French tracing-paper above it, the phenomena were very beautiful, and such as supported the view taken. By placing the screen at different distances, it could be adapted to the

focal length due to the curvature at different parts of the surface of fluid, so that by observing the luminous figure produced and its transitions as the screen was moved nearer or further, the general form of the surface could be deduced. Each heap with a certain distance of screen gave a star

Fig. 24.

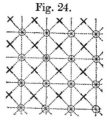

of light ⊕, fig. 24, which twinkled, *i. e.* appeared and disappeared alternately, as the heap rose and fell. At the corners × equidistant from these, fainter starred lights appeared ; and by putting the screen nearer to or further from the surface, lines of light, in two or even four directions, appeared intersecting the luminous centres and apparently permanent, whilst circumstances remained unchanged. These effects could be magnified to almost any scale (72).

97. When heaps of similar magnitude were produced, with diluted ink on glass (75), and white paper or an illuminated screen looked at through them, a chequered appearance was observed. In one position, lines of a certain intensity separated the heaps from each other, but the square places representing the heaps looked generally lighter. In another position, when but little reflected light came from the surface of the heaps, their places could be perceived as dark, from the greater depth of ink there. By care, another position could be found in which the whole surface looked like an alternate arrangement of light and dark chequers, fig. 25, not steady, but with

Fig. 25.

a quivering motion, which further attention could trace as due to a rapid alternation in which the light spaces became dark and the dark light, simultaneously. When, instead of glass, a bright tin plate was used under the diluted ink, the chequered spaces and their alternations could be seen still more beautifully.

98. It was in consequence of these effects that very large arrangements were made (72), giving heaps that were two inches and a half wide each*; and now it was evident, by ordinary

---

* This estimate is given in accordance with the mode of estimating the former and smaller heaps, as if the heaps were formed simultaneously ; but it is evident that if only half the number exist at once, each heap will have twice the width or four times the area of those which can be formed if all exist together.

inspection, that the heaps were not stationary, but rose and fell; and also that there were two sets regularly and alternately arranged, the one set rising as the other descended.

99. Sand gave no indications of arrangement with these large heaps (86); but when some coarse sawdust was soaked, so as to sink in water, and then distributed in the fluid, its motions were beautifully illustrative of the whole philosophy of the phenomena. It was immediately washed away from under the rising and falling heaps, and collected in the places equidistant between these spots, as the sand did in the former experiments (86), and by its vibratory motion to and fro, it showed distinctly how the water oscillated from one heap towards another, as the heaps sunk and rose.

100. When milk (75) was used instead of water for these large arrangements in a dark room, and a candle was placed beneath, the appearances also were very beautiful, resembling in character those described (97).

101. Each heap (identified by its locality) recurs or is re-formed in two complete vibrations of the sustaining surface*; but as there are two sets of heaps, a set occurs for each vibration. The maximum and minimum of height for the heaps appears to be alternately, almost immediately after the supporting plate has begun to descend in one complete vibration.

102. Many of these results are beautifully confirmed by the appearances produced, when regular crispations have been sustained for a short time with mercury, on which a certain degree of film has been allowed to form (77). On examining the film afterwards in one light, lines could be seen on it, coinciding with the intervals of the heaps in one direction; in another light, lines coinciding with the other direction came into sight, whilst the first disappeared; and in a third light, both sets of lines could be seen cutting out the square places where the heaps had existed: in these spaces the film was minutely wrinkled and bagged, as if it had there been distended; at the lines it was only a little wrinkled, giving the appearance of texture; and at the crossing of the lines themselves, it was

---

* A vibration is here considered as the motion of the plate, from the time that it leaves its extreme position until it returns to it, and not the time of its return to the intermediate position.

quite free from mark, and fully distended.   All these are natural consequences, if the film be considered as a flexible but
inelastic envelope formed over the whole surface whilst the
heaps were rising and falling.

103. The mode of action by which these heaps are formed
is now very evident, and is analogous in some points to that by
which the currents and the involving heaps already described
are produced.   The plate in rising tends to lift the overlying
fluid, and in falling to recede from it; and the force which it
is competent to communicate to the fluid can, in consequence
of the physical qualities of the latter, be transferred from particle to particle in any direction.   The heaps are at their maximum elevation just after the plate begins to recede from them;
before it has completed its motion downwards, the pressure of
the atmosphere and that part of the force of the plate which
through cohesion is communicated to them, has acted, and by the
time the plate has begun to return, it meets them endowed with
momentum in the opposite direction, in consequence of which
they do not rise as a heap, but expand laterally, all the forces
in action combining to raise a similar set of heaps, at exactly
intermediate distances, which attain their maximum height just
after the plate again begins to recede; these therefore undergo
a similar process of demolition, being resolved into exact dupli-
cates of the first heaps.   Thus the two sets oscillate with each
vibration of the plate, and the action is sustained so long as
the plate moves with a certain degree of force; much of that
force being occupied in sustaining this oscillation of the fluid
against the resistance offered by the cohesion of the fluid, the
air, the friction on the plate, and other causes.   Fig. 26.

104. A natural reason now appears for the
quadrangular and right-angled arrangement
which is assumed, when the crispation is most
perfect.   The hexagon, the square, and the
equilateral triangle are the only regular figures
that can fill an area perfectly.   The square and
triangle are the only figures that can allow of
one half alternating symmetrically with the
other, in conformity with what takes place between the two reciprocating sets of heaps,
fig. 26; and of these two the boundary lines

between squares are of shorter extent than those between equilateral triangles of equal area. It is evident therefore that one of these two will be finally assumed, and that that will be the square arrangement; because then the fluid will offer the least resistance in its undulations to the motions of the plate, or will pass most readily to those positions into which the forces it receives from the plate conspire to impel it.

105. All the phenomena observed and described may, as it appears to me, be now comprehended. The fluid may be considered as a pendulum vibrating to and fro under a given impulse; the various circumstances of specific gravity, cohesion, friction, intensity of vibrating force, &c. determining the extent of oscillation, or, what is the same thing, the number of heaps in a given interval. When the number of vibrations in a given time is increased, these heaps are more numerous, because the oscillation, to be more rapid, must occur in a shorter space. The necessity of a certain depth of fluid (73) is evident, and also the reason why, by varying the depth (82), the lateral extent of the heaps is changed. The arrangement of the sand and lycopodium, by the crispations, and the occurrence of the latter at centres of vibration, and only upon surfaces vibrating normally, are all evident consequences. The permanency of the lateral extension of the heaps, when the velocity of the vibrating plate varies, is a very marked effect; and it is probable that the investigation of these phenomena may hereafter importantly facilitate inquiries into the undulations of fluids, their physical qualities, and the transmission of forces through them.

106. As to the origin or determination of crispations, no difficulty can arise; the smallest possible difference in almost any circumstance, at any one part, would, whilst the plate is vibrating, cause an elevation or depression in the fluid there; the smallest atom of dust falling on the surface, or the smallest elevation in the plate, or the smallest particle in the fluid of different specific gravity to the liquid itself, might produce this first effect; this would, by each vibration of the plate, be increased in amount, and also by each vibration extended the breadth of a heap, in at least four directions: so that in less than a second a large surface would be affected, even under the improbable supposition that only one point should at first be disturbed.

107. I have thought it unnecessary to dwell upon the explanation of the circular linear heaps (83. 93. 110) produced on long or circular plates by feeble vibration. They are explicable upon the same principles, account being at the same time taken of the arrangement and proportion of vibrating force in the various parts of the plates.

108. The heaps which constitute crispation (as the word has been used in this paper) are in form, quality, and motion of their parts, the same with what are called stationary undulations; and if the mercury in a small circular basin be tapped at the middle, stationary undulations, resembling the ring-like heaps (83. 110), will be obtained; or if a rectangular frame be made to beat at equal intervals of time on mercury or water, heaps like those of the crispations, arranged quadrangularly at angles of 45° to the frame, will be produced. These effects are in fact the same with those described, but are produced by a cause differing altogether. The first are the result of two progressing and opposed undulations, the second of four: but the heaps of crispations are produced by the power impressed on the fluid by the vibrating plate; are due to vibrations of that fluid occurring in twice the time of the vibrations of the plate; and have no dependence on progressive undulations, originating laterally, as many of the phenomena described prove. Thus, when the edges were beveled (72. 110), or covered with cloth, or wet sawdust, so that waves reaching the side should be destroyed, or when the limits of the water or plates were round (91) or irregular, still the heaps were produced, and their arrangement square. When the round plate (93) was used, regular crispations were still produced, though, as the water extended over the nodal line, and was there perfectly undisturbed, no progressing and opposed undulations could originate to produce them. Vellum stretched over a ring, and rendered concave by the pressure of the exciting rod, produced the same effect.

109. When a plate of tin, rendered very slightly concave, was attached to a lath (69), so as to have equality of vibratory motion in all its parts, and a little dilute alkali (which would wet the surface) put into it, the crispations formed in the middle, but ceased towards the sides, where, though well-wetted, there was not depth enough of water, and whence also no

waves could be reflected to produce stationary undulations in the ordinary manner.

110. When a similar arrangement was made with mercury on a concave tin plate, the effects were still more beautiful and convincing. The centre portion was covered with one regular group of quadrangular crispations; at some distance from the centre, and where the mercury was less in depth, these passed into concentric, ring-like heaps, of which there were a great many; and outside of these there was a part wet with mercury, but with too little fluid to give either lines or heaps. Here there could be no reflected waves; or, if that were thought possible, those waves could not have formed both the circular rings and the square crispation. When this plate was vibrated, the mercury spread in all directions up the side; a natural consequence of the production of powerful oscillations at the middle, which would extend their force laterally, but quite against their being due to the opposition and crossing of waves originating at the sides.

111. A limited depth of fluid is by no means necessary to produce crispations on the surface (73). A circular glass basin about five inches in diameter and four inches deep was attached to a lath (69), filled with water and vibrated, the exciting rod being applied at the side (71). The surface of the water was immediately covered with the most regular crispations, *i. e.* heaps arranged quadrangularly. On taking out part of the water and filling it up with oil, the oil assumed the same superficies. On putting an inch in depth of mercury under the water, the mercury became crispated. The experiment was finally made with water fourteen inches in depth. Particles at a very moderate depth in the water seemed to have no motion except the general motion of the fluid, and the whole of the lower part of the water may be considered as performing the part of a solid mass upon which the superficial undulating portion reposed. In fact, it matters not to the fluid, what is beneath, provided it has sufficient cohesion, is uniform in relation to the surface fluid, and can transmit the vibrations to it in an undisturbed manner*.

* I have seen the water in a pail placed in a barrow, and that on the head of an upright cask in a brewer's van passing over stones, exhibit these elevations.

112. The beautiful action thus produced at the limits of two immiscible fluids, differing in density or some other circumstances, by which the denser was enabled most readily to accommodate itself to rapid, regular and alternating displacements of its support when that support was horizontal, suggested an inquiry into the probable arrangement of the fluid when the displacements were lateral or even superficial.

113. On arranging the long plate (67. 81) vertically, so that the lower extremity dipped about one-third of an inch into water, fig. 27, and causing it to vibrate by applying the rod at ×, or by tapping the plate with the finger, undulations of a peculiar character were observed: those passing from the plate towards the sides of the basin were scarcely visible though the plate vibrated strongly, but in place of such appeared others, in the production of which the mechanical force of the vibrating plate exerted upon the fluid was principally employed. These were apparently permanent elevations, at regular intervals, strongest at the plate, projecting directly out from it over the surface of the water, like the teeth of a coarse comb gradually diminishing in height, and extending half or three-quarters of an inch in length. These varied in commencing at the glass, or having intervening ridges, or in height, or in length, or in number, or in breaking up into violently agitated pimples and drops, &c. according as the plate dipped more or less into the water, or vibrated more or less violently, or subdivided whilst vibrating into parts, or changed in other circumstances. But when the plate (sixteen or seventeen inches long) dipped about one-sixth of an inch, then four of these linear heaps occupied as nearly as possible the same space as four heaps formed with the same plate in the former way (83) and accompanied with the same sound.

114. By fixing a wooden lath (69) perpendicularly downwards in a vice, plates of any size or form could be attached to its lower end and immersed more or less in water; and by varying the immersion of the plate, or the length of the lath, or the place against which the exciting rod (71) was applied, the vibrations could be varied in rapidity to any extent.

115. On using a piece of board at the extremity of the lath, eight inches long and three inches deep, with pieces of tin

plate four inches by five, fixed on at the
ends in a perpendicular position to prevent lateral disturbance at those parts,
very regular and beautiful ridges were
obtained of any desired width, fig. 28.

Fig. 28.

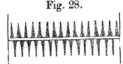

These ridges, as before, formed only on the wood, and were
parallel to the direction of its vibration. They occurred on
each side of the vibrating plane with equal regularity, force
and magnitude, but seemed to have no connexion, for sometimes they corresponded in position, and at other times not;
the one set shifting a little, without the others being displaced.

116. It could now be observed that the ridges on either side
the vibrating plane consisted of two alternating sets; the one
set rising as the other fell. For each fro and to motion of the
plane, or one complete vibration, one of the sets appeared, so
that in two complete vibrations the cycle of changes was complete. Pieces of cork and lycopodium powder showed that
there was no important current setting in the direction of the
ridges; towards the heads of the ridges pieces of cork oscillated from one ridge towards its neighbour, and back again.
The lycopodium sometimes seemed to move on the ridges from
the wood, and between them to it; but the motion was irregular, and there was no general current outwards or inwards.
There was not so much disturbance as amongst the heaps (90).

117. A very simple arrangement exhibits these ripples beautifully. If an oval or circular pan, fifteen or eighteen inches
in diameter, be filled with water, and a piece of lath (69) twelve
or fifteen inches long be held in it, edge upwards, so as to
bear against the sides of the pan as supporting points, and cut
the surface of the water, then on being vibrated horizontally
by the glass rod and wet finger, the phenomenon immediately
appears with ripples an inch or more in length. When the
upper edge of the lath was an inch below the surface, the ripples could be produced. When the vessel had a glass bottom,
the luminous figures produced by a light beneath and a screen
above, were very beautiful (96). Glass, metal and other plates
could thus be easily experimented with.

118. These ripple-like stationary undulations are perfectly
analogous as to cause, arrangement and action with the heaps
and crispations already explained, *i. e.* they are the results of

that vibrating motion in directions perpendicular to the force applied (105), by which the water can most readily accommodate itself to rapid, regular, and alternating changes in bulk in the immediate neighbourhood of the oscillating parts.

119. From this view of the effect it was evident that similar phenomena would be produced if a substance were made to vibrate in contact with and normally to the surface of a fluid, or indeed in any other direction. A lath was therefore fixed horizontally in a vice by one end, so that the other could vibrate vertically; a cork was cemented to the under surface of the free end, and a basin of water placed beneath with its surface just touching the cork; on vibrating the lath by means of the glass rod and fingers(67), a beautiful and regular star of ridges two, three, or even four inches n length, was formed round the cork, fig. 29. These ridges were more or less numerous according to the number of vibrations, &c. As the water was raised, and more of the cylinder immersed, the ridges diminished in strength, and at last disappeared: when the cylinder of cork just touched the surface, they were most powerfully developed. This is a necessary consequence of the dependence of the ridges upon the portion of water which is vertically displaced and restored at each vibration. When that, being partial in relation to the whole surface, is at or near the surface, the ridges are freely formed in the immediate vicinity; when at a greater depth (being always at the bottom of the cork), the displacement is diffused over a larger mass and surface, each particle moves through less space and with less velocity, and consequently the vibrations must be stronger or the ridges be weaker or disappear altogether. The refraction of a light through this star produces a very beautiful figure on a screen.

Fig. 29.

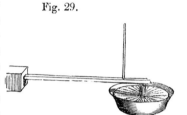

120. A heavy tuning-fork vibrating, but not too strongly, if placed with the end of one limb either vertical, inclined, or in any other position, just touching the surface of water, ink, milk, &c. (75), shows the effect very well for a moment. It also shows the ridges on mercury, but the motion and resistance of so dense a body quickly bring the fork to rest. It formed

ridges in hot oil, but not in cold oil (76). With cold oil a very inclined fork produced a curious pump-like action, throwing up four streams, easily explained when witnessed, but not so closely connected with the present phenomena as to require more notice here.

121. There is a well-known effect of crispation produced when a large glass full of water is made to sound by passing the wet finger round the edges. The glass divides into four vibrating parts opposite to which the crispations are strongest, and there are four nodal points considered in relation to a horizontal section, at equal distances from each other, the finger always touching at one of them. If the vessel is a large glass jar, and soft sounds are produced, the surface of the water exhibits the ridges at the centres of vibration; as the sound is rendered louder, these extend all round the glass, and at last break up at the centres of vibration into irregular crispations; but both the ridges and crispations are effects of the kind already described, and require no further explanation.

122. There are some other effects, one of which I wish here briefly to notice, as connected more or less with the vibratory phenomena that have been described. If, during a strong steady wind, a smooth flat sandy shore, with enough water on it, either from the receding tide or from the shingles above, to cover it thoroughly, but not to form waves, be observed in a place where the wind is not broken by pits or stones, stationary undulations will be seen over the whole of the wet surface, forming ridges like those already described, and each several inches long. These are not waves of the ordinary kind; they are accurately parallel to the course of the wind; they are of uniform width whatever the extent of surface, varying in width only as the force of the wind and the depth of the stratum of water vary. They may be seen at the windward side of the pools on the sand, but break up so soon as waves appear. If the waves be quelled by putting some oil on the water to windward, these ripples then appear on those parts. They are often seen (but so confused that their nature could not be gathered from such observations) on pavements, roads, and roofs when sudden gusts of wind occur with rain. The character of these ripples, and their identity with stationary undulations, may be ascertained by exerting the eye and the mind to resolve

them into two series of ordinary advancing waves moving directly across the course of the wind in opposite directions. But as such series could not be caused by the wind exerted in a manner similar to that by which ordinary waves are produced (the direction being entirely opposed to such an idea), I think the effect is due to the water acquiring an oscillatory condition similar to those described, probably influenced in some way by the elastic nature of the air itself (124) and analogous to the vibration of the strings of the Æolian harp, or even to the vibration of the columns of air in the organ-pipe and other instruments with embouchures.

These ridges were strong enough to arrange the sand beneath, where ordinary waves had not been powerful enough to give form to the surface.

123. All the phenomena as yet described are such as take place at the *surfaces* of those fluids in common language considered as inelastic, and in which the elasticity they possess performs no necessary part; nor is it possible that they could be produced within their mass. But on extending the reasoning, it does not seem at all improbable that analogous effects should take place in gases and vapour, their elasticity supplying that condition necessary for vibration which in liquids is found in an abrupt termination of the mass by an unconfined surface.

124. If this be so, then a plate vibrating in the atmosphere may have the air immediately in contact with it separated into numerous portions, forming two alternating sets like the heaps described (95); the one denser, and the other rarer than the ordinary atmosphere; these sets alternating with each other by their alternate expansion and condensation with each vibration of the plate.

125. With the hope of discovering some effect of this kind, a flat circular tin plate had a raised edge of tin three quarters of an inch high fixed on all round, and the plate was then attached to a lath (69), a little lycopodium put on to it, and vibrated powerfully, so that the powder should form a mere cloud in the air, which, in consequence of the raised edge and the equal velocity (70) of all parts of the plate, had no tendency to collect. It was seen immediately that in place of a uniform cloud a misty honeycomb appearance was produced, the whole being in a quivering condition; and on exerting the attention to

perceive waves travelling as it were across the cloud in opposite directions, they could be most distinctly traced. This is exactly the appearance that would be produced by a dusty atmosphere lying upon the surface of a plate and divided into a number of alternate portions rapidly expanding and contracting simultaneously.

126. The spaces were very many times too small to represent the interval through which the air by its elasticity would vibrate laterally once for two vibrations of the plate, in analogy with the phenomena of liquids; and this forms a strong objection to its being an effect of that kind. But it does not seem impossible that the air may have vibrated in subdivisions like a string or a long column of air; and the air itself also being laden with particles of lycopodium would have its motions rendered more sluggish thereby. I have not had time to extend these experiments, but it is probable that a few, well-chosen, would decide at once whether these appearances of the particles in the air are due to real lateral vibrations of the atmosphere, or merely to the direct action of the vibrating plate upon the particles.

127. If the atmosphere vibrates laterally in the manner supposed, the effect is probably not limited to the immediate vicinity of the plate, but extends to some distance. The vertical plates intersecting the surface of water and vibrating in a horizontal plane (117) produced ripples proceeding directly out from them five or six inches long; whilst the waves parallel to the vibrating plate were hardly sensible; and something analogous to this may take place in the atmosphere. If so, it would seem likely that these vibrations occurring conjointly with those producing sound, would have an important influence upon its production and qualities, upon its apparent direction, and many other of its phenomena.

128. Then by analogy these views extend to the undulatory theory of light, and especially to that theory as modified by M. Fresnel. That philosopher, in his profound investigations of the phenomena of light, especially when polarized, has conceived it necessary to admit that the vibrations of the ether take place transversely to the ray of light, or to the direction of the wave causing its phenomena. "In fact we may conceive direct light to be an assemblage, or rather a rapid succession,

of an infinity of systems of waves polarized (*i. e.* vibrating transversely) in all azimuths, and so, that there is as much polarized light in any one plane as in a plane perpendicular to it." Herschel says that Fresnel supposes the eye to be affected *only* by such vibrating motions of the etherial molecules as are performed in planes perpendicular to the direction of the rays. Now the effects in question seem to indicate how the direct vibration of the luminous body may communicate transversal vibration in every azimuth to the molecules of the ether, and so account for that condition of it which is required to explain the phenomena.

129. When the star of ridges formed by a vibrating cylinder (119) upon the surface of water is witnessed instead of the series of circular waves that might be expected, it seems like the instant production of the phenomena of radiation by means of vibratory action. Whether the contiguous rarefied and condensed portions which I have supposed in air, gases, vapour and the ether, are arranged radially like the ridges in the experiment just quoted, or whether rare and dense alternate in the direction of the radii as well as laterally, is a question which may perhaps deserve investigation by experiment or calculation.

Royal Institution, July 30th, 1831.

Notice of a Means of preparing the Organs of Respiration, so as considerably to extend the Time of holding the Breath; with Remarks on its Application in Cases in which it is required to enter an irrespirable Atmosphere, and on the Precautions necessary to be observed in such Cases*.

To the Editors of the Philosophical Magazine and Journal.

Gentlemen,—There are many facts which present themselves to observant men, which, though seen by them to be curious, interesting, and new to the world, are not considered worthy of distinct publication. I have often felt this conclusion to be objectionable, and am convinced that it is better to publish such facts, and even known facts under new forms, provided it be done briefly, clearly, and with no more pretension

* Philosophical Magazine, 1833, vol. iii. p. 241.

than the phenomena fairly deserve. It is this feeling which makes me send for your acceptance or rejection an account of an effect, new to me, and to all to whom I have mentioned it, and which seems to have some valuable applications.

At one of the scientific meetings at the apartments of His Royal Highness the President of the Royal Society, whilst speaking of certain men who, by means of peculiar apparatus for breathing, could walk about at the bottom of waters, and also of the pearl fishers, Sir Graves C. Haughton described to me an observation he had made, by the application of which a man could hold his breath about twice as long as under ordinary circumstances. It is as follows:—If a person inspire deeply, he will be able immediately after to hold breath for a time, varying with his health, and also very much with the state of exertion or repose in which he may be at the instant. A man, during an active walk, may not be able to cease from breathing for more than half a minute, who, after a period of rest on a chair or in bed, may refrain for a minute or a minute and a half, or even two minutes. But if that person will prepare himself by breathing in a manner deep, hard and quick (as he would naturally do after running), and, ceasing that operation with his lungs full of air, will then hold his breath as long as he is able, he will find that the time during which he can remain without breathing will be double, or even more than double the former, other circumstances being the same. I hope that I have here stated Sir Graves C. Haughton's communication to me correctly; at all events, whilst confirming his observation by personal experience, I found the results to be as above.

Whilst thus preparing myself, I always find that certain feelings come on, resembling in a slight degree those produced by breathing a small dose of nitrous oxide; slight dizziness and confusion in the head are at last produced ; but on ceasing to breathe, the feeling gradually goes off, no inconvenience results from it either at the time or afterwards, and I can hold my breath comfortably for a minute and a quarter, or a minute and a half, walking briskly about in the mean time.

Now this effect may be rendered exceedingly valuable. There are many occasions on which a person who can hold breath for a minute or two minutes, might save the life of another. If, in a brewer's fermenting vat, or an opened cesspool, one man sinks senseless and helpless, from breathing the unsuspected noxious atmosphere within, another man of cool mind would by means of this mode of preparation, which requires nothing but what is always at hand, have abundant time, in most cases, to descend by the ladder or the bucket, and rescue the sufferer without any risk on his own part. If a chamber were on fire, the difference in the help which could be given to any one within it by a person thus prepared, and another who goes in, perhaps, with lungs partially exhausted, and who, if he inhale any portion of the empyreumatic vapours of the atmosphere, is stimulated to inspire more rapidly, and is therefore urged to instant retreat into fresh air, is so great, that no one who has noticed what can be done in a minute or in two minutes of time can doubt the value of the preparation under such circumstances, even though from want of practice and from hurry and alarm it may be very imperfectly made. In cases of drowning, also, a diver may find his powers of giving aid wonderfully increased by taking advantage of Sir Graves Haughton's fact.

I have myself had occasion to go more than once or twice into places with atmospheres rendered bad by carbonic acid, sulphuretted hydrogen or combustion; and I feel how much I should have valued at such times the knowledge of the fact above stated. Hoping, therefore, that it may be useful, I will add one or two precautions to be borne in mind by those who desire to apply it.

Avoid all unnecessary action; for activity exhausts the air in the lungs of its vital principle more quickly, and charges it with bad matter. Go collectedly, coolly and quietly to the spot where help is required: do no more than is needful, leaving what can be done by those who are in a safe atmosphere (as the hauling up of a senseless body, for example,) for them to do.

Take the precautions usual in cases of danger in *addition* to the one now recommended. Thus, in a case of choke-damp, as in a brewer's vat, hold the head as high as may be; in a case of fire in a room, keep it as low down as possible.

If a rope is at hand, by all means let it be fastened to the person who is *giving* help, that he may be succoured if he should venture too far. It is astonishing how many deaths happen in succession in cesspools, and similar cases, for want of this precaution.

It is hardly needful to say, do not try to breathe the air of the place where help is required. Yet many persons fall in consequence of forgetting this precaution. If the temptation to breathe be at all given way to, the *necessity* increases, and the helper himself is greatly endangered. Resist the tendency, and retreat in time.

Be careful to commence giving aid with the lungs *full* of air, not *empty*. It may seem folly to urge this precaution, but I have found so many persons who, on trying the experiment on which the whole is based, have concluded the preparation by closing the mouth and nostrils *after an expiration*, that I am sure the precaution requires to be borne in mind.

I have thought it quite needless to refer to the manner in which the preparation enables a person to increase so considerably the time during which he may suspend the operation of breathing. It consists, of course, chiefly in laying up for the time, in the cells of the lungs, a store of that vital principle which is so essential to life. Those who are not aware of the state of the air in the lungs during ordinary respiration, and its great difference from that of the atmosphere, may obtain a clearer notion from the following experiment. Fill a pint or quart jar with water over the pneumatic trough, and with a piece of tube and a forced expiration throw the air from the lungs in their *ordinary state* into the jar; it will be found that a lighted taper put into that air will be immediately extinguished.

A very curious fact connected with the time of holding the breath was observed by Mr. Brunel, jun., and has, I think, never been published. After the river had broken into the tunnel at Rotherhithe, Mr. Brunel descended with a companion (Mr. Gravatt, I think,) in a diving-bell, to examine the place: at the depth of about 30 feet of water, the bell touched the bottom of the river, and was over the hole; covering it, but too large to pass into it. Mr. Brunel, after attaching a rope to himself, inspired deeply, and sunk, or was lowered through the water, in the hole, that he might feel the frames with his feet, and gain further knowledge, if possible, of the nature of the leak. He remained so long beneath without giving any signal, that his companion, alarmed, drew him up before he desired; and then it was found that either of them could remain about twice as long under water, going into it from the diving-bell at that depth, as they could under ordinary circumstances.

This was supposed to be accounted for, at the time, by the circumstance that at the depth of 30 feet the atmosphere was of double pressure, and that the lungs, therefore, held twice as much air as they could do under common circumstances. It is, however, quite evident that another advantageous circumstance must have occurred, and that the air in the lungs was also better in quality than it would have been at the surface of the river, as well as denser; for supposing the deterioration by breathing to continue the same for the same time, it is clear that every inspiration passed into the lungs twice as much pure air as would have entered under common circumstances: the injured air must, therefore, have been removed more rapidly, and the quality of that at any one time in the lungs must have risen in consequence. When to this is added the effect of double quantity, it fully accounts for the increased time of holding the breath; and had the effect of the mode of preparation now described been also added, it is probable that the time would have appeared astonishingly increased.

I am, Gentlemen, yours, &c.,

M. Faraday.

On the Ventilation of Lighthouse Lamps; the points necessary to be observed, and the manner in which these have been or may be attained*.

The author states that the fuel used in lighthouses for the production of light is almost universally oil, burnt in lamps of the Argand or Fresnel construction; and from the nature and use of the buildings, it very often happens that a large quantity of oil is burnt in a short time, in a small chamber exposed to low temperature from without, the principal walls of the chamber being only the glass through which the light shines; and that these chambers being in very exposed situations, it is essential that the air within should not be subject to winds or partial draughts, which might interfere with the steady burning of the lamps.

* Proceedings of the Institution of Civil Engineers, June 27, 1843, p. 206.

If the chamber or lantern be not perfectly ventilated, the substances produced by combustion are diffused through the air, so that in winter or damp weather the water condenses on the cold glass windows, which, if the light be a fixed one, greatly impairs its brilliancy and efficiency, or, if the light be a revolving one, tends to confound the bright and dark periods together. The extent to which this may go, may be conceived, when it is considered that some lighthouses burn as much as twenty, or more, pints of oil in one winter's night, in a space of 12 or 14 feet diameter, and from 8 to 10 feet high, and that each pint of oil produces more than a pint of water; or, from this fact, that the ice on the glass within, derived from this source, has been found in some instances an eighth, and even a sixth of an inch in thickness, and required to be scraped off with knives.

The carbonic acid makes the air unwholesome, but it is easily removed by any arrangement which carries off the water as vapour. One pound of oil in combustion produces about $1·06$ pound of water and $2·86$ pounds of carbonic acid.

The author's plan is to ventilate the lamps themselves by fit flues, and then the air inside the lantern will always be as pure as the external air, yet with closed doors and windows, a calm lantern, and a bright glass.

In lighthouses there are certain conditions to which the ventilating arrangement must itself submit, and if these are not conformed with, the plan would be discarded, however perfect its own particular effect might be. These conditions are chiefly, that it should not alter the burning of the oil, or charring of the wicks,—that it should not interfere with the cleaning, trimming, and practice of the lamps and reflectors,—that it should not obstruct the light from the reflectors,—that it should not, in any sudden gust or tempest, cause a downward blast or impulse on the flame of the lamp,—that, if thrown out of action suddenly, it should not alter the burning; and, added to these, that it should perform its own ventilating functions perfectly.

Lighthouses have either one large central lamp, the outer wick of which is sometimes $3\frac{3}{4}$ inches in diameter, or many single Argand burners, each with its own parabolic reflector. The former is a fixed lamp; the latter are frequently in motion. The former requires the simplest ventilating system, which may be thus described :—

The ventilating pipe or chimney is a copper tube, 4 inches in diameter, not, however, in one length, but divided into three or four pieces : the lower end of each of these pieces, for about $1\frac{1}{2}$ inch, is opened out into a conical form about $5\frac{1}{2}$ inches in diameter at the lowest part. When the chimney is put together, the upper end of the bottom piece is inserted about $\frac{1}{2}$ an inch into the cone of the next piece above, and fixed there by three ties or pins, so that the two pieces are firmly held together ; but there is still plenty of air-way, from the surrounding atmosphere into the chimney between them. The same arrangement holds good with each succeeding piece. When the ventilating chimney is fixed in its place, it is adjusted, so that the lamp-chimney enters about $\frac{1}{2}$ an inch into the lower cone, and the top of the ventilating chimney enters into the cowl or head of the lantern.

With this arrangement, it is found that the action of the ventilating flue is to carry up every portion of the products of combustion into the cowl; none passes out of the flue into the air of the lantern by the cone apertures, but a portion of the air passes from the lantern by these into the flue, and so the lantern itself is in some degree ventilated.

The important use of these cone apertures is, that when a sudden gust, or eddy of wind, strikes into the cowl of the lantern, it does not have any effect in disturbing or altering the flame. It is found that the wind may blow suddenly in at the cowl, yet the effect never reach the lamp. The upper, or the second, or the third, or even the fourth portion of the ventilating flue might be entirely closed, yet without altering the flame. The cone junctions in no way interfere with the tube in carrying up all the products of combustion; but if any downward current occurs, they dispose of the whole of it into the room, without ever affecting the flame. The ventilating flue is, in fact, a tube which, as regards the lamp, can carry everything *up*, but conveys nothing *down*.

In lighthouses with many separate lamps and reflectors, the case is more difficult and the arrangement more complicated, yet the conditions before referred to are more imperatively called for, because any departure from them was found to have greater influence in producing harm. The object has been attained thus :—A system of gathering pipes has been applied to the lamps which may be considered as having the different beginnings at each lamp, and being fixed to the frame which supports the lamps, is made to converge together and to the axis of the frame by curved lines. The object is to bring the tubes together behind the reflectors, as soon as convenient, joining two or more into one, like a system of veins, so that one ventilating flue may at last carry off the whole of the lamp products. It is found that a pipe $\frac{7}{8}$ths of an inch in diameter is large enough for one lamp; and where, by junction, two or more pipes have become one, if the one pipe has a sectional area, proportionate to the number of lamps which it governs, the desired effect is obtained.

Each of the pipes, $\frac{7}{8}$ths of an inch in diameter, passes downwards through the aperture in the reflector over the lamp, and dips an inch into the lamp-glasses; it is able to gather and carry off all the products of combustion, though, perhaps, still 2 inches from the top of the flame, and therefore not interfering in any respect with it, nor coming as a shade between it and any part of the reflector: the flame and reflector are as free in their relation to each other as they were before. Neither does this tube hide from the observer or mariner, a part of the reflector larger than about $1\frac{1}{2}$ square inch of surface, and it allows of a compensation to two or three times the amount; for, when 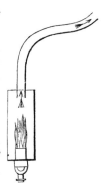 in its place, all the rest of the aperture over the lamp which is left open and inefficient in the ordinary service, may be made effectual reflecting surface, simply by filling it up with a loose, fitly formed, reflecting plate.

At this termination of the ventilating flue an important adjustment is effected. If the tube dip about an inch into the lamp-glass, the draught up it is such that not only do all the products of combustion enter the tube, but air passes down between the top edge of the lamp-glass and the tube, going, finally, up the latter with the smoke. In this case, however, an evil is produced, for the wick is charred too rapidly; but if the ventilating flue descends until only level with the top of the lamp-glass, the whole of the burnt air does not usually go up it, but some passes out into the chamber, and at such times the charring of the wick is not hastened. Here, therefore, there is an adjusting power, and it was found by the trials made, that when the tube dipped about $\frac{1}{2}$ an inch into the lamp-glass, it left the burning of the lamp unaltered, and yet carried off all the products of combustion.

The power already referred to, of dividing a chimney into separate and independent parts, and yet enabling it to act perfectly as a whole, as shown in the single central chimney, was easily applicable in the case of several lamps, and gave a double advantage; for it not only protected the lamps from any influence of down draught, but it easily admitted of the rotation of the system of gathering flues, fixed to the frame sustaining the lamps and reflectors in a revolving lighthouse, and of the delivery of the burnt air, &c., from its upper extremity into the upper immoveable portion of the flue. This capability in a revolving light is essential, for in all, the support of the frame-work is of such a nature, as to require that the upper part of the flue should be a fixture.

The author explains, that it is as an officer of the Trinity House, and under its instructions, that he entered into the consideration of this subject; that, as to the central chimney, its action has been both proved and approved, and that all the central lights are ordered to be furnished with them; that as respects the application to separate and revolving lamps, the experiment has been made under the direction of the Trinity House on a face of six lamps, being a full-sized copy of the Tynemouth revolving light, and, so far to the satisfaction of the Deputy Master and Brethren, that the plan is to be applied immediately to two lighthouses which suffer most from condensation on the glass: he believes it will be with full success.

Thoughts on Ray-vibrations.*
To Richard Phillips, Esq.

DEAR SIR,—At your request I will endeavour to convey to you a notion of that which I ventured to say at the close of the last Friday-evening Meeting, incidental to the account I gave of Wheatstone's electro-magnetic chronoscope; but from first to

* Philosophical Magazine, 1846, vol. xxviii. p. 345.

last understand that I merely threw out as matter for speculation, the vague impressions of my mind, for I gave nothing as
the result of sufficient consideration, or as the settled conviction,
or even probable conclusion at which I had arrived.

The point intended to be set forth for the consideration of
the hearers was, whether it was not possible that the vibrations
which in a certain theory are assumed to account for radiation
and radiant phenomena may not occur in the lines of force
which connect particles and consequently masses of matter
together; a notion which, as far as it is admitted, will dispense
with the ether which, in another view, is supposed to be the
medium in which these vibrations take place.

You are aware of the speculation* which I some time since
uttered respecting that view of the nature of matter which
considers its ultimate atoms as centres of force, and not as so
many little bodies surrounded by forces, the bodies being
considered in the abstract as independent of the forces and capable of existing without them.   In the latter view, these little
particles have a definite form and a certain limited size; in the
former view such is not the case, for that which represents size
may be considered as extending to any distance to which the
lines of force of the particle extend : the particle indeed is supposed to exist only by these forces, and where they are it is.
The consideration of matter under this view gradually led me
to look at the lines of force as being perhaps the seat of the
vibrations of radiant phenomena.

Another consideration bearing conjointly on the hypothetical
view both of matter and radiation, arises from the comparison
of the velocities with which the radiant action and certain powers
of matter are transmitted. The velocity of light through space
is about 190,000 miles in a second; the velocity of electricity
is, by the experiments of Wheatstone, shown to be as great as
this, if not greater: the light is supposed to be transmitted by
vibrations through an ether which is, so to speak, destitute of
gravitation, but infinite in elasticity ; the electricity is transmitted
through a small metallic wire, and is often viewed as transmitted
by vibrations also.   That the electric transference depends on
the forces or powers of the matter of the wire can hardly be
doubted, when we consider the different conductibility of the

* Philosophical Magazine, 1844, vol. xxiv. p. 136—or Experimental Researches in Electricity, vol. ii. p. 284.

various metallic and other bodies; the means of affecting it by heat or cold; the way in which conducting bodies by combination enter into the constitution of non-conducting substances, and the contrary; and the actual existence of one elementary body, carbon, both in the conducting and non-conducting state. The power of electric conduction (being a transmission of force equal in velocity to that of light) appears to be tied up in and dependent upon the properties of the matter, and is, as it were, existent in them.

I suppose we may compare together the matter of the ether and ordinary matter (as, for instance, the copper of the wire through which the electricity is conducted), and consider them as alike in their essential constitution; *i. e.* either as both composed of little nuclei, considered in the abstract as matter, and of force or power associated with these nuclei, or else both consisting of mere centres of force, according to Boscovich's theory and the view put forth in my speculation; for there is no reason to assume that the nuclei are more requisite in the one case than in the other. It is true that the copper gravitates and the ether does not, and that therefore the copper is ponderable and the ether is not; but that cannot indicate the presence of nuclei in the copper more than in the ether, for of all the powers of matter gravitation is the one in which the force extends to the greatest possible distance from the supposed nucleus, being infinite in relation to the size of the latter, and reducing that nucleus to a mere centre of force. The smallest atom of matter on the earth acts directly on the smallest atom of matter in the sun, though they are 95,000,000 of miles apart; further, atoms, which, to our knowledge, are at least nineteen times that distance, and indeed, in cometary masses, far more, are in a similar way tied together by the lines of force extending from and belonging to each. What is there in the condition of the particles of the supposed ether, if there be even only *one* such particle between us and the sun, that can in subtilty and extent compare to this?

Let us not be confused by the *ponderability* and *gravitation* of heavy matter, as if they proved the presence of the abstract nuclei; these are due not to the nuclei, if they exist at all, but to the force superadded to them; and, if the *ether* particles be without this force, which according to the assumption is the case, then they are more material, in the abstract sense, than the matter of this our globe; for matter, according to the assumption, being made up of nuclei and force, the ether particles have in this respect proportionately more of the nucleus and less of the force.

On the other hand, the infinite elasticity assumed as belonging to the particles of the ether, is as striking and positive a force of it as gravity is of ponderable particles, and produces in its way effects as great; in witness whereof we have all the varieties of radiant agency as exhibited in luminous, calorific, and actinic phenomena.

Perhaps I am in error in thinking the idea generally formed of the ether is that its nuclei are almost infinitely small, and that such force as it has, namely its elasticity, is almost infinitely intense. But if such be the received notion, what then is left in the ether but force or centres of force? As gravitation and solidity do not belong to it, perhaps many may admit this conclusion; but what is gravitation and solidity? certainly not the weight and contact of the abstract nuclei. The one is the consequence of an *attractive* force, which can act at distances as great as the mind of man can estimate or conceive; and the other is the consequence of a *repulsive* force, which forbids for ever the contact or touch of any two nuclei; so that these powers or properties should not in any degree lead those persons who conceive of the ether as a thing consisting of force only, to think any otherways of ponderable matter, except that it has more, and other, *forces* associated with it than the ether has.

In experimental philosophy, we can, by the phenomena presented, recognize various kinds of lines of force; thus there are the lines of gravitating force, those of electro-static induction, those of magnetic action, and others partaking of a dynamic character might be perhaps included. The lines of electric and magnetic action are by many considered as exerted through space like the lines of gravitating force. For my own part, I incline to believe that when there are intervening particles of matter (being themselves only centres of force), they take part in carrying on the force through the line, but that when there are none, the line proceeds through space*. Whatever the

* Experimental Researches in Electricity, pars. 1161, 1613, 1663, 1710, 1729, 1735, 2443.

view adopted respecting them may be, we can, at all events, affect these lines of force in a manner which may be conceived as partaking of the nature of a shake or lateral vibration. For suppose two bodies, A B, distant from each other and under mutual action, and therefore connected by lines of force, and let us fix our attention upon one resultant of force having an invariable direction as regards space ; if one of the bodies move in the least degree right or left, or if its power be shifted for a moment within the mass (neither of these cases being difficult to realize if A and B be either electric or magnetic bodies), then an effect equivalent to a lateral disturbance will take place in the resultant upon which we are fixing our attention ; for, either it will increase in force whilst the neighbouring resultants are diminishing, or it will fall in force as they are increasing.

It may be asked, what lines of force are there in nature, which are fitted to convey such an action, and supply for the vibrating theory the place of the ether ? I do not pretend to answer this question with any confidence ; all I can say is, that I do not perceive in any part of space, whether (to use the common phrase) vacant or filled with matter, anything but forces and the lines in which they are exerted. The lines of weight or gravitating force are, certainly, extensive enough to answer in this respect any demand made upon them by radiant phenomena ; and so, probably, are the lines of magnetic force : and then, who can forget that Mossotti has shown that gravitation, aggregation, electric force, and electro-chemical action may all have one common connexion or origin ; and so, in their actions at a distance, may have in common that infinite scope which some of these actions are known to possess ?

The view which I am so bold as to put forth considers, therefore, radiation as a high species of vibration in the lines of force which are known to connect particles and also masses of matter together. It endeavours to dismiss the ether, but not the vibrations. The kind of vibration which, I believe, can alone account for the wonderful, varied, and beautiful phenomena of polarization, is not the same as that which occurs on the surface of disturbed water, or the waves of sound in gases or liquids, for the vibrations in these cases are direct, or to and from the centre of action, whereas the former are lateral. It seems to me, that the resultant of two or more lines of force

is in an apt condition for that action which may be considered as
equivalent to a *lateral* vibration; whereas a uniform medium,
like the ether, does not appear apt, or more apt than air or water.

The occurrence of a change at one end of a line of force easily
suggests a consequent change at the other. The propagation
of light, and therefore probably of all radiant action, occupies
time; and, that a vibration of the line of force should account
for the phenomena of radiation, it is necessary that such vibration should occupy time also. I am not aware whether there
are any data by which it has been, or could be ascertained,
whether such a power as gravitation acts without occupying
time, or whether lines of force being already in existence, such
a lateral disturbance of them at one end as I have suggested
above, would require time, or must of necessity be felt instantly
at the other end.

As to that condition of the lines of force which represents
the assumed high elasticity of the ether, it cannot in this respect
be deficient: the question here seems rather to be, whether
the lines are sluggish enough in their action to render them
equivalent to the ether in respect of the time known experimentally to be occupied in the transmission of radiant force.

The ether is assumed as pervading all bodies as well as space:
in the view now set forth, it is the forces of the atomic centres
which pervade (and make) all bodies, and also penetrate all
space. As regards space, the difference is, that the ether
presents successive parts or centres of action, and the present
supposition only lines of action; as regards matter, the difference
is, that the ether lies between the particles, and so carries on
the vibrations; whilst as respects the supposition, it is by the
lines of force between the centres of the particles that the
vibration is continued. As to the difference in intensity of
action within matter under the two views, I suppose it will be
very difficult to draw any conclusion; for when we take the
simplest state of common matter and that which most nearly
causes it to approximate to the condition of the ether, namely
the state of rare gas, how soon do we find in its elasticity and
the mutual repulsion of its particles, a departure from the law,
that the action is inversely as the square of the distance!

And now, my dear Phillips, I must conclude. I do not think
I should have allowed these notions to have escaped from me,

had I not been led unawares, and without previous consideration, by the circumstances of the evening on which I had to appear suddenly and occupy the place of another. Now that I have put them on paper, I feel that I ought to have kept them much longer for study, consideration, and, perhaps, final rejection; and it is only because they are sure to go abroad in one way or another, in consequence of their utterance on that evening, that I give them a shape, if shape it may be called, in this reply to your inquiry. One thing is certain, that any hypothetical view of radiation which is likely to be received or retained as satisfactory, must not much longer comprehend alone certain phenomena of light, but must include those of heat and of actinic influence also, and even the conjoined phenomena of sensible heat and chemical power produced by them. In this respect, a view, which is in some degree founded upon the ordinary forces of matter, may perhaps find a little consideration amongst the other views that will probably arise. I think it likely that I have made many mistakes in the preceding pages, for even to myself, my ideas on this point appear only as the shadow of a speculation, or as one of those impressions on the mind which are allowable for a time as guides to thought and research. He who labours in experimental inquiries knows how numerous these are, and how often their apparent fitness and beauty vanish before the progress and development of real natural truth.

I am, my dear Phillips, ever truly yours,

Royal Institution, April 15, 1846.   M. Faraday.

---

*On Certain conditions of Freezing Water. A Discourse, &c.**
[Royal Institution, Friday Evening, June 7, 1850.]

THE chief object of the discourse was the great, various, and extraordinary forms of affinity which exist between the particles of water. Having experimentally illustrated the combining power of water, and shown how this attraction passes from a physical to a chemical force, Mr. Faraday confined the rest of his discourse to ice, as being that condition of water in which its particles are allowed to associate with each other without the intervention of foreign matter. Such ice as is now imported

* Athenæum, 1850, p. 640. The report is by the author.

into this country under the name of the Wenham Lake ice (though it is chiefly supplied from Norway) may be regarded as one of the purest natural substances. Mr. Faraday first showed how entirely colouring matter, salts and alkalies are expelled in freezing[*]. A solution of sulphate of indigo, diluted sulphuric acid, and diluted ammonia were partially frozen in glass test-tubes: as soon as the operation had been carried on long enough to produce an icy lining of each tube, the unfrozen liquid was poured out and the ice dislodged. This ice was found in every instance perfectly colourless, and, when dissolved, perfectly free from acid or alkali, although the unfrozen liquid exhibited in the first experiment a more intense blue colour, in the second a stronger acid, and in the third a more powerful alkaline reaction than the liquor which was put into the freezing mixture. Mr. Faraday also devised a method for making this ice perfectly clear and transparent as well as colourless. By continually stirring the liquid, while freezing, with a feather, he brushed away globules of air as fast as they were dislodged from the freezing fluid, and thus prevented their becoming imbedded in the ice. Having noticed the rapidity with which water absorbs air as soon as it is thawed, Mr. Faraday called attention to the importance of this natural arrangement to aquatic plants and animals, to whose life air is as indispensable as to those which live on land. Mr. Faraday then referred to Mr. Donny's discovery, that water, when deprived of air, does not boil till it reaches the temperature of 270°, and that at that degree of heat it explodes. He mentioned that he suggested to Mr. Donny that ice when placed in oil (so as to prevent its receiving any air from the atmosphere on thawing) would probably explode on reaching a sufficient temperature. This experiment had been successfully tried by Mr. Donny, and was as successfully repeated on this occasion. Mr. Faraday then invited attention to the extraordinary property of ice in solidifying water which is in contact with it. Two pieces of moist ice will consolidate into one. Hence the property of damp snow to become compacted into a snowball—an effect which cannot be produced on dry, hard-frozen snow. Mr. Faraday suggested, and illustrated by a diagram, that a film of water must possess the property of freezing when placed between two

[*] See ice from solution of chlorine, p. 82.

sets of icy particles, though it will not be affected by a single set of particles.   Certain solid substances, as flannel, will also freeze to an icy surface, though other substances, as gold-leaf, cannot be made to do so.   In this freezing action latent heat becomes sensible heat; the contiguous particles must therefore be raised in temperature while the freezing water is between them.   It follows from hence, that, by virtue of the solidifying power at points of contact, the same mass may be freezing and thawing at the same moment, and even that the freezing process in the inside may be a thawing process on the outside.   Mr. Faraday then referred to Mr. Thomson's memoirs on the effect of pressure on the freezing-point.   Mr. Thomson has shown that immense pressure will prevent water from freezing at $32°$— ice naturally occupies a greater volume than that of the water which forms it; and we may conceive that when ice is pressed the tendency is to give it both the water bulk and state.

In conclusion, Mr. Faraday noticed briefly, and chiefly by way of suggestion, the molecular condition of ice as presenting many curious results, and called attention to the strangeness of striæ being found in a body of such uniform composition as pure water frozen into ice.

On Ice of Irregular Fusibility*.

MY DEAR TYNDALL,

HAVE the following remarks, made in reference to the irregular fusibility of ice, to which you drew my attention, any interest to you, or by an occasional bearing on such cases, any value in themselves?   Deal with them as you like.

Imagine a portion of the water of a lake about to freeze, the surface S being in contact with an atmosphere considerably below $32°$, the previous action of which has been to lower the temperature of the whole mass of water, so that the portion below the line M is at

$32°$ .... S ....
$33°$ b ————————
$34°$ c ————————
$35°$ d ————————
$36°$ e ————————
$37°$ f ————————
$38°$ g ————————
$39°$ h ————————
$40°$ i ....M...........
$40°$ ————————
$40°$ ————————

* From Tyndall's paper " On some Physical Properties of Ice," Phil. Trans. 1858, p. 228.

40°, or the maximum density, and the part above at progressive temperatures from 40° upwards to 32°; each stratum keeping its place by its relative specific gravity to the rest, and having therefore, in that respect, no tendency to form currents either upwards or downwards. Now generally, if the surface became ice, the water below would go on freezing by the cold conducted downwards through the ice; but the successive series of temperatures from 32° to 40° would always exist in a layer of water contained between the ice and the dense water at 40° below M. If the water were *pure*, no action of the cold would tend to change the places of the particles of the water or cause currents; because the lower the cold descended, the more firmly would any given particle tend to retain its place above those beneath it: a particle at *e*, for instance, at 36° Fahr., would, when the cold had frozen what was above it, be cooled sooner and more than any of the particles beneath, and so always retain its upper place as respects them.

But now, suppose the water to contain a trace of saline matter in solution. As the water at 32° froze, either at the surface or against the bottom of the previously-formed ice, these salts would be expelled; for the ice first formed (and that *always* formed, if the proper care be taken to displace the excluded salts) is perfectly free from them, and PURE. The salts so excluded would pass into the layer of water beneath, and there produce two effects: they would make that layer of greater specific gravity than before, and so give it a tendency to sink into the warmer under layer; but they would also make it require a lower temperature than 32° for congelation; this it would acquire from the cold ice above, and by that it would become lighter and float, tending to remain uppermost; for it has already been shown that the diminution of temperature below 32° in sea water and solution of salts, is accompanied by the same enlargement of bulk as between 32° and 40° with pure water. The stratum of water, therefore, below the ice, would not of necessity sink because it contained a little more salt than the stratum immediately below it; and *certainly would not* if the increase of gravity conferred by the salts was less than the decrease by lowering of temperature. An approximation of the strata between the freezing place and the layer at 40° would occur, *i. e.* the distance between these temperatures would be less, but the water particles would keep their respective places.

When water freezes, it does not appear that this process is continuous, for many of the characters of the ice seem to show that it is intermittent; *i. e.* either a film of ice is formed, and then the process stops until the heat evolved by solidification has been conducted away upwards, and the next stratum of water has been sufficiently cooled to freeze in turn; or else the freezing being, so to speak, continuous, still is not continued at the same constant rate, but, as it were, by intermittent pulsations. Now it may well be, when a layer next the previously-formed ice, and containing an undue proportion of salts, has been cooled down to its required temperature for freezing (which would be below 32°), that on freezing, the congelation will pervade at once a certain thickness of the water, excluding the salts from the larger portion of ice formed, but including them as a weak solution within its interstices. The next increment of cold conducted from the ice above would freeze up these salts in the ice containing them, at the same time that a layer of pure ice was formed beneath it. Thus a layer of ice fusible at a lower temperature than the ice either above or below it might be produced; and by a repetition of the process many such layers might be formed.

It does not follow necessarily that the layers would be perfectly exact in their disposition. Very slight circumstances tending to disturb the regularity of the water-molecules would be sufficient, probably, to disturb the layers more or less. Ice contains *no* air, and the exclusion of a minute bubble of air from the water in the act of freezing might disturb the direction and progress of the congelation, and cause accumulation of the extra saline liquid in one spot rather than another. So might the tendency to the formation of little currents, either arising from the separation of the saline water from the forming ice, or from the elevation of temperature in different degrees at those places where the congelation was going on at different rates.

The effect would not depend upon the quantity of salts contained in the freezing water, though its degree would. The proportion of salts necessary to be added to pure water to lower its freezing-point 1° Fahr. may be very sensible to chemical tests, but the proportion required to make the difference $\frac{1}{100}$th or $\frac{1}{1000}$th of a degree would be far less : and if we suppose that only $\frac{1}{20}$th of a piece of ice is brought into the condition of melting before the rest of the mass, and that the salts in that proportion were originally in the *whole* of the water, then its quantity there may be so small as to escape detection except by very careful analysis.   However, it would be desirable to examine the water chemically which is produced by ice distinguished by having in its interior much, that liquefies before the rest.

It is easy to make ice perfectly free from air, and, as I believe, from salts, by a process I formerly described[*].   It would be interesting to see if such ice had within it portions melting at a lower temperature than the general mass.   I think it ought not.

Ever truly yours,

Royal Institution, Dec. 9, 1857.   M. FARADAY.

On Regelation.

THE subject of regelation has of late years acquired very great interest through the experimental investigations of Tyndall, J. Thomson, Forbes and others, and in its present state will perhaps justify a few additional remarks on my part as to the cause.   On the first observation of the effect eight years ago, I attributed it to the greater tendency which a particle of fluid water had to assume the solid state, when in contact with ice on two or more sides, above that it had when in contact on one side only[†].   Since then Mr. Thomson has shown that pressure lowers the freezing-point of water[‡], and has pointed out how such an effect occurring at the places where two masses of ice press against each other, may lead first to fusion and then union of the ice at those places; and so he explains the fact of regelation.   Prof. J. D. Forbes[§] does not think that pressure causes regelation in this manner, though it favours it by moulding the touching surfaces to each other.   He admits Person's view of the gradual liquefaction of ice[||], and assumes that ice must be essentially colder than ice-cold water, *i. e.* the water in contact with it.

[*] P. 373.   [†] Pp. 373, 374.
[‡] Belfast Society Proceedings, December 2, 1857.
[§] Royal Society Edinburgh Proceedings, April 19, 1858.
[||] *Comptes Rendus*, 1850, xxx. 526.

I find no difficulty in thinking it would be easy to arrange a mixture of water and snow in such a manner that it might be kept for hours and days without any transition of heat either to or from it; but I find great difficulty in thinking that the particles of snow, small as they may be made, would remain for the whole of the time at a lower temperature by $0°·3$ F. than the particles of water intermingled with them:—still admitting for the present the possibility that Prof. Forbes's view may be correct, and also the truthfulness of Mr. Thomson's principle, and its possible action in regelation, I wish to say a few words on the other principle already referred to, which was originally assumed by myself, which in relation with the mechanical theory of heat, has been adopted by Dr. Tyndall, and which, after all, may be the sole cause of the effect.

The principle I have in view being more distinctly expressed is this:—In all uniform bodies possessing cohesion, *i. e.* being in either the solid or the liquid state, particles which are surrounded by other particles having the like state with themselves tend to preserve that state, even though subject to variations of temperature, either of elevation or depression, which, if the particles were not so surrounded, would cause them instantly to change their condition. As water is the substance in which regelation occurs, I will illustrate the principle by the phenomena which it presents. Water may be cooled many degrees below $32°$ Fahr.* and yet retain its liquid state, for as far as we know any length of time, without solidification; yet, introduce a piece of the same chemical substance, ice, at a higher temperature, and the cold water freezes and becomes warm. It is certainly not the change of temperature which causes the freezing, for the ice introduced is warmer than the water. I assume that it is the difference in the condition of cohesion existing on the different sides of the changing particles which sets them free and causes the change. The cold water particles would willingly, as to temperature, have solidified without the ice, but were held fluid by the cohesion with them of other like fluid particles on all sides.

* Water may be cooled to $22°$ F. It is probable that if it were perfectly freed from air it would remain fluid at a much lower temperature, for the air is excluded at the freezing-point, and the occurrence of this exclusion would break cohesion.

In the other direction, Donny's experiments have taught us that the cohesion amongst the particles of water is so great that it will support a column of the fluid four or more feet high when there is no other power to sustain it; or will cause it to resist conversion into the state of vapour at temperatures so much higher than its ordinary boiling- or condensing-point, that explosion will occur when the continuity, and therefore the cohesion, is destroyed. The water may be exalted to the temperature of 270° Fahr. at the ordinary pressure of the atmosphere, and remain as water; but the introduction of the smallest particle of air or steam will cause it at once to burst into vapour, and at the same time its temperature falls.

This ability which water has to retain by cohesion its liquid state, refusing to solidify when below the freezing-point, or to become vapour when above the boiling-point, it has in common with many other substances. Acetic acid, sulphur, phosphorus, many metals, many solutions, may be cooled below the congealing temperature prior to the solidification of the first portions; many other substances, such as alcohol, sulphuric acid, ether, camphine, &c., boil with bumping, or boil with different degrees of facility in vessels of different substances*. The conclusion, that these differences are due to a certain range of cohesion in the case of each body, seems to me both simple and natural; this cohesion enabling the substances to withstand a change of temperature, which, without the cohesion, ought to have caused a change of state. The effect of extraneous matters as nuclei also appears to me to be simple; for though when introduced, as into cooled or heated water, their particles may exert a cohesive force (so to say) upon the particles of the fluid, the force so exerted in the first instance is rarely equal to the force exerted between the water particles themselves. Extraneous substances require preparation before their adhesion to fluid is at a maximum: glass will permit water to boil in contact with it at 212°, or by preparation will remain in contact with it at 270° Fahr., as in Donny's experiment. It will also remain in contact with water at 22° Fahr. without causing its solidification, and yet an ordinary piece of glass will set it off at once.

Enough has been said, I think, to show that water particles

* Marcet.

surrounded by water tend to retain their fluid state in both directions at temperatures which are abundantly sufficient to make it equally retain the solid or the vaporous state when either of them is conferred upon it.   There is nothing against the assumption that ice has the like kind of power, *i. e.* the power of retaining its solid state at temperatures higher than the temperature of ice against water.   Nevertheless, the fact is more difficult to show ; still some experiments may be quoted in favour of the view.   If hydrated crystals of sulphate of soda, carbonate of soda, phosphate of soda, &c.*, be carefully prepared in clean basins, by spontaneous evaporation of the water they will retain their form unbroken, and their hydrated state undisturbed, through the high temperatures of a whole summer ; though if broken or scratched even in winter, they will commence to effloresce at the place where the cohesion, and with it the balance of force was disturbed, and will from thence change progressively throughout the whole mass †. As regelation concerns the condition of water, there is perhaps no occasion to go further.   Such facts as the following, however, concern the extension of the principle and illustrate the power of cohesion, especially in cases where it is coming into activity.   Camphor in bottles, or iodide of cyanogen in proper glass vessels produce crystals sometimes an inch or two in length, which grow by the deposition of solid matter on them from an atmosphere unable to deposit like solid matter upon the surrounding glass, except at a lower temperature.   Crystals in solution grow by the deposition of solid matter on them which does not deposit elsewhere in the solution :—many such like cases may be produced.

Returning to the particular case of regelation, it is seen that water can remain fluid at temperatures below that at which ice forms, by virtue of the cohesion of its particles, and in so far the change is rendered independent of a given temperature. Next, I rest on the fact that ice has the same property as camphor, sulphur, phosphorus, metals, &c., which cause the

---

* Philosophical Transactions, 1834, p. 74 ; or Exp. Res. Electricity, vol. i. p. 191, note.

† Such a case shows combined solid water at a temperature ready to separate and change into vapour, yet not changing, because, as far as we can see, the undisturbed cohesion holds all together.

deposition of solid particles upon them from the surrounding fluid, that would not have been so deposited without the presence of the previous solid portions; a fact sufficiently proved by the growth of fine crystals of ice in ice-cold water. This effect was admirably shown in Mr. Harrison's freezing apparatus, where beautiful thin crystals of ice, six, eight, and ten inches long, would form in the surrounding fluid; and these crystals, which could not be colder than the surrounding fluid, exhibited the phenomena of regelation when purposely brought in contact with each other.

The next point may be considered as an assumption: it is that many particles in a given state exert a greater sum of their peculiar cohesive force upon a given particle of the like substance in another state than few can do; and that as a consequence a water particle with ice on one side and water on the other, is not so apt to become solid as with ice on both sides; also that a particle of ice at the surface of a mass in water is not so apt to remain ice as when, being within the mass, there is ice on all sides, temperature remaining the same. If that be admitted, then regelation is sufficiently accounted for. Difference of temperature above or below that of the changing points of water is not alone sufficient to cause change of state, the change being independent of temperature throughout a large range. At such times the particles appear to be governed by cohesion. Cohesion resolves itself into the force exerted on one particle by its neighbours, and this force seems to me to be sufficient, under the circumstances, to account for regelation.

Supposing this to be the true view of the state of things, then a particle of ice within ice can exist at a temperature higher than a like particle of ice on its surface in contact with water; and though it does not appear at present how a higher temperature could be communicated to the interior of a mass of freezing ice than that existing over its surface, still there may be principles of action in radiation, and even in conduction and liquefaction, producing that effect. Assuming, however, that a piece of freezing ice is in such a state, then, if it were to be pulverized, it ought to produce a mixed mass of ice and water colder than the ice was before. Such seems to be the result in one of Prof. Forbes's experiments, in which ice rapidly pounded showed a temperature of $0°·3$ Fahr. below the temperature of snow in a thawing state. The experiment, however, would require much consideration in every point of view, and much care before it could be considered as telling anything beyond the temperature of ice-cold water.

On the other hand, if a spherical cup of ice could be prepared containing water within, to which no heat could pass except by conduction through the ice itself, that water ought to be a little colder than the ice cup around it:—also if a mixture of snow and water were pressed together, the temperature should rise whenever regelation occurred, being an effect in the contrary direction to that which Prof. J. Thomson contemplates; and such a mixture, as a whole, ought to be warmer than the water in the ice sphere mentioned above. No doubt nice experiment will hereafter enable us to criticise such results as these, and separating the true from the untrue, will establish the correct theory of regelation.

September, 1858.

On Table-turning*.

To the Editor of the Times.

Sir,—I have recently been engaged in the investigation of table-turning. I should be sorry that you should suppose I thought this necessary on my own account, for my conclusion respecting its nature was soon arrived at, and is not changed; but I have been so often misquoted, and applications to me for an opinion are so numerous, that I hoped, if I enabled myself by experiment to give a strong one, you would consent to convey it to all persons interested in the matter. The effect produced by table-turners has been referred to electricity, to magnetism, to attraction, to some unknown or hitherto unrecognized physical power able to affect inanimate bodies—to the revolution of the earth, and even to diabolical or supernatural agency. The natural philosopher can investigate all these supposed causes but the last; that must, to him, be too much connected with credulity or superstition to require any attention on his part. The investigation would be too long in description to obtain a place in your columns. I therefore purpose asking admission

* Times, June 30, 1853.

for that into the 'Athenæum' of next Saturday, and propose here to give the general result.   Believing that the first cause assigned—namely, a quasi involuntary muscular action (for the effect is with many subject to the wish or will)—was the true cause; the first point was to prevent the mind of the turner having an undue influence over the effects produced in relation to the nature of the substances employed.   A bundle of plates, consisting of sand-paper, millboard, glue, glass, plastic clay, tinfoil, cardboard, gutta-percha, vulcanized caoutchouc, wood, and resinous cement, was therefore made up and tied together, and being placed on a table, under the hand of a turner, did not prevent the transmission of the power; the table turned or moved exactly as if the bundle had been away, to the full satisfaction of all present.   The experiment was repeated, with various substances and persons, and at various times, with constant success; and henceforth no objection could be taken to the use of these substances in the construction of apparatus.   The next point was to determine the place and source of motion, *i. e.* whether the table moved the hand, or the hand moved the table; and for this purpose indicators were constructed.   One of these consisted of a light lever, having its fulcrum on the table, its short arm attached to a pin fixed on a cardboard, which could slip on the surface of the table, and its long arm projecting as an index of motion. It is evident that if the experimenter willed the table to move towards the left, and it did so move before the hands, placed at the time on the cardboard, then the index would move to the left also, the fulcrum going with the table.   If the hands involuntarily moved towards the left without the table, the index would go towards the right; and, if neither table nor hands moved, the index would itself remain immoveable.   The result was, that when the parties saw the index it remained very steady; when it was hidden from them, or they looked away from it, it wavered about, though they believed that they always pressed directly downwards; and, when the table did not move, there was still a resultant of hand force in the direction in which it was wished the table should move, which, however, was exercised quite unwittingly by the party operating.   This resultant it is which, in the course of the waiting time, while the fingers and hands become stiff, numb, and insensible by continued pressure, grows up to an amount sufficient to move the table or the substances pressed upon. But the most valuable effect of this test-apparatus (which was afterwards made more perfect and independent of the table) is the corrective power it possesses over the mind of the table-turner. As soon as the index is placed before the most earnest, and they perceive—as in my presence they have always done—that it tells truly whether they are pressing downwards only or obliquely, then all effects of table-turning cease, even though the parties persevere, earnestly desiring motion, till they become weary and worn out. No prompting or checking of the hands is needed—the power is gone; and this only because the parties are made conscious of what they are really doing mechanically, and so are unable unwittingly to deceive themselves. I know that some may say that it is the cardboard next the fingers which moves first, and that it both drags the table, and also the table-turner with it. All I have to reply is, that the cardboard may in practice be reduced to a thin sheet of paper weighing only a few grains, or to a piece of goldbeaters' skin, or even the end of the lever, and (in principle) to the very cuticle of the finger itself. Then the results that follow are too absurd to be admitted: the table becomes an incumbrance, and a person holding out the fingers in the air, either naked or tipped with goldbeaters' skin or cardboard, ought to be drawn about the room, &c.; but I refrain from considering imaginary yet consequent results which have nothing philosophical or real in them. I have been happy thus far in meeting with the most honourable and candid though most sanguine persons, and I believe the mental check which I propose will be available in the hands of all who desire truly to investigate the philosophy of the subject, and, being content to resign expectation, wish only to be led by the facts and the truth of nature. As I am unable, even at present, to answer all the letters that come to me regarding this matter, perhaps you will allow me to prevent any increase by saying that my apparatus may be seen at the shop of the philosophical instrument maker—Newman, 122 Regent-street. Permit me to say, before concluding, that I have been greatly startled by the revelation which this purely physical subject has made of the condition of the public mind. No doubt there are many persons who have formed a right judgment or used a cautious reserve, for I know several such, and public communications have shown it to be so; but their number is almost as nothing to the great body who have believed and borne testimony, as I think, in the cause of error. I do not here refer to the distinction of those who agree with me and those who differ. By the great body, I mean such as reject all consideration of the equality of cause and effect, who refer the results to electricity and magnetism, yet know nothing of the laws of these forces,—or to attraction, yet show no phenomena of pure attractive power,—or to the rotation of the earth, as if the earth revolved round the leg of a table,—or to some unrecognized physical force, without inquiring whether the known forces are not sufficient,—or who even refer them to diabolical or supernatural agency, rather than suspend their judgment, or acknowledge to themselves that they are not learned enough in these matters to decide on the nature of the action. I think the system of education that could leave the mental condition of the public body in the state in which this subject has found it, must have been greatly deficient in some very important principle.

I am, Sir, your very obedient Servant,

Royal Institution, June 28, 1853.   M. FARADAY.

---

*Experimental Investigation of Table-Moving**.*

THE object which I had in view in this inquiry was not to satisfy myself, for my conclusion had been formed already on the evidence of those who had turned tables; but that I might be enabled to give a strong opinion, founded on facts, to the many who applied to me for it. Yet the proof which I sought for, and the method followed in the inquiry, were precisely of the same nature as those which I should adopt in any other physical investigation. The parties with whom I have worked were very honourable, very clear in their intentions, successful table-movers, very desirous of succeeding in establishing the existence of a peculiar power, thoroughly candid, and very effectual. It is with me a clear point that the table moves when the parties, though they strongly wish it, do not intend, and do not believe that they move it by ordinary mechanical power. They say, the table draws their hands;

* Athenæum, July 2, 1853.

that it moves first, and they have to follow it,—that sometimes it even moves from under their hands. With some the table will move to the right or left according as they wish or will it, —with others the direction of the first motion is uncertain:— but all agree that the table moves the hands and not the hands the table. Though I believe the parties do not intend to move the table, but obtain the result by a *quasi* involuntary action, still I had no doubt of the influence of expectation upon their minds, and through that upon the success or failure of their efforts. The first point, therefore, was to remove all objections due to expectation, having relation to the substances which I might desire to use :—so, plates of the most different bodies, electrically speaking,—namely, sand-paper, millboard, glue, glass, moist clay, tinfoil, cardboard, gutta percha, vulcanized rubber, wood, &c.,—were made into a bundle and placed on a table under the hands of a turner. The table turned. Other bundles of other plates were submitted to different persons at other times,—and the tables turned. Henceforth, therefore, these substances may be used in the construction of apparatus. Neither during their use nor at other times could the slightest trace of electrical or magnetic effects be obtained. At the same trials it was readily ascertained that one person could produce the effect; and that the motion was not necessarily circular, but might be in a straight line. No form of experiment or mode of observation that I could devise gave me the slightest indication of any peculiar natural force. No attractions, or repulsions, or signs of tangential power, appeared,— nor anything which could be referred to other than the mere mechanical pressure exerted inadvertently by the turner. I therefore proceeded to analyse this pressure, or that part of it exerted in a horizontal direction:—doing so, in the first instance, unawares to the party. A soft cement, consisting of wax and turpentine, or wax and pomatum, was prepared. Four or five pieces of smooth slippery cardboard were attached one over the other by little pellets of the cement, and the lower of these to a piece of sand-paper resting on the table; the edges of these sheets overlapped slightly, and on the under surface a pencil line was drawn over the laps so as to indicate position. The upper cardboard was larger than the rest, so as to cover the whole from sight. Then the table-turner placed the hands upon the upper card,—and we waited for the result. Now, the cement was strong enough to offer considerable resistance to mechanical motion, and also to retain the cards in any new position which they might acquire,—and yet weak enough to give way slowly to a continued force. When at last the tables, cards, and hands all moved to the left together, and so a true result was obtained, I took up the pack. On examination it was easy to see by the displacement of the parts of the line, that the hand had moved further than the table, and that the latter had lagged behind ;—that the hand, in fact, had pushed the upper card to the left, and that the under cards and the table had followed and been dragged by it. In other similar cases when the table had not moved, still the upper card was found to have moved, showing that the hand had carried it in the expected direction. It was evident, therefore, that the table had not drawn the hand and person round, nor had it moved simultaneously with the hand. The hand had left all things under it behind, and the table evidently tended continually to keep the hand back.

The next step was to arrange an index which should show whether the table moved first, or the hand moved before the table, or both moved or remained at rest together. At first this was done by placing an upright pin fixed on a leaden foot upon the table, and using that as the fulcrum of a light lever. The latter was made of a slip of foolscap paper, and the short arm, about $\frac{1}{4}$ of an inch in length, was attached to a pin proceeding from the edge of a slipping card placed on the table, and prepared to receive the hands of the table-turner. The other arm, of $11\frac{1}{2}$ inches long, served for the index of motion. A coin laid on the table marked the normal position of the card and index. At first the slipping card was attached to the table by the soft cement, and the index was either screened from the turner, or the latter looked away: then, before the table moved, the index showed that the hand was giving a resultant pressure in the expected direction. The effect was never carried far enough to move the table, for the motion of the index corrected the judgment of the experimenter, who became aware that, inadvertently, a side force had been exerted. The card was now set free from the table, *i. e.* the cement was removed. This, of course, could not interfere with any of the results expected by the table-turner, for both the bundle of plates spoken of and single cards had been freely moved on the tables before; but now that the index was there, witnessing to the eye, and through it to the mind, of the table-turner, not the slightest tendency to motion either of the card or of the table occurred. Indeed, whether the card was left free or was attached to the table, all motion or tendency to motion was gone. In one particular case there was relative motion between the table and the hands: I believe that the hands moved in one direction; the table-turner was persuaded that the table moved from under the hand in the other direction :—a gauge, standing upon the floor, and pointing to the table, was therefore set up on that and some future occasions,—and then, neither motion of the hand nor of the table occurred.

A more perfect lever apparatus was then constructed in the following manner :—Two thin boards, $9\frac{1}{2}$ inches by 7 inches, were provided; a board, 9 inches by 5 inches, was glued to the middle of the underside of one of these (to be called the table-board), so as to raise the edges free from the table; being placed on the table, near and parallel to its side, an upright pin was fixed close to the further edge of the board, at the middle, to serve as the fulcrum for the indicating lever. Then four glass rods, 7 inches long and $\frac{1}{4}$ in diameter, were placed as rollers on different parts of this table-board, and the upper board placed on them; the rods permitted any required amount of pressure on the boards, with a free motion of the upper on the lower to the right and left. At the part corresponding to the pin in the lower board, a piece was cut out of the upper board, and a pin attached there, which, being bent downwards, entered the hole in the end of the short arm of the index lever: this part of the lever was of cardboard; the indicating prolongation was a straight hay-stalk 15 inches long. In order to restrain the motion of the upper board on the lower, two vulcanized rubber rings were passed round both, at the parts not resting on the table: these, whilst they tied the boards together, acted also as springs,—and whilst they allowed the first feeblest tendency to motion to be seen by the index, exerted, before the upper board had moved a quarter of an inch, sufficient power in pulling the upper board back from either side, to resist a strong lateral action of the hand. All being thus arranged, except that the lever was away, the two boards were tied together with string, running parallel to the vulcanized rubber springs, so as to be immoveable in relation to each other. They were then placed on the table, and a table-turner sat down to them :—the table very shortly moved in due order, showing that the apparatus offered no impediment to the action. A like apparatus, with metal rollers, produced the same result under the hands of another person. The index was now put into its place and the string loosened, so that the springs should come into play. It was soon seen, with the party that could will the motion in either direction (from whom the index was purposely hidden), that the hands were gradually creeping up in the direction before agreed upon, though the party certainly thought they were pressing downwards only. When shown that it was so, they were truly surprised; but when they lifted up their hands and immediately saw the index return to its normal position, they were convinced. When they looked at the index and could see for themselves whether they were pressing truly downwards, or obliquely so as to produce a resultant in the right- or left-handed direction, then such an effect never took place. Several tried, for a long while together, and with the best will in the world; but no motion, right or left, of the table, or hand, or anything else occurred.—[Then occurs a passage from the 'Times,' already printed at pp. 383, 384.]

Another form of index was applied thus :—a circular hole was cut in the middle of the upper board, and a piece of cartridge paper pasted under it on the lower surface of the board; a thin slice of cork was fixed on the upper surface of the lower board corresponding to the cartridge paper; the interval between them might be a quarter of an inch or less. A needle was fixed into the end of one of the index hay-stalks, and when all was in place the needle point was passed through the cartridge paper and pressed slightly into the cork beneath, so as to stand upright: then any motion of the hand, or hand-board, was instantly rendered evident by the deflection of the perpendicular hay-stalk to the right or left.

I think the apparatus I have described may be useful to many who really wish to know the truth of nature, and would prefer that truth to a mistaken conclusion; desired, perhaps only because it seems to be new or strange. Persons do not know how difficult it is to press directly downward, or in any given direction against a fixed obstacle, or even to *know only* whether they are doing so or not; unless they have some indicator, which, by visible motion or otherwise, shall instruct them: and this is more especially the case when the muscles of the fingers and hand have been cramped and rendered either tingling, or insensible, or cold by long-continued pressure. If a finger be pressed constantly into the corner of a window-frame for ten minutes or more, and then, continuing the pressure, the mind be directed to judge whether the force at a given moment is all horizontal, or all downward, or how much is in one direction and how much in the other, it will find great difficulty in deciding; and will at last become altogether uncertain: at least such is my case. I know that a similar result occurs with others; for I have had two boards arranged, separated, not by rollers, but by plugs of vulcanized rubber, and with the vertical index: when a person with his hands on the upper board is requested to press only downwards, and the index is hidden from his sight, it moves to the right, to the left, to him and from him, and in all horizontal directions; so utterly unable is he strictly to fulfil his intention without a visible and correcting indicator. Now, such is the use of the instrument with the horizontal index and rollers: the mind is instructed, and the involuntary or *quasi* involuntary motion is checked in the commencement, and therefore never rises up to the degree needful to move the table, or even permanently the index itself. No one can suppose that looking at the index can in any way interfere with the transfer of electricity or any other power from the hand to the board under it or to the table. If the board tends to move, it may do so, the index does not confine it; and if the table tends to move, there is no reason why it should not. If both were influenced by any power to move together, they may do so,—as they did indeed when the apparatus was tied, and the mind and muscles left unwatched and unchecked.

I must bring this long description to a close. I am a little ashamed of it, for I think, in the present age, and in this part of the world, it ought not to have been required. Nevertheless, I hope it may be useful. There are many whom I do not expect to convince ; but I may be allowed to say that I cannot undertake to answer such objections as may be made.   I state my own convictions as an experimental philosopher, and find it no more necessary to enter into controversy on this point than on any other in science, as the nature of matter, or inertia, or the magnetization of light, on which I may differ from others.   The world will decide sooner or later in all such cases, and I have no doubt very soon and correctly in the present instance.   Those who may wish to see the particular construction of the test apparatus which I have employed, may have the opportunity at Mr. Newman's, 122 Regent Street. Further, I may say, I have sought earnestly for cases of lifting by attraction, and indications of attraction in any form, but have gained no traces of such effects.   Finally, I beg to direct attention to the discourse delivered by Dr. Carpenter at the Royal Institution on the 12th of March, 1852, entitled, "On the Influence of Suggestion in modifying and directing Muscular Movement, independently of Volition;" which, especially in the latter part, should be considered in reference to table-moving by all who are interested in the subject.

Royal Institution, June 27.   M. FARADAY.

THE BAKERIAN LECTURE.—*Experimental Relations of Gold (and other Metals) to Light*.*

[Received November 15, 1856,—Read February 5, 1857.]

THAT wonderful production of the human mind, the undulatory theory of light, with the phenomena for which it strives to account, seems to me, who am only an experimentalist, to stand midway between what we may conceive to be the coarser mechanical actions of matter, with their explanatory philosophy, and that other branch, which includes, or should include, the physical idea of forces acting at a distance; and admitting for the time the existence of the ether, I have often struggled to perceive how far that medium might account for or mingle in with such actions generally; and to what extent experimental trials might be devised, which, with their results and consequences, might contradict, confirm, enlarge, or modify the idea we form of it, always with the hope that the corrected or instructed idea would

* Philosophical Transactions, 1857, p. 145.

approach more and more to the truth of nature, and in the fulness of time coincide with it.

The phenomena of light itself are, however, the best and closest tests at present of the undulatory theory; and if that theory is hereafter to extend to and include other actions, the most effectual means of enabling it to do so will be to render its application to its own special phenomena clear and sufficient. At present the most instructed persons are, I suppose, very far from perceiving the full and close coincidence between all the facts of light and the physical account of them which the theory supplies. If perfect, the theory would be able to give a reason for every physical affection of light; whilst it does not do so, the affections are in turn fitted to develope the theory, to extend and enlarge it if true, or if in error to correct it or replace it by a better. Hence my plea for the possible utility of experiments and considerations such as those I am about to advance.

Light has a relation to the matter which it meets with in its course, and is affected by it, being reflected, deflected, transmitted, refracted, absorbed, &c. by particles very minute in their dimensions. The theory supposes the light to consist of undulations, which, though they are in one sense continually progressive, are at the same time, as regards the particles of the ether, to and fro transversely. The number of progressive alternations or waves in an inch is considered as known, being from 37,600 to 59,880, and the number which passes to the eye in a second of time is known also, being from 458 to 727 billions; but the extent of the lateral excursion of the particles of the ether, either separately or conjointly, is not known, though both it and the velocity are probably very small compared to the extent of the wave and the velocity of its propagation. Colour is identified with the number of waves. Whether reflexion, refraction, &c., have any relation to the extent of the lateral vibration, or whether they are dependent in part upon some physical action of the medium unknown to and unsuspected by us, are points which I understand to be as yet undetermined.

Conceiving it very possible that some experimental evidence of value might result from the introduction into a ray of separate particles having great power of action on light, the particles being at the same time very small as compared to the wavelengths, I sought amongst the metals for such. Gold seemed especially fitted for experiments of this nature, because of its comparative opacity amongst bodies, and yet possession of a real transparency; because of its development of colour both in the reflected and transmitted ray; because of the state of tenuity and division which it permitted with the preservation of its integrity as a metallic body; because of its supposed simplicity of character; and because known phenomena appeared to indicate that a mere variation in the size of its particles gave rise to a variety of resultant colours. Besides, the waves of light are so large compared to the dimensions of the particles of gold which in various conditions can be subjected to a ray, that it seemed probable the particles might come into effective relations to the much smaller vibrations of the ether particles; in which case, if reflexion, refraction, absorption, &c., depended upon such relations, there was reason to expect that these functions would change sensibly by the substitution of different sized particles of this metal for each other. At one time I hoped that I had altered one coloured ray into another by means of gold, which would have been equivalent to a change in the number of undulations; and though I have not confirmed that result as yet, still those I have obtained seem to me to present a useful experimental entrance into certain physical investigations respecting the nature and action of a ray of light. I do not pretend that they are of great value in their present state, but they are very suggestive, and they may save much trouble to any experimentalist inclined to pursue and extend this line of investigation.

Gold-leaf—effect of heat, pressure, &c.

Beaten gold-leaf is known in films estimated at the $\frac{1}{282000}$th of an inch in thickness; they are translucent, transmitting green light, reflecting yellow, and absorbing a portion. These leaves consist of an alloy in the proportions of 12 silver and 6 copper to 462 of pure gold. 2000 leaves $3\frac{3}{5}$ths of an inch square are estimated to weigh 384 grains. Such gold-leaf is no doubt full of holes, but having, in conjunction with Mr. W. De la Rue, examined it in the microscope with very high powers (up to 700 linear), we are satisfied that it is truly transparent where the gold is continuous, and that the light transmitted is green. By the use of the balance Mr. De la Rue found that the leaf employed was on the average $\frac{1}{278000}$th of an inch thick.

Employing polarized light and an arrangement of sulphate of lime plates, it was found that other rays than the green could be transmitted by the gold-leaf. The yellow rays appeared to be those which were first stopped or thrown back. Latterly I have obtained some pure gold-leaf beaten by Marshall, of which 2000 leaves weighed 408 grains, or 0·2 of a grain per leaf: its reflected colour is orange-yellow, and its transmitted colour a warm green. Gold alloy containing 25 per cent. of silver produces pale gold-leaf, which transmits a blue purple light, and extinguishes much more than the ordinary gold-leaf.

So a leaf of beaten gold occupies in average thickness no more than from $\frac{1}{3}$th to $\frac{1}{5}$th part of a single wave of light. By chemical means, the film may be attenuated to such a degree as to transmit a ray so luminous as to approach to white, and that in parts which have every appearance of being continuous in the microscope, when viewed with a power of 700. For this purpose it may be laid upon a solution of chlorine, or, better still, of the cyanide of potassium*. If a clean plate of glass be breathed upon and then brought carefully upon a leaf of gold, the latter will adhere to it; if distilled water be immediately applied at the edge of the leaf, it will pass between the glass and gold, and the latter will be perfectly stretched; if the water be then drained out, the gold-leaf will be left well extended, smooth, and adhering to the glass. If, after the water is poured off, a weak solution of cyanide be introduced beneath the gold, the latter will gradually become thinner and thinner; but at any moment the process may be stopped, the cyanide washed away by water, and the attenuated gold film left on the glass. If towards the end a washing be made with alcohol, and then with alcohol containing a little varnish, the gold film will be left cemented to the glass†.

* The chlorine leaves a film of chloride of silver behind, the cyanide leaves only metal.

† Air-voltaic circles are formed in these cases, and the gold is dissolved almost exclusively under their influence. When one piece of gold-leaf was placed on the surface of a solution of cyanide of potassium, and another, moistened on both sides, was placed under the surface, both dissolved; but twelve minutes sufficed for the solution of the first, whilst above twelve hours were required for the submerged piece. In weaker solutions, and with silver also, the same results were obtained; from sixty to a hundredfold as much time being required for the disappearance of the submerged metal as for that which, floating, was in contact both with the air and the solvent. An action

In this manner the leaf may be obtained so thin, that I think 50 or even 100 might be included in a single progressive undulation of light.   But the character of the effect on light is not changed, the light transmitted is green, as before; and though that green tint is due to a condition of the gold induced by pressure, it as yet remains unchanged through all these varieties of thickness and of proportion to the progressive or the lateral undulation.

Gold-leaf, either fine or common, examined in the microscope, appears as a most irregular thing. It is everywhere closely mottled or striated, according as a part at the middle or the edge of a leaf is selected, minute portions which are close to other parts being four or five times as thick as the latter, if the proportion of light which passes through may be accepted as an indication.   Yet this irregular plate does not cause any sensible distortion of an object seen through it, that object being the line of light reflected from a fine wire in the focus of a moderate microscope. Nor perhaps was any distortion due to consecutive convexities and concavities to be expected; for when the thicker parts of the leaf were examined they seemed to be accumulated plications of the gold, the leaf appearing as a most irregular and crumpled object, with dark veins running across both the thicker and thinner parts, and from one to the other. Yet in the best microscope, and with the highest power, the leaf seemed to be continuous, the occurrence of the smallest sensible hole making that continuity at other parts apparent, and every part possessing its proper green colour.   How such a film can act as a plate on polarized light in the manner it does, is one of the queries suggested by the phenomena which requires solution.

When gold-leaf is laid upon glass and its temperature raised considerably without disturbance, either by the blowpipe or an ordinary Argand gas-burner, it seems to disappear, *i. e.* the lustre passes away, the light transmitted is abundant and nearly white, and the place appears of a pale brown colour. One would think that much of the metal was dissipated, but all is there, and if the heat has been very high (which is not necessary for the best results), the microscope shows it in minute

of this kind has probably much to do with the *formation* of the films to be described hereafter.

globular portions. A comparatively low heat, however, and one unable to cause separation of the particles, is known to alter the molecular condition of gold, and the gold-beater finds important advantage in the annealing effect of a temperature that does not hurt the skins or leaves between which he beats the metal.

It might be supposed that the annealed metal, in contracting from the constrained and attenuated state produced by beating, drew up, leaving spaces through which white light could pass, and becoming itself almost insensible through the smallness of its quantity; and if gold-leaf unattached to glass be heated carefully with oil in a tube, it does shrink up considerably even before it loses its green colour, which finally happens. But if the gold-leaf laid upon glass plates by water only be carefully dried, then introduced into a bath of oil and raised to a temperature as high as the oil can bear for five or six hours, and then suffered to cool, the plates, when taken out and washed, first in camphine, and then in alcohol, present specimens of gold which has lost its green colour, transmits far more light than before and reflects less, whilst yet the film remains in form and other conditions apparently quite unchanged. Being now examined in the microscope, it presents exactly the forms and appearance of the original leaf, except in colour; the same irregularities appear, the same continuity, and if the destruction of the green colour has not been complete, it will be seen that it is the thicker folds and parts of the mottled mass that retains the original state longest.

This change does not depend upon the substance in contact with which the gold is heated*. If the leaf be laid upon mica, rock-crystal, silver or platinum, the same result occurs; the surrounding medium also may change, and be air, oil or carbonic acid, without causing alteration. Nor has the gold disappeared; a piece of leaf, altered in one part and not in another, was divided into four equal parts, and the gold on each converted by chlorine gas into crystallized chloride of gold; the same amount was found in each division.

When the gold-leaf is laid by water on plates of rock-crystal,

---

* The disappearance of gold-leaf as metal, when mingled with lime, alumina and other bodies, and then heated, has been already observed; and referred to oxidation (J. A. Buchner). See Gmelin's ' Chemistry,' vi. p. 206, " Purple oxide of gold."

and then gradually heated in a muffle not higher than is necessary, an excellent result is obtained. The gold is then of a
uniform pale brown colour by common observation; but when
examined by a lens, and an oblique light, all the mottle of the
original leaf appears. It adheres but very slightly to the rock-
crystal, and yet can bear the application of the pressure now
to be described.

When gold rendered colourless by annealing is subjected to
pressure, it again becomes of a green colour. I find a convex
surface of agate or rock-crystal having a radius of from a quarter
to half an inch very good for this purpose, the metal having
very little tendency to adhere to this substance. The greening
is necessarily very imperfect, and if examined by a lens it will be
evident that the thinner parts of the film are rarely reached by
the pressure, it being taken off by the thicker corrugations;
but when reached they acquire a good green colour, and the
effect is abundantly shown in the thicker parts. At the same
time that the green colour is thus reproduced, the quantity of
light transmitted is diminished, and the quantity of light
reflected is increased. When the gold-leaf has been heated on
glass in a muffle, it generally adheres so well as to bear streak-
ing with the convex rock-crystal, and then the production of
the reflecting surface and the green transmission is very striking.
In other forms of gold film, to be described hereafter, the
greening effect of pressure (which is general to gold) is still
more strikingly manifested, and can be produced with the
touch of a card or a finger. In these cases, and even with gold-
leaf, the green colour reproduced can be again taken away by
heat, to appear again by renewed pressure.

As to the essential cause of this change of colour, more investigation is required to decide what that may be. As already
mentioned, it might be thought that the gold-leaf had run up
into separate particles. If it were so, the change of colour by
division is not the less remarkable, and the case would fall into
those brought together under the head of gold fluids. On the
whole, I incline to this opinion; but the appearance in the microscope, the occurrence of thin films of gold acting altogether
like plates and yet not transmitting a green ray until they are
pressed, and their action on a polarized ray of light, throw
doubts in the way of such a conclusion.

It may be thought that the beating has conferred a uniform strained condition upon the gold, a difference in quality in one direction which annealing takes away; but when the gold is examined by polarized light, there is no evidence as yet of such a condition, for the green and the colourless gold present like results; and there is a little difficulty in admitting that such an irregular corrugated film as gold-leaf appears to be can possess any general compression in one direction only, especially when it is considered that it is beaten amongst tissues softer than itself, and made up with it into considerable masses. The greening effect of pressure occurs with the deposited particles of electric discharges, and here it appears either amongst the larger particles near the line of the discharge, or amongst the far finer ones at a considerable distance. Such results do not suggest a dependence upon either the size of the particles or their quantity, but rather upon the relative dimensions of the particles in the direction of the ray and transverse to that direction. One may imagine that spherical or other particles, which, being disposed in a plane, transmit ruby rays or violet rays, acquire the power, when they are flattened, of transmitting green rays, and such a thought sends the mind at once from the wave of light to the direction and extent of the vibrations of the ether. For it does not seem likely that pressure can produce its peculiar result by affecting the relation of the dimension of the particle to the length-dimension of a progressive undulation of light, the latter being so very much greater than the former; but the relation to the dimension of the direct or lateral vibration of the particles of the ether may be greatly affected, that being probably very small and much nearer to, if not even less than, the size of the particles of gold.

Silver-leaf, as usually obtained by beating, is so opake, as perfectly to exclude the light of the sun. When this is laid by water on plates of rock-crystal and heated in a muffle, it begins to change at a temperature lower than that required for gold, and becomes very translucent, losing at the same time its reflective power: it looks very like the film of chloride produced when a leaf of silver is placed in chlorine gas. When examined by a lens or an ordinary microscope, the leaf seems to be as continuous as in its original state; the finest hole, or the finest line drawn by a needle point, appears only to prove the continuity of the metallic film up to the very edges of these real apertures. When pressure is applied to this translucent film, the compressed metal becomes either opake or of a very dark purple colour, and resumes its high reflective power. If a higher heat than that necessary for this first change be applied, then the leaf, viewed in the microscope, assumes a mottled appearance, as if a retraction into separate parts had occurred. At a still higher temperature this effect is increased; but the heat, whether applied in the muffle or by a blowpipe, which is necessary to fuse the metal and make it run together in globules, is very much higher than that which causes the first change of the silver: the latter is, in fact, below such a red heat as is just visible in the dark. Whatever the degree of heat applied, the metal remains as metallic silver during the whole time. When many silver leaves were laid loosely one upon another, rolled up into a loose coil, introduced into a glass tube, and the whole placed in a muffle and heated carefully for three or four hours to so low a degree that the glass tube had not been softened or deformed, it was found that the silver-leaf had sunk together a little and shaped itself in some degree upon the glass, touching by points here and there, but not adhering to it. But it was changed, so that the light of a candle could be seen through forty thicknesses: it had not run together, though it adhered where one part touched another. It did not look like metal, unless one thought of it as divided dead metal, and it even appeared too unsubstantial and translucent for that; but when pressed together, it clung and adhered like clean silver, and resumed all its metallic characters.

When the silver is much heated, there is no doubt that the leaf runs up into particles more or less separate. But the question still remains as to the first effect of heat, whether it merely causes a retraction of the particles, or really changes the optical and physical nature of the metal from the beaten or pressed state to another from which pressure can return it back again to its more splendid condition. It seems just possible that the leaf may consist of an infinity of parts resulting from replications, foldings and scales, all laid parallel by the beating which has produced them, and that the first action of heat is to cause these to open out from each other; but that supposi- tion leaves many of the facts either imperfectly explained or untouched. The Arts do not seem to furnish any process which can instruct us as to this condition, for all the operations of polishing, burnishing, &c. applied to gold, silver and other metals, are just as much fitted to produce the required state under one view as under the other.

To return to gold: it is clear that that metal, reduced to small dimensions by mere mechanical means, can appear of two colours by transmitted light, whatever the cause of the difference may be. The occurrence of these two states may prepare one's mind for the other differences with respect to colour, and the action of the metallic particles on light, which have yet to be described.

Many leaves of gold, when examined by a lens and transmitted light, present the appearance of red parts; these parts are small, and often in curved lines, as if a fine hair had been there during the beating. At first I thought the gold was absolutely red in these parts, but am inclined to believe that in the greatest number of cases the tint is subjective, being the result of the contrast between the white light transmitted through bruised parts, and the green light of the neighbouring continuous parts. Nevertheless, some of these places, when seen in the microscope, appeared to have a red colour of their own, that is, to transmit a true red light. As I believe that gold in a certain state of division can transmit a ruby light, I am not prepared to say that gold-leaf may not, in some cases, where the effect of pressure in a particular direction has been removed, do the same.

Many of the prepared films of gold were so thin as to have their reflective power considerably reduced, and that in parts which, under the microscope and in other ways, appeared to be quite continuous: this agrees with the transmission of all the rays already mentioned, but it seems to imply that a certain thickness is necessary for full reflexion; therefore, that more than one particle in depth is concerned in the act, and that the division of gold into separate particles by processes to be described, may bring them within or under the degree necessary for ordinary reflexion.

As particles of pure gold will be found hereafter to adhere by contact, so the process of beating may be considered as one which tends to weld gold together in all directions, and especially in that transverse to the blow,—a point favourable to continuity in that direction, both as it tends to preserve and even reproduce it.

If a polarized ray be received on an analyser so that no light passes, and a plate of annealed glass, either thick or thin, be interposed vertically across the ray, and no difference is observed on looking through the analyser, the image of the source of light does not appear; but if the plate be inclined until it makes an angle of from 30° to 45°, or thereabouts, with the ray, the light appears, provided the inclination of the glass is not in the plane of polarization or at right angles to it, the effect being a maximum if the inclination be in a plane making an angle of 45° with that of polarization. This effect, which is common to all uncrystallized transparent bodies, is also produced by leaf-gold, and is one of the best proofs of the true transparency of this metal according to the ordinary meaning of the term. In like manner, if a leaf of gold be held obliquely across an ordinary ray of light, it partly polarizes it, as Mr. De la Rue first pointed out to me. Here again the condition of true transparency is established, for it acts like a plate of glass or water or air. But the relations of gold and the metals in different conditions to polarized light shall be given altogether at the close of this paper.

Deflagrations of Gold (and other metals)—heat—pressure, &c.

Gold wire deflagrated by explosions of a Leyden battery produces a divided condition, very different to that presented by gold leaves. Here the metal is separated into particles, and no pressure in any direction, either regular or irregular, has been exerted upon them in the act of division. When the deflagrations have been made near surfaces of glass, rock-crystal, topaz, fluor-spar, card-board, &c., the particles as they are caught are kept separate from each other and in place, and generally those which remain in the line of the discharge have been heated by the passage of the electricity. The deposits consist of particles of various sizes, those at the outer parts of the result being too small to be recognized by the highest powers of the microscope. Beside making these deflagrations over different substances, as described above, I made them in different atmospheres, namely, in oxygen and hydrogen, to compare with air; but the general effects, the colours produced, and the order of the colours, were precisely the same in all the cases. These deposits were insoluble in nitric acid and in hydrochloric acid, but in the mixed acids or in chlorine solution were soluble, exactly in the manner of gold. There is no reason to doubt that they consisted of metallic gold in a state of extreme division.

Now as to the effects on light, *i. e.* as to the coloured rays reflected or transmitted by these deposited particles, and first, of those in the line of the discharge where the wire had been. Here the mica was found abraded much, the glass less, and the rock-crystal and topaz least. Where abraded, the gold adhered; in all the other parts it could be removed with the slightest touch. The gold deposited in this central place was metallic and golden by reflected light, and of a fine ruby colour by transmitted light. On each side of this line the deposit had a dark colour, but when particularly examined gave a strong golden metallic reflexion, and by transmission a fine violet colour, partaking of green and ruby in different parts, and sometimes passing altogether into green. Beyond this, on each side, where the tints became paler and where the particles appeared to be finer, the transmitted tint became ruby or violet-ruby, and this tint was especially seen when the deposit was caught on a card. As to the reflected light, even at these faintest parts it is golden and metallic. This is easily observed by wiping off a sharp line across the deposit on glass in the very faintest part, and then causing the sun's rays collected in the focus of a small lens to travel to and fro across that edge; the presence of the metallic gold on the unwiped part is at once evident by the high illumination produced there. It is evident that all the colours described are produced by one and the same substance, namely gold, the only apparent difference being the state of division and different degrees of the application of heat. The thickest parts of these deposits are so discontinuous, that they cannot conduct the electricity of a battery of two or three pairs of plates, *i. e.* of a battery unable to produce a spark among the particles.

When any of these deposits of divided gold are heated to dull redness, a remarkable change occurs. The portions which before were violet, blue, or green by transmitted light, now change to a ruby, still preserving their metallic reflecting power, and this ruby is in character quite like that which is presented in the arts by glass tinged by gold. This change is often far better shown in the more distant and thinner parts of the deposit, than in those nearest to the line of discharge; for near the latter place, where the deposit is most abundant, the metal appears to run up into globules, as with gold-leaf, and so disappears as a film. I believe that the ruby character of the deposit *in* the line of discharge, is caused by the same action of heat produced at the moment by the electricity passing there. In the distant parts, the deposit, rubified by after-heat, is not imbedded or fused into the glass, rock-crystal, topaz, &c., but is easily removed by a touch of the finger, though in parts of the glass plate which are made very hot, it will adhere.

If the agate pressure before spoken of, in respect of gold-leaf, be applied to ruby parts not too dense, places will easily be found where this pressure increases the reflective power considerably, and where at the same time it converts the transmitted ray from ruby to green; making the gold, as I believe, then accord in condition with beaten gold-leaf. On the other hand, if parts of the *unheated* electric deposit, where they are purple-grey, and so thin as to be scarcely visible without care, be in like manner pressed, they will acquire the reflective power, and then transmit the green ray; and I think I am justified by my experiments in stating, that fine gold particles, so loosely deposited that they will wipe off by a light touch of the finger, and possessing one conjoint structure, can in one state transmit light of a *blue-grey* colour, or can by heat be made to transmit light of a *ruby* colour, or can by pressure from either of the former states be made to transmit light of a *green* colour; all these changes being due to modifications of the gold, as gold, and independent of the presence of the bodies upon which for the time the gold is supported; for I ought to have said, if I have not said so, that these changes happen with all the deposits upon glass, mica, rock-crystal, and topaz, and whatever the atmosphere in which they were formed.

When gold is deflagrated by the voltaic battery near glass (I have employed sovereigns laid on glass for the terminals), a deposit of metallic gold in fine particles is produced. The densest parts have a dark slate-violet colour passing into violet

2 D 2

and ruby-violet in the outer thinner portions; a ruby tint is presented occasionally where the heat of the discharge has acted on the deposit. The deposited gold was easily removed by wiping, except actually at the spot where the discharge had passed. When these deposits were heated to dull redness they changed and acquired a ruby tint, which was very fine at the outer and thinner parts. The portions nearer the place of discharge presented ruby-violet and then violet tints, suggesting that accumulation of that which presented a fine ruby tint would, by stopping more and more light, transmit a ruby-violet or violet ray only. Pressure with the agate surface had a like effect as before, both with the heated and the unheated portions, *i. e.* with the violet and the ruby particles; but the effect was not altogether so good, and the tint of the transmitted ray was rather a green violet than a pure green. Still the difference produced by the pressure was very remarkable. The unheated particles at the surface, away from the glass, presented by reflexion almost a black; being heated, they became much more golden and metallic in appearance.

I prepared an apparatus by which many of the common metals could be deflagrated in hydrogen by the Leyden battery, and being caught upon glass plates could be examined as to reflexion, transmission, colour, &c., whilst in the hydrogen and in the metallic, yet divided state. The following are briefly the results; which should be considered in connexion with those obtained by employing polarized light. *Copper*: a fine deposit presenting by reflexion a purplish red metallic lustre, and by transmission a green colour, dark in the thicker parts, but always green; agate pressure increased the reflexion where it was not bright, and a little diminished the transmission, rendering the green deeper, but not changing its character as in the case of gold. *Tin* gave a beautiful bright white reflexion, and by transmission various shades of light and dark brown; agate pressure diminished the transmission and increased the reflexion in places before dull or dead; the effect appeared to be due simply to the lateral expansion of the separate particles filling up the space. *Iron* presented a fine steel grey, or slate metallic reflexion and a dark brown transmission; agate pressure gave the same effect as with tin, but no change of colour. *Lead*: a bright white reflexion, the transmission a dark smoky brown, agate pressure appeared to change this brown towards blue. *Zinc*: the reflexion bright white and metallic; the transmission a dark smoky colour with portions of blue-grey, brown-grey and pale brown; agate pressure tended to change the blue-grey to brown. *Palladium*: the reflexion fine metallic and dark grey; the transmitted light, where most abundant, sepia-brown; agate pressure converted the tint in the thinner places from brown towards blue-grey. *Platinum*: the reflexion white, bright and metallic; the transmission brown or warm grey with no other colours; agate pressure increased the reflexion and diminished the transmission as with tin. *Aluminium*: the reflexion metallic and white, very beautiful; the transmitted light was dark brown, bluish brown, and occasionally in the thinner parts orange; agate pressure caused but little change.

Films of Gold (and other metals) by Phosphorus, Hydrogen, &c.—effect of heat—pressure.

The reduction of gold from its solution by phosphorus is well known. If fifteen or twenty drops of a strong solution of gold, equal to about $1\frac{1}{2}$ grain of metal, be added to two or three pints of water, contained in a large capsule or dish, if four or five minute particles of phosphorus be scattered over the surface, and the whole be covered and left in quietness for twenty-four or thirty-six hours, then the surface will be found covered with a pellicle of gold, thicker at the parts near the pieces of phosphorus, and possessing there the full metallic golden reflective power of the metal; but passing by gradation into parts, further from the phosphorus, where the film will be scarcely sensible except upon close inspection. If plates of glass be introduced into the fluid under the pellicle, and raised gradually, the pellicle will be raised on them; it may then be deposited on the surface of pure distilled water to wash it; may be raised again on the glass; the water allowed to drain away, and the whole suffered to dry. In this way the pellicle remains attached to the glass, and is in a very convenient condition for preservation and examination.

If phosphorus be dissolved in two or three times its bulk of sulphide of carbon, and a few drops of the fluid be placed on the bottom of a dry basin, vapour of the phosphorus will soon rise up and bring the atmosphere in the basin to a reducing state. If a plate of glass large enough to cover the basin have six or eight drops of a strong neutral solution of chloride of gold placed on it, and this be spread about by a glass stirrer, so as to form a flowing layer on the surface, the glass may then be inverted and placed over the dish. So arranged, the gold solution will keep its place, but will have a film of metal reduced on its under surface. The plate being taken off after twenty, thirty, or forty minutes, and turned with the gold solution upwards, may then gradually be depressed in an inclined position into a large basin of pure water, one edge entering first, and the gold film will be left floating. After sufficient washing it may be taken up in portions on smaller plates of glass, dried, and kept for use. Mr. Warren De la Rue taught me how to make and deal with these films: they may by attention be obtained very uniform, of very different degrees of thickness, from almost perfect transparency to complete opacity, and by successive application of the same collecting glass plate may be superposed with great facility.

These films may be examined either on the water or on the glass. When thick, their reflective power is as a gold plate, full and metallic; as they are thinner, they lose reflective power, and they may be obtained so thin as to present no metallic appearance, all the coloured rays of light then passing freely through them. As to the transmitted light, the thinner films generally present one kind of colour; it appears as a feeble grey-violet, which increases in character as the film becomes thicker and sometimes approaches a violet; a greenish violet also appears; and the likeness of the grey-violet tint of these films to the stains produced by a solution of gold on the skin or other organic reducing substance, or the stain produced on common pottery, cannot be mistaken. Superposition of several grey-violet films does not produce a green tint, but only a diminution of light without change of colour. In those specimens made by particles of phosphorus floating on the solution of gold, very fine green tints occur at the thicker and golden parts of the film. The colour of the gold here may depend in some degree on the *manner* in which these films are formed: the thicker parts are not produced altogether by the successive addition of reduced gold from the portion of fluid immediately beneath them. When a particle of phosphorus is placed on pure water, it immediately throws out a film which appears to cover the whole of the surface; in a little while the film thickens around the particle and is easily distinguished by its high reflective power. It is this film which reduces the gold in solution, being itself consumed in the action; the result is a continued extension from the phosphorus outwards, which, after it has covered the solution with a thin film of gold, continues to cause a compression of the parts around the phosphorus and an accumulation there, rendering the gold at a distance of half an inch from the phosphorus so thick, that it is brilliant by reflexion and nearly opake by transmission; whilst near to the phosphorus the forming film is so thin as to be observed only on careful examination, and is still travelling outwards and compressing the surrounding parts more and more. The phosphorus is very slowly consumed; a particle not weighing $\frac{1}{100}$th of a grain will remain for four or five days on the surface of water before it disappears.

Though the particles of these films adhere together strongly, as may be seen by their stiffness on water, still the films cannot be considered as continuous. If they were, those made by vapour of phosphorus could not thicken during their formation, neither could they dry on glass in the short time found sufficient for that purpose. Experimentally also, I find that vapours and gases can pass through them. Very thin films without folds did not sensibly conduct the electricity of a single pair of Grove's plates; thicker films did conduct; yet with these proofs that these films could not be considered as continuous, they acted as thin plates upon light, producing the concentric rings of colours round the phosphorus at their first formation, though their thickness then could scarcely be the $\frac{1}{100}$th, perhaps not the $\frac{1}{500}$th of a wave undulation of light. Platinum, palladium, and rhodium produced films, showing these concentric rings very well.

Many of these films of gold, both thick and thin, which, being of a grey colour originally, were laid on a solution of cyanide of potassium to dissolve slowly, changed colour as they dissolved and became green; if change occurred, it was always towards green. On the other hand, when laid on a solution of chlorine, the change during solution was towards an amethyst or ruby tint. The films were not acted upon by pure nitric, or hydrochloric, or sulphuric acids, or solutions of potassa or brine. They dissolved in damp chlorine gas, not changing in colour during the solution. I believe them to consist of pure gold.

When these gold films were heated to dull redness they changed. The reflexion, though not much altered, was a little more metallic and golden than before; more light was transmitted after the heating and the colour had altered from greenish to violet, or from grey-green to ruby or amethyst; and now two or three films superposed often gave a very ruby colour. This action is like that of heat on the particles separated by electric explosions. If not overheated, the particles were not fused to the glass, but could be easily wiped off. Whenever these heated particles were pressed by the convex agate, they changed in character and transmitted green light. Heat took away this character of the gold, the heat of boiling oil, if continued, being sufficient; but on applying pressure at the same spot, the power of transmitting green light was restored to the particles. In many cases, where the gold adhered sufficiently to the glass to bear a light drawing touch from the finger or a card, such touch altered the light transmitted from amethystine to green; so small is the pressure required when the particles are most favourably disposed.

Heating injured the conducting power for electricity of these films, no doubt by retraction of the particles, though there was no such evident appearance in these cases, as in the un-attached gold-leaf of the particles running up into globules.

A given film, examined very carefully in the microscope by transmitted lamp-light, with an aperture of 90° and power of 700 linear, presented the following appearances. The un-heated part was of a grey colour, and by careful observation was seen to be slightly granular. By very close observation this grey part was often resolvable into a mixture of green and amethystine striæ, it being the compound effect of these which in general produce the grey sensation in the eye. When a part of such a film was heated, the transmitted colour was changed from grey to purple, as before described, and the part thus heated was evidently more granular than before. This dif-ference was confirmed in other cases. That the heated part should thus run up, seems to show that many of the particles

must have been touching though they did not form a continuous film; and on the other hand, the difference between
the effect here and with unattached gold-leaf, shows that the
degree of continuity as a film must be very small. When these
heated films were greened by agate pressure, or the drawing
pressure of a card, the green parts remained granulated,
apparently in the same degree as when purple. The green
was not subjective or an effect of interference, but a positive
colour belonging to the gold in that condition. Every touch of
the agate was beautifully distinct as a written mark. The parts
thus greened and the purple parts appeared to transmit about
the same amount of light. Though the film appeared granulated, no impression was made upon the mind that the
individual particles of which the film consisted were in any
degree rendered sensible to the eye.

The unheated gold films when pressed by agate often
indicated an improved reflective power, and the light transmitted was also modified; generally it was less, and occasionally
tended towards a green tint; but the effect of pressure was
by no means so evident as in particles which had been heated.

Films of some other metals were reduced by phosphorus in
like manner, the results in all these cases being of course
much affected by the strength of the solution and the time of
action; they are briefly as follows:—*Palladium*: a weak solution of the chloride gave fine films, apparently very continuous
and stiff; the reflexion was strong and metallic, of a dark grey
colour; the transmission presented every shade of Indian ink.
Platinum chloride gave traces of a film excessively thin, and
very slow in formation. *Rhodium* chloride in three or four
hours gave a beautiful film of metal in concentric rings, varying
in reflecting and transmitting power over light and also in
colour; those which reflected well, transmitted little light;
and those which transmitted, reflected little light; one migh
have thought there was no metal in some of the rings betwee
other rings that reflected brilliantly, but the metal was the r
of transmitting thickness; the transmitted colour of rhodium
varied from brown to blue. *Silver*: a solution of the nitrate
gave films showing the concentric rings; the light transmitted
by the thinner parts was of a warm brown, or sepia tint; the
film becomes very loose and mossy in the thicker parts and is

wanting in adherence; pressure brings out the full metallic lustre in every part, and in the thin places converts the colour from brown to blue, being in that respect like the result with pale gold-leaf, in which the silver present dominates over the colour of the gold. I do not think there is phosphorus combined with this silver; I did not find any, and considering the surface action on metals which float as films between air and water, it seems improbable that it should be there.

Hydrogen was employed to reduce some of the metals, their solutions being placed in an atmosphere of the gas. The action differed considerably from that of phosphorus, as might be expected. *Gold* produced a very thin film, too thin to be washed; it had a faint metallic reflexion, and transmitted a slate-blue colour like the former films. *Platinum* chloride was acted on at once; minute spots appeared here and there on the surface; these enlarged, became rough and corrugated at the middle, though brilliant at the edges, and at last formed an irregular coat over the fluid; at the part where the film was flat and brilliant, it resembled that produced by the electric explosion, and by transmission gave a dark grey colour. *Iridium* required much time, and formed a crust from centres like the platinum. *Palladium* gave an instant action, but most of the reduced metal sank in a finely divided state; a film may be obtained, but it has very little adhesion. *Rhodium* is reduced, but the film consists of floating particles, having so little adhesion that it cannot be gathered up. *Silver* is reduced, but the film is very thin and has no tenacity.

A copper film of very beautiful character may be obtained as follows in all varieties of thickness. Let a little oxide of copper be dissolved in olive-oil to form a bath, and having immersed some plates of glass, for which purpose microscope plates 3 × 1 inches are very convenient, let the whole be heated up to the decomposing temperature of the oil; being left to cool, and the plates then drained and washed successively in camphine and alcohol, they will be found covered with a film of copper, having the proper metallic lustre and colour by reflexion; and by transmission, presenting a green colour, which, though generally inclining to olive, is in the thinner films often more beautiful than the green presented by pressed gold.

Diffused particles of gold—production—proportionate size—
colour—aggregation and other changes.

Agents competent to reduce gold from its solution are very numerous, and may be applied in many different ways, leaving it either in films, or in an excessively subdivided condition. Phosphorus is a very favourable agent when the latter object is in view. If a piece of this substance be placed under the surface of a moderately strong solution of chloride of gold, the reduced metal adheres to the phosphorus as a granular crystalline crust. If the solution be weak and the phosphorus clean, part of the gold is reduced in exceedingly fine particles, which becoming diffused, produce a beautiful ruby fluid.

This ruby fluid is well obtained by pouring a weak solution of gold over the phosphorus which has been employed to produce films, and allowing it to stand for twenty-four or forty-eight hours ; but in that case all floating particles of phosphorus should be removed. If a stronger solution of gold be employed, the ruby fluid is formed, but it soon becomes turbid and tends to produce a deposit. When the gold is in such proportion that it remains in considerable excess, still the ruby formation is not prevented, and being formed it mingles unchanged with the excess of gold in solution. If an exceedingly weak solution of gold be employed the production of ruby appears to be imperfect and retarded. The nearer the solution is to neutrality at the commencement the better ; when a little hydrochloric acid was added the effect was not so good, and the colour of the fluid was more violet than ruby.

If a pint or two of the weak solution of gold before described be put into a *very clean* glass bottle, a drop of the solution of phosphorus in sulphide of carbon added, and the whole well shaken together, it immediately changes in appearance, becomes red, and being left for six or twelve hours, forms the ruby fluid required ; too much sulphide and phosphorus should not be added, for the reduced gold then tends to clot about the portions which sink to the bottom.

Though the sulphide of carbon is present in such processes and very useful in giving division to the phosphorus, still it is not essential. A piece of clean phosphorus in a bottle of the gold solution gradually produces the ruby fluid at the bottom, but the action is very slow. If the phosphorus be attached to the side of the bottle, but always beneath the surface of the solution, the streams of ruby fluid may be seen moving both upwards and downwards against the side of the glass, and forming films in the vicinity of the phosphorus perfect in their golden reflexion, and yet transmitting light of ruby, violet, and other tints, thus giving, first a proof that the particles are gold, and then connecting the present condition of the gold with that of the films already described. On the other hand, the phosphorus may be excluded and the sulphide of carbon employed alone; for when it and the solution of gold are shaken together, the gold is reduced and the ruby fluid formed; but it soon changes to purple or violet.

A quick and ready mode of producing the ruby fluid, is to put a quart of the weak solution of gold (containing about 0·6 of a grain of metal) into a clean bottle, to add a little solution of phosphorus in ether, and then to shake it well for a few moments: a beautiful ruby or amethystine fluid is immediately produced, which will increase in depth of tint by a little time. Generally, however, the preparations made with phosphorus dissolved in sulphide of carbon, are more ruby than those where ether is the phosphorus solvent. The process of reduction appears to consist in a transfer of the chlorine from the gold to the phosphorus, and the formation of phosphoric or phosphorous acids and hydrochloric acid, by the further action of the water.

The fluids produced may easily be tested for any gold yet remaining unreduced, by trial of a portion with solution of protochloride of tin. If any be found, it is easily reduced by the addition of a little more of the phosphorus in solution. After all the gold is separated as solid particles, the fluid may be considered in its perfected state. Occasionally it may smell of phosphorus in excess, even after it has been poured off from the deposited particles of it and the sulphide. In that case it is easy to deprive it of this excess by agitation in a bottle with air. When kept in closed vessels mouldiness often occurs. If this be in groups it is collected with facility at the end of a splinter of wood and removed, or the whole fluid may be poured through a wet plug of cotton in the neck of a funnel, the reduced gold passing freely. All the vessels used in these operations must be very clean; though of glass, they should not be supposed in proper condition after wiping, but should be soaked in water, and after that rinsed with distilled water. A glass supposed to be clean, and even a new bottle, is quite able to change the character of a given gold fluid.

Fluids thus prepared may differ much in appearance. Those from the basins, or from the stronger solutions of gold, are often evidently turbid, looking brown or violet in different lights. Those prepared with weaker solutions and in bottles, are frequently more amethystine or ruby in colour and apparently clear. The latter, when in their finest state, often remain unchanged for many months, and have all the appearance of solutions. But they never are such, containing in fact no dissolved, but only diffused gold. The particles are easily rendered evident, by gathering the rays of the sun (or a lamp) into a cone by a lens, and sending the part of the cone near the focus into the fluid; the cone becomes visible, and though the illuminated particles cannot be distinguished because of their minuteness, yet the light they reflect is golden in character, and seen to be abundant in proportion to the quantity of solid gold present. Portions of fluid so dilute as to show no trace of gold, by colour or appearance, can have the presence of the diffused solid particles rendered evident by the sun in this way. When the preparation is deep in tint, then common observation by reflected light shows the suspended particles, for they produce a turbidness and degree of opacity which is sufficiently evident. Such a preparation contained in a pint bottle will seem of a dull pale-brown colour, and nearly opake by reflexion, and yet by transmission appear to be a fine ruby, either clear or only slightly opalescent.

That the ruby and amethystine fluids hold the particles in suspension only, is also shown by the deposit which occurs when they are left at rest. If the gold be comparatively abundant, a part will soon settle, *i. e.* in twenty-four or forty-eight hours; but if the preparation be left for six or eight months, a part will still remain suspended. Even in these portions, however, the diffused state of the gold is evident; for where, as in some cases, the top to the depth of half an inch or more has become clear, it is seen that the ruby portion below is as a cloud sinking from it; and in the part which has apparently been cleared from colour by the settling of the particles, the lens and cone of light still show the few, or rather the fine diffused particles yet in suspension, though the proto-chloride of tin can show no gold in solution. The mould or mucus before spoken of, often collects the larger, heavier particles, and becomes of a dark blue colour; it may then be taken out by a splinter of wood, and being shaken in water, disengages the particles, which issue from it in clouds like the sporules from a ripe puff-ball.

A gradual change goes on amongst the particles diffused through these fluids, especially in the cases where the gold is comparatively abundant. It appears to consist of an aggregation. Fluids, at first clear or almost clear to ordinary observation, become turbid; being left to stand for a few days, a deposit falls. If the supernatant fluid be separated and left to stand, another deposit may be obtained. This process may be repeated, and whilst the deposition goes on, the particles in the fluid still seem to aggregate; it is only when the fluid is deprived of much gold that the process appears to stop. Even after the fluid has attained a fine marked ruby tint, if allowed to stand for months in a place of equable temperature, the colouring particles will appear in floating clouds, and probably the aggregation is then still going on. That the particles of gold when they touch each other do in many cases adhere together with facility, is shown in many experiments. In order to test this matter mechanically, I gave much agitation to a dense ruby fluid, but did not find it cause any sensible change in the character. When gold particles of a much larger size were agitated in water, they did cohere together, and the fluid, which required a certain time for settling at the beginning of the experiment, settled in a much shorter time at the termination.

If these fluids be examined generally their appearances differ not merely under different circumstances, but also under the same circumstances, though they always consist of a colourless liquid and diffused particles of gold. A certain fluid in a bottle or glass, looked at from the front, *i. e.* the illuminated side by general daylight, may appear hazy and amethystine, whilst in

bright sunlight it will appear light brown and almost opake.
From behind, the same fluid may appear of a pure blue in both
lights, whilst from the side it may appear amethystine or ruby.
These differences result from the mixture of reflected and
transmitted lights, both derived from the particles, the former
appearing in greatest abundance from the front or side, and the
latter from behind. The former is seen by common observation
in a purer state if a black background be placed behind the
fluid; when a white background is there, much of the transmitted light from that source comes to the eye, and the appearance is greatly altered. A mode of observing the former
by a strong ray of light and a lens has been already described;
but even in that case some effects of transmitted light are
observed if the focus is thrown deep into the fluid; and it is
only the particles near the surface, whether illuminated by the
base or the apex of the cone, which give the nearly pure effect
of reflexion. In order to observe the transmitted ray in an
unmingled state, a glass tube closed at one end was surrounded
with a tube of black paper longer than itself, and with the
black surface inwards. When a fluid (or the particles in it)
was to be examined, it was put into this tube, and a surface of
white paper illuminated by daylight or the sun, regarded
through it, other light being excluded from the eye; or the
tube was sometimes interposed between the eye and the sky,
and sometimes the rays of the sun itself were reflected up to
the eye through it. In speaking hereafter of the tints of the
light transmitted by the particles (which will of course vary
with the proportion of different rays in the original beam of
light), a pure white original light is to be understood, but
occasionally differently-tinted papers were employed with this
tube as sources of different coloured lights.

The very oblique angle at which reflected light comes to
the eye from the diffused particles, is well seen when the lens
cone, or a direct ray of the sun, is passed into the fluid and
observed from different positions; it is only when the eye is
behind and nearly in the line of the ray, that the unmixed
transmitted ray is observed. In the dark tube I think that no
reflected light arrives at the eye: for if half an inch in depth
of water be introduced, white light passes; if a drop of the
washed deposit, to be hereafter described, be introduced, the

light transmitted is either blue or ruby, or of other intermediate tint, according to the character of the deposit; but
if water be then added until the column is six inches or
more in length, the quantity of light transmitted does not
sensibly alter, nor its tint; a fact, which I think excludes the
idea of any light being reflected from particle to particle, and
finally to the eye.

If a given ruby-tinted fluid, containing no gold in solution,
be allowed to stand for a few days, a deposit will fall from
which the fluid may be removed by a siphon; being now
allowed to stand for a week, a second deposit will be produced;
if the fluid be again removed and allowed to stand for some
months, another deposit will be obtained, and the fluid will
probably be of a bright ruby; if it be now allowed to stand for
several months, it will still yield a deposit, looking, however,
more like a ruby fluid than a collection of fine particles at the
bottom of the fluid, whilst traces of yet finer particles of gold
in suspension may be obtained by the lens. All these deposits
may be washed with water and will settle again; the coarser
are not much affected, but the finer are, and tend to aggregate;
nevertheless specimens often occur, especially after boiling,
which tend to preserve their fine character after washing, if
the water be very clean and pure.

The colour of these particles whilst under, or diffused
through water, is by common reflected light brown, paler and
richer, sometimes tending to yellow, and sometimes to red.
The same difference is shown when illuminated by sunlight.
Everything tends to show that the light reflected is very
bright considering the size of the particles, and therefore of
the reflecting surfaces; yet comparing by the cone of light a
ruby fluid when first prepared and before it has become very
sensibly turbid, with the same fluid after the evident turbidity
is produced, in both of which cases I believe the gold to be in
solid metallic particles, though of different sizes, it would seem
that more light is transmitted and absorbed and less reflected
by the finer particles than by the coarser set, the same quantity
of gold being in the same space. I believe that there may be
particles so fine as to reflect very little light indeed, that
function being almost gone. Occasionally some of the fluids
containing the very finest particles in suspension, when illu-

minated by the sun's rays and a lens, appeared to give a fine green reflexion, but whether this is a true colour as compared to white light, or only the effect of contrast with the bright ruby in the other parts of the fluid, I am not prepared to say.

When the deposits were examined in the dark tube by transmitted light, being first diffused in more or less water to give them the form of fluid, those first deposited, and therefore presumed to be the heavier and larger, transmitted a pure blue light. The second and the third had the same character, perhaps the fourth, if the subdivision into portions had been numerous; then came some which transmitted an amethystine ray from the white of paper; and others followed progressing to the finest, which transmitted a rich ruby tint. It is probable that many of these deposits were mixtures of particles having different characters, and this is perhaps the reason that in some cases, when the fluids were contained in round-bottomed flasks, the lens-like deposit was ruby at the edges, though deep violet in the middle, the former having settled last; but as a pure blue deposit could be obtained, and also one transmitting a pure ruby ray, and as a comparatively pure intermediate preparation transmitting a ruby violet, or amethystine ray, was obtained, it is probable that all gradations from blue to ruby exist; for the production of which I can see no reason to imagine any other variation than the existence of particles of intermediate sizes or proportions.

When light other than white was passed through the fluids, then of course other tints were produced, yet some of these were unexpected. A fluid of a pure blue colour, whilst in the dark tube, would in an open glass and by reflected light appear of a strong ruby-violet tint. Dropping some of the wet deposit into pure water, the striæ which it formed would in one part be ruby in colour and in another violet: these effects were referable to the light reflected from the solid particles back through the fluid to the eye, but it seemed redder than any which light reflected from gold was likely to produce. However, upon regarding the surface of dull gold-leaf, or the thick wet deposit of gold, or the hand, it was found that the red rays easily passed through the blue fluid and formed a ruby-violet tint. Prevost showed in old times, how much the red and warm rays are reflected by gold, in preference to the others contained in white light.

The supernatant fluid in specimens that had stood long and deposited, was always ruby; yet because it showed no dissolved gold, because it showed the illuminated cone by the lens, and because by standing ruby clouds settled in it, there was every reason to believe that the gold was there in separated particles, and that such specimens afforded cases of extreme division, which by long standing would form deposits of the finest kind.

Those fluids which on standing gave abundance of deposits, transmitting blue light, consisted in the first instance of particles transmitting a ruby light, and in these cases it would seem that the particles at their first separation were always competent to transmit this ruby light; and if the preparation were not too rich in gold, the ruby condition appeared to be retained, the division being then most extreme. But purple or amethystine fluids could be procured, which, containing no colouring particles other than suspended gold, still retained them in suspension for many months together, so that they must have been as light or as finely divided as those in the ruby fluids. When the phosphorous ether was employed for the reduction of the gold, such fluids occurred; also when the solution of the phosphorus in sulphide of carbon was used, provided the solution of gold had a very little chloride of sodium contained in it. They appear to show that the mere degree of division is not the only circumstance which determines the aptitude to transmit in preference this or that ray of light.

Considering the fluids as owing their properties to diffused particles, it may be observed, that many of them which in small quantities in the dark tube transmit an amethystine light, send forward a ruby light when the quantity is increased; and this appears to be the general progression. I have not found any which by increase in quantity tended to transmit the blue rays in preference to the red.

Elevation of temperature had an effect upon these fluids which is advantageous in their preparation. On boiling an apparently clear ruby fluid for some time, its colour passed a little towards amethystine, and on boiling a like amethystine fluid, its tint passed towards blue. The separation of the gold particles was also facilitated, for now they would settle in three or four days from a fluid which, prior to this operation, would not have deposited them in an equal degree for weeks. In the case of the ruby fluids the colour often became more rosy and luminous, and by reflected light the fluid seemed to have become more turbid, as if the particles had gained in reflective power; in fact the boiling often appeared to confer a sort of permanency on the particles in their new state. When settled, they formed collections looking like little lenses of a deep ruby or violet colour, at the bottom of the flasks containing the fluid; when all was shaken up the original fluid was reproduced, and then, by rest, the gold re-settled. This effect could be obtained repeatedly. The particles could fall together within a certain limit, but many weeks did not bring them nearer or into contact; for they remained free to be diffused by agitation. The space they occupied in this lens-like form must have been a hundred-fold or even a thousandfold, more than that, which they would have filled as solid gold. Whether the particles be considered as mutually repulsive, or else as molecules of gold with associated envelopes of water, they evidently differ in their physical condition, for the time, from those particles which by the application of salt or other substances are rendered mutually adhesive, and so fall and clot together.

In preparing some of these fluids, I made the solution of gold hot and boiling before adding the solution of phosphorus. The phenomena were the same in kind as before: but when the phosphorus was dissolved in sulphide of carbon, the gold soon fell as a dark flocculent deposit; when it was dissolved in ether a more permanent turbid ruby fluid was obtained, which, if it does not go on changing in aggregation, may give a good ruby deposit.

The particles in these fluids are remarkable for a set of physical alterations occasioned by bodies in small quantities, which do not act chemically on the gold, or change its intrinsic nature; for through all of them it seems to remain gold in a fine state of division. They occur most readily where the particles are finest, *i. e.* in the ruby fluids, and so readily that it is difficult to avoid them; they are often occasioned by the contact of vessels which are supposed to be perfectly clean. An idea of their nature may be obtained in the following manner. Place a layer of ruby fluid in a clean white plate, dip the tip of a glass rod in a solution of common salt and touch the ruby fluid; in a few moments the fluid will become blue or violet-blue, and sometimes almost colourless: by mingling up the neighbouring parts of the fluid, it will be seen how large a portion of it can be affected by a small quantity of the salt. By leaving the whole quiet, it will be found that the changed gold tends to deposit far more readily than when in the ruby state. If the experiment be made with a body of fluid in a glass, twelve or twenty-four hours will suffice to separate gold, which in the ruby state has remained suspended for six months.

The fluid changed by common salt or otherwise, when most altered, is of a violet-blue, or deep blue. Any tint, however, between this and the ruby may be obtained, and, as it appears to me, in either of two ways; for the intermediate fluid may be a mixture of ruby and violet fluids, or, as is often the case, all the gold in the fluid may be in the state producing the intermediate colour; but as the fluid may in all cases be carried on to the final violet-blue state, I will, for brevity sake, describe that only in a particular manner. The violet or blue fluid, when examined by the sun's rays and a lens, always gives evidence showing that the gold has not been redissolved, but is still in solid separate particles; and this is confirmed by the non-action of protochloride of tin, which, in properly prepared fluids, gives no indication of dissolved gold. When a ruby solution is rendered blue by common salt, the separation of the gold as a precipitate is greatly hastened; thus when a glass jar containing about half a pint of the ruby fluid had a few drops of brine added and stirred into the lower part, the lower half of the fluid became blue whilst the upper remained ruby; in that state the cone of sun's rays was beautifully developed in both parts. On standing for four hours the lower part became paler, a dark deposit of gold fell, and then the cone was feebly luminous there, though as bright as ever in the ruby above. In three days no cone was visible in the lower fluid; a fine cone appeared in the upper. After many days, the salt diffused gradually through the whole, first turning the gold it came in contact with blue, and then causing its precipitation.

Such results would seem to show that this blue gold is aggregated gold, *i. e.* gold in larger particles than before, since they precipitate through the fluid in a time which is as nothing to that required by the particles of the ruby fluid from which they are obtained. But that the blue particles are always merely larger particles does not seem admissible for a moment, inasmuch as violet or blue fluids may be obtained in which the particles will remain in suspension as long as in the ruby fluids; there is probably some physical change in the condition of the particles, caused by the presence of the salt and such affecting media, which is not a change of the gold as gold, but rather a change of the relation of the surface of the particles to the surrounding medium.

When salt is added in such quantity as to produce its effect in a short time, it is seen that the gold reflexion of the particles is quickly diminished, so that either as a general turbidness or by the cone of rays it becomes less visible; at last the metal contracts into masses, which are comparatively so few and separate, that when shaken up in the fluid, they confer little or no colour or character, either by reflected or transmitted light. In these cases no re-solution of the metal is effected, for neither the salt nor hydrochloric acid, when used in like manner, has any power to redissolve the gold. The same aggregating effect is shown with all the fluids whatever their colour, and also with the deposits that settle down from them. When salt is added to the solution of gold *before* the phosphorus, and therefore before the reduction of the gold, the fluid first produced is always ruby; but it becomes violet, purple, or blue, with a facility in proportion to the quantity of salt present. If that be but small, the ruby will remain for many days unchanged in colour, and the violet-ruby for many weeks, before the gold will be deposited, the degree of dilution or concentration always having its own particular effect, as before described; the more finely divided preparations, *i. e.* the ruby and amethystine, appear to be more permanent than when the salt is added after the separation of the gold.

Many other bodies besides salt have like action on the particles of gold. A ruby fluid is changed to or towards blue by solutions of chlorides of calcium, strontium, manganese; sulphates of magnesia, manganese, lime; nitrates of potassa, soda, baryta, magnesia, manganese; acetates of potassa, soda, and lime; these effect the change freely: the sulphate of soda, phosphates of soda and potassa, chlorate of potassa, and acetate of ammonia acted feebly. Sulphuric and hydrochloric acids produce the change, but show no tendency to dissolve the gold. Nitric acid acts in the same manner, but not so strongly: it often causes re-solution of the gold after some time, because of the hydrochloric acid which remains in the fluid.

Amongst the alkalies, potash produces a similar action in a weak degree. So also does soda. Lime-water produces a change in the same direction, but the gold quickly precipitates associated with the lime.

Ammonia causes the ruby fluid to assume a violet tint; the deposit is slow of formation and often ruby in colour; the alkali apparently retards the action of common salt.

Chlorine or nitromuriatic acid turns the ruby fluid blue or violet-blue before they dissolve the gold.

Solution of sulphuretted hydrogen changes the ruby slowly to purple, and finally to deep blue. Ether, alcohol, camphine, sulphide of carbon, gum, sugar and glycerine cause little or no change in the fluids; but glycerine added to the dense deposits causes serious condensation and alteration of them, so that it could not be employed as a medium for the suspension of particles in the microscope.

All endeavours to convert the violet gold back into ruby were either failures, or very imperfect in their results. A violet fluid will, upon long standing, yield a deposit and a supernatant ruby fluid, but this I believe to be a partial separation of a mixture of violet and ruby gold, by the settlement of the blue or violet gold from ruby gold, which remains longer in suspension. Mucus, which often forms in portions of these fluids that have been exposed to the air, appears sometimes to render a fluid more ruby, but this it does by gathering up the larger violet particles; it often becomes dark blue or even black by the particles of gold adhering to it, many of which may be shaken out by agitation in water; but I never saw it become ruby-coloured as a filter can, and I think that in these cases it is the gathering out of the blue or violet particles which makes the fluid left appear more ruby in tint. I have treated blue or violet fluid with phosphorus in various ways, but saw no appearance of a return in any degree towards ruby. Sometimes the fluids possess a tendency to re-solution of the gold, a condition which may often be given by addition of a very little nitric acid, but in these cases the gold does not become ruby before solution.

It would rather appear that the finer ruby particles dissolve first, for the tint of the fluid, if ruby-violet at the commencement, changes towards blue. One effect only seemed to show the possibility of a reversion. Filtering-paper rendered ruby by a ruby fluid was washed and dried; being wetted by solution of caustic potash, it did not change; but being heated in a tube with the alkali, it became of a grey-blue tint; pouring off the alkali, washing the paper, and then adding dilute sulphuric or nitric acid to it, there was no change; but on boiling the paper in the mixed acids there was a return, and when the paper was washed and dried it approached considerably to the original ruby state. Again, potash added to it rendered it blue, which by washing with water, and especially with a little nitric acid, was much restored towards ruby. These changes may be due to an affection of the surface, or that which may be considered the surface of the particles.

The state of division of these particles must be extreme; they have not as yet been seen by any power of the microscope. Whether those that are ruby have their colour dependent upon a particular degree of division, or generally upon their being under a certain size, or whether it is consequent in part upon some other condition of the particles, is doubtful; for judging of their magnitude by the time occupied in their descent through the fluid, it would appear that violet and blue fluids occur giving violet deposits, which still consist of particles so small as to require a time equally long with the ruby particles for their deposition, and indeed in some specimens to remain undeposited in any time which has yet occurred since their formation. These deposits, when they occur, look like clear solutions in the fluid, even under the highest power of the microscope.

I endeavoured to obtain an idea of the quantity of gold in a given ruby fluid, and for this purpose selected a plate of gold ruby glass, of good full colour, to serve as a standard, and compared different fluids with it, varying their depth, until the light from white paper, transmitted through them, was apparently equal to that transmitted by the standard glass. Then known quantities of these ruby fluids were evaporated to dryness, the gold converted into chloride, and compared by reduction on glass and otherwise with solutions of gold of known strengths. A portion of chloride of gold, containing

0·7 of a grain of metal, was made up to 70 cubic inches by the
addition of distilled water and converted into ruby fluid: on
the sixth day it was compared with the ruby glass standard,
and with a depth of 1·4 inch was found equal to it; there was
just one hundredth of a grain of gold diffused through a cubic
inch of fluid. In another comparison, some gold leaves were
dissolved and converted into ruby fluid, and compared; the
result was a fluid, of which 1·5 inch in depth equalled the
standard, a leaf of gold being contained in 27 cubic inches of
the fluid. Hence looking through a depth of 2·7 inches, the
quantity of gold interposed between the light and the eye
would equal that contained in the thickness of a leaf of
gold. Though the leaf is green and the fluid ruby, yet it is
easy to perceive that more light is transmitted by the latter
than the former; but inasmuch as it appears that ruby fluids
may exist containing particles of very different sizes (or that
settle at least with very different degrees of rapidity), so it is
probable that the degree of colour, and the quantity of gold
present, may not be always in the same proportion. I need
hardly say that mere dilution does not alter the tint sensibly,
i. e. if a deep ruby fluid be put into a cylindrical vessel, and
the eye look through it along the axis of the vessel, dilution
of the fluid to eight or ten times its volume does not sensibly
alter the light transmitted. From these considerations, it
would appear that one volume of gold is present in the ruby
fluid in about 750,600 volumes of water; and that whatever
the state of division to which the gold may be reduced, still
the proportion of the solid particles to the amount of space
through which they are dispersed, must be of that extreme
proportion. This accords perfectly with their invisibility in
the microscope; with the manner of their separation from the
dissolved state; with the length of time during which they can
remain diffused; and with their appearance when illuminated
by the cone of the sun's rays.

The deposits, when not fixed upon glass or paper, are much
changed by drying; they cannot be again wetted to the same
degree as before, or be again diffused; and the light reflected
or refracted is as to colour much altered, as might be expected.
Whilst diffused through water, they seem to be physical asso-
ciations of metallic centres with enveloping films of water, and

as they sink together will lie for months at the bottom of the
fluid without uniting or coming nearer to each other, or without
being taken up by metallic mercury put into the same vessel.
This is consistent with what we know of the manner in which
gold and platinum can be thoroughly wetted if cleaned in water,
and of the difference which occurs when they are dried and
become invested with air. I endeavoured to transfer the gold
particles unchanged into other media, for the purpose of noting
any alteration in the action on light. By decanting the water
very closely, and then carefully adding alcohol with agitation,
I could diffuse them through that fluid; they still possessed a
blue colour when looked through in the dark tube, but seemed
much condensed or aggregated, for the fluid was obscure, not
clear, and the particles soon subsided. I could not transfer
them from alcohol to camphine; they refused association with
the latter fluid, retaining a film of alcohol or water, and adhering by it to the glass of the vessel; but when the camphine
was removed, a partial diffusion of them in fresh alcohol could
be effected, and gave the colour as before. All these transfers,
however, injured the particles as to their condition of division.
In one case I obtained a ruby film on a white plate; on pouring off the water and allowing parts to become dry, these became violet, seen by the light going through them to the plate
and back again to the eye. I could not wet these places with
water; a thin feebly reflecting surface remained between it and
them. Using alcohol, the parts already dry remained violet,
when wetted by it; but wetting other parts with alcohol before
they were dry from water, they remained rosy, became bluish
when dry from the alcohol, and became rosy again when re-
wetted by it.

It will be necessary to speak briefly of the reduction of gold
into a divided state by some other chemical agents than those
already described[*]. If a drop of solution of protosulphate of
iron be introduced to, and instantly agitated with, a weak
neutral solution of chloride of gold in such proportion that
the latter shall be in excess, the fluid becomes of a blue-grey
colour by transmission and brown by reflexion; and a deposit is
formed of a green colour by transmitted light, greatly resembling

---

[*] See Gmelin's 'Chemistry,' vi. p. 219, "Terchloride of gold," for numerous references in relation to changes of these kinds.

the colour of beaten or pressed metal. It is not, however, pure gold, but an association of it and oxide of iron. Hydrochloric or other acids remove the iron and reduce the gold to a dark, dense, insoluble set of particles, in very small quantity apparently, yet containing all that was present in the bulky green deposit. If the solution of gold be made slightly acid beforehand, then the change and precipitation is to appearance much less; the reflexion by the particles is feeble, but of a pale brown colour: the general transmitted light is amethystine; in the dark tube the tint is blue; the particles are much condensed and settle quickly, but occasionally leave a good ruby film on the side of the glass, which has all the characters of the ruby films and particles before described. The loose gold particles quickly adhere together. Hence it appears that the green precipitate often obtained by protosulphate of iron is not pure gold in a divided state; and that when care is taken to produce such pure divided gold, it presents the appearances of divided gold obtained by other means, the gold being competent to produce the ruby, amethystine, and blue colours by transmission. Usually the gold rapidly contracts and becomes almost insensible, and yet the test of protochloride of tin will show *that* all has been separated from solution; it then forms a *striking* contrast to the depth of colour presented by the same solution of gold precipitated by phosphorus, and most impressively directs attention to the molecular condition of the metal in the latter state.

A very small quantity of *protochloride of tin*, added to a dilute solution of gold, gave, first the ruby fluid, showing diffused particles by the cone of rays; this gradually became purple, and if the gold were in sufficient quantity, a precipitate soon began to fall, being the purple of Cassius. If the chloride of tin were in larger quantity, a more bulky precipitate fell and more quickly. Acid very much reduced this in quantity, dissolving out oxide of tin, and leaving little else than finely-divided gold, which, when diffused and examined in the dark tube, transmitted a blue colour. I believe the purple of Cassius to be essentially finely-divided gold, associated with more or less of oxide of tin.

Tartaric acid being added to a weak solution of gold gradually reduced it. The amethystine tint produced by diffused particles first appeared, and then a blue deposit of larger particles; whilst the side and bottom of the glass became covered by an adhering film of finer particles, presenting the perfect ruby tint of gold.

Ether added to a weak solution of gold gradually reduced it; the fluid was brown by reflected light, fine blue by transmitted light, and gave a good cone by the sun's rays and lens. The blue colour was not deep, though all the gold had been separated from solution; the preparation closely resembled that made with protosulphate of iron and a little acid.

A weak solution of gold, mingled with a little sugar, being heated, yielded a very characteristic decomposition. The gold was reduced into diffused particles, which rendered the fluid of a ruby-amethystine colour, and which, upon standing for twenty-four hours, gave signs of separation by settling as on former occasions. A little glycerine with solution of gold reduces it at common temperatures, producing a fluid, brown by reflexion, blue by transmission, giving a fine cone of rays by its suspended particles. Heat quickens the action, and causes a blue deposit.

Organic tissues often reduce solutions of gold, light if present assisting the action; and they afford valuable evidence in aid of the solution of the question relative to the condition of the metal in the divided state. If the skin be touched with a solution of gold, it soon becomes stained of a dull purple colour. If a piece of the large gut of an ox be soaked first in water, then in a solution of gold, and be afterwards taken out and allowed to dry, either exposed to light or not, the inner membrane will become so stained, that though of a dull purple colour by common observation, a transmitted ray will show it to be generally a very fine ruby, equal to that of ruby-coloured glass, or the gold fluids already described, though perhaps in places of a beautiful violet hue. The character of the particles which are here located and not allowed to diffuse and aggregate, as in the fluids, will be resumed when dealing with the whole question of the metallic nature of the particles of the variously divided gold.

Chloride of gold is reducible by heat alone. If a drop of solution of chloride of gold be evaporated in a watch-glass, or on a plate of rock-crystal, and then heated over a spirit-lamp until the gold is reduced, it will generally be found that the vapour has carried a portion of gold on to the neighbouring part of the glass, and that this part, when placed over a sheet of white paper, has the ruby tint. With the rock-crystal both ruby and blue parts are produced; and when the ruby parts are subjected to rock-crystal pressure, they become beautifully green. In the arts also glass is oftentimes coloured ruby by gold; I think that glass in this state derives its colour from diffused divided gold; and if either the ruby glass or the watch-glass be examined by a lens and the cone of rays, it will be seen that the colours are not due to any gold dissolved, but to solid and diffused particles. There is nothing in any of the appearances or characters, or in the processes resorted to to obtain the several effects, that point at any physical difference in the nature of the results; and without saying that gold cannot produce a ruby colour whilst in combination or solution, I think that in all these cases the ruby tint is due simply to the presence of diffused finely-divided gold.

Metallic character of the divided gold.

Hitherto it may seem that I have assumed the various preparations of gold, whether ruby, green, violet, or blue in colour, to consist of that substance in a metallic divided state. I will now put together the reasons which caused me to draw that conclusion. With regard to *gold-leaf* no question respecting its metallic nature can arise, but it offers evidence reaching to the other preparations. The green colour conferred by pressure, and the removal of this colour by heat, evidently belong to it as a metal; these effects are very striking and important as regards the action on light; and where they recur with other forms of gold, may be accepted as proof that the gold is in the metallic state. Although I do not attach equal importance to the fact already described, that gold-leaf frequently presents fine parts that appear to be ruby in colour, I am not as yet satisfied that they are not in themselves ruby; and if they should be so, it will be another proof by analogy of the metallic nature of other kinds of preparations eminently ruby.

The *deflagrations* of gold wire by the Leyden discharge can be nothing but divided gold. They are the same whatever the atmosphere surrounding them at the time, or whatever the substance on which they are deposited. They have all the chemical reactions of gold, being, though so finely divided, insoluble in the fluids that refuse to act on the massive metal, and soluble in those that dissolve it, producing the same result. Heat makes these divided particles assume a ruby tint, yet such heat is not likely to take away their metallic character, and when heated they still act with chemical agents as gold. Pressure then confers the green colour, which heat takes away, and pressure reconfers. All these changes occur with particles attached to the substances which support them by the slightest possible mechanical force, just enough indeed to prevent their coalescence and to keep them apart and in place, and yet offering no resistance to any chemical action of test agents, as the acids, &c., not allowing any supposition of chemical action between them and the body supporting them. Still this gold, unexceptionable as to metallic state, presents different colours when viewed by transmitted light. Ruby, green, violet, blue, &c. occur, and the mere degree of division appears to be the determining cause of many of these colours. The deflagrations by the voltaic battery lead to the same conclusion.

The *gold films* produced by phosphorus have every character belonging to the metallic state. When thick, they are in colour, lustre, weight, &c. equal to gold-leaf; but in the unpressed state, their transmitted colour is generally grey, or violet-grey. The progression of their lustre and colour is gradual from the thickest to the thinnest, and the same is generally true, if thick films are gradually thinned and dissolved whilst floating on solvents; the thick and the thin films must both be accepted as having the same amount of evidence for their metallic nature. When subjected to chemical agents, both the thick and the thin films have the same relations as pure metallic gold. These relations are not changed by the action of heat, yet heat shows the same peculiar effect that it had with preparations of gold obtained by beating, or by electric deflagrations. The remarkable and characteristic effect of pressure is here reproduced, and sometimes with extraordinary results; since from the favourable manner in which the particles are occasionally divided and then held in place on the glass, the mere touch of a finger or card is enough to produce the result. Yet with gold thus proved to be metallic, colours including grey, grey-violet, green, purple, ruby, especially by heat, and green again by pressure, and by thinning of grey films, may be obtained by transmitted light, almost all of them at pleasure.

It may be thought that the *fluid preparations* present more difficulty to the admission, that they are simply cases of pure gold in a divided state; yet I have come to that conclusion, and believe that the differently-coloured fluids and particles are quite analogous to those that occur in the deflagrations and the films. In the first place they are produced as the films are, except that the particles are separated under the surface and out of the contact of the air; still, when produced in sufficient quantity against the side of the containing vessel to form an adhering film, that film has every character of lustre, colour, &c. in the parts differing in thickness, that a film formed at the surface has. Whilst the particles are diffused through the fluid it is difficult to deal with them by tests and reagents; for their absolute quantity is very small, and their physical characters are very changeable, chiefly as I believe by aggregation; still there are some expedients which enable one to submit even the finest of them to proof. In several cases particles from ruby and amethystine fluids adhere to the sides of the bottles or flasks in which the fluids had been preserved, and the process of boiling seemed to favour such a result; the adhesion was so strong, that when the fluid contents were removed and the bottles well-washed, the glass remained tinged of a ruby or of a violet colour. These films, in which the fine particles were fixed mechanically apart and in place, were then submitted to the action of various chemical agents. Drying and access of air did not cause any marked alterations in them. Strong nitric acid produced no change, nor hydrochloric acid, nor sulphuric acid. Neither did a solution of chloride of sodium, even up to brine, cause any alteration in the colour or any other character of the deposit. A little solution of chlorine or of nitromuriatic acid dissolved them at once, producing the ordinary solutions of gold. I can see no other mode of accounting for these effects (which are in strong contrast with what happens when ruby fluid is acted on by these agents), than to suppose that the gold particles, being in a high state of division, were retained in that state for the time by their adhesion to the glass. Of course chemical change was free to occur, but not a change dependent upon their mutual aggregation; yet they were not held by any special chemical attraction to, or combination with, the glass; for a touch with a card, a feather, or the finger, was sufficient to remove them at once; and if rubbed off with a point of wood, they coated it with brilliant metallic gold.

Again, though these particles are so finely divided that they pass easily through ordinary filters, still a close filter catches some; and if a ruby fluid be passed through again and again, the paper at last becomes of a rosy hue, because of the gold which adheres to it; being then well-washed, and, if needful, dried, the gold is again ready for experiment. Such gold paper, placed across the middle of the dark tube and examined by transmitted light, was of the same ruby tint as when looked through in the open air. It was unaffected by salt or brine, though these, added to the rosy fluid which had passed the filter, instantly changed it to violet-blue. Portions of the paper were put into separate glasses with brine, solutions of hydrochloric, nitric and sulphuric acids, ammonia, potassa, soda and sulphuretted hydrogen, but no change occurred with any of them in two days. On the other hand, a very dilute solution of chlorine immediately turned the ruby to blue, and then gradually dissolved the gold. A piece of the ruby paper immersed in a strong solution of cyanide of potassium suffered a very slow action, if any, and remained unaltered in colour; being brought out into the air, the gold very gradually dissolved, becoming first blue. A portion of the ruby paper was dried and heated in oil until the oil and the paper began to change their hue; the gold had not altered in its colour or character. Another portion was heated in the vapour of alcohol and also of ether until the paper began to alter; the gold remained unaltered. A blue fluid being passed oftentimes through a filter gave a blue paper, which, being washed and tried in the same manner, was found to contain particles unchanged by the simple acids or alkalies, or by heat or vapours, but dissolving, as gold would do, in chlorine or nitromuriatic acid. These tests are, I think, sufficient to prove the metallic nature and permanence of the gold as it exists in the ruby, amethystine, violet, and other coloured fluids.

The production by such different agents as phosphorus, sulphide of carbon, ether, sugar, glycerine, gelatine, tartaric acid, protosulphate of iron and protochloride of tin, of gold fluids all more or less red or ruby at the commencement, and all passing through the same order of changes, is again a proof that only gold was separated ; no single one or common compound of gold, as an oxide or a phosphide, could be expected in all these cases. Many of the processes, very different as to the substances employed to reduce the gold, left good ruby films adhering to the glass vessels used, presenting all the characters of the gold described already : this was the case with phosphorus, sugar, tartaric acid, protosulphate of iron, and some other bodies.

Again, the high reflective power of these particles (unalterable by acids and salts), when illuminated by the sun's rays and a lens, and the colour of the light reflected, is in favour of their metallic character.   So also is their aggregation, and their refusal to return from blue, violet or amethystine to ruby ; for the cohesive and adhering force of the gold particles and their metallic nature and perfect cleanliness is against such a reverse change. Particles transmitting blue light could be obtained in such quantity as to admit of their being washed and dried in a tube, and being so prepared they presented every character of gold : when heated, no oxygen, water, phosphorus, acid of phosphorus, nor any other substance was evolved from them : they changed a little, as the film when heated changed, becoming more reflective and of a pale brown colour, and contracted into aggregated porous masses of pure ordinary gold.

Gold is reduced from its solution by organic tissues ; and *stained gut* has been quoted as a case. I have a very fine specimen which by transmitted light is as pure a ruby as gold-stained glass, and I believe that the gold has been simply reduced and diffused through the tissue. The preparation stood all the trials that had been applied to the ruby films on glass or the gold deposit on filtering-paper. Portions of it remained soaking in water, solution of chloride of sodium and dilute sulphuric acid for weeks, but these caused no change from ruby to blue, such as could be effected on loose ruby particles. Strong hydrochloric acid caused no change as long as the tissue held together; but as that became loose the gold flowed out into the acid in ruby-amethystine streams, finally changing to blue. Caustic potassa caused no change for days whilst the tissue kept together, but on mixing all up by pressure the loosened gold became at last blue. Strong nitric acid caused no change of colour until, by altering the tissue, the gold particles first flowed out in ruby and amethystine streams, and then were gradually changed to the condition of common aggregated gold. All these effects, and the actions on light, accord with the idea that the stain was simply due to diffused particles of finely-divided gold; and I am satisfied that all such stains upon the skin, or other organic matter, are of exactly the same nature.

As to the gold in ruby glass, I think a little consideration is sufficient to satisfy one that it is in the metallic condition. The action of heat tends to separate gold from its state of combination, and when so separated from the chloride, either upon the surface of glass, rock-crystal, topaz, or other inactive bodies, a ruby film of particles is frequently obtained. The sunlight and lens show that in ruby glass the gold is in separated and diffused particles. The parity of the gold glass, with the ruby-gold deflagrations and fluids described, is very great. These considerations, with the sufficiency of the assigned cause to produce the ruby tint, are strong reasons, in the absence of any to the contrary, to induce the belief that finely divided metallic gold is the source of the ruby colour.

When a pure, clean, stiff jelly is prepared, and mixed, whilst warm and fluid, with a little dilute chloride of gold, as if to prepare a ruby fluid, it gelatinizes when cold, and if left for two or three days may become a ruby jelly; sometimes, however, the gold in the jelly changes but little or changes to blue, or it may happen that it is reduced on the surface as a film, brilliant and metallic by reflected light, and blue-grey by transmitted light. I have not yet ascertained the circumstances determining one or the other state. If a trace of phosphorus in sulphide of carbon be added to the solution of gold in a dilute state, and some salt be added to the warm jelly, and the latter be then mixed gradually and with agitation with the gold

2 F

solution, a ruby jelly is generally produced. In such ruby jelly the reduced particles of gold preserve their state and relative place, and the tint does not pass to blue, even though a considerable proportion of salt be present. Such jelly will remain in the air for weeks before it decays, and has every character, in colour and appearance, of gold ruby glass. It is hardly possible to examine the series of ruby glass, ruby membrane, ruby jelly cold and gelatinous, ruby jelly warm and fluid, and the ruby fluids, to consider their production, and then to conclude that the cause of their common ruby colour is not the same in all.

When the warm ruby jelly is poured into a capsule or on to a plate, allowed to gelatinize and then left in the air, it gradually becomes dry. When dry, some of these jellies remain ruby; others will probably be of an amethystine violet colour, or perhaps almost blue. When one of the latter is moistened with water, and has absorbed that fluid, it becomes gelatinous, and whilst in that state resumes its first ruby colour; but on being suffered to dry again, it returns to its amethystine or blue colour. This change will occur for any number of times, as often as the jelly is wetted and dried. Here the gold remains in the same metallic state through this great change of colour, the association or the absence of water being the cause: and the effect strengthens in my mind the thought before expressed, that in the ruby fluids the deposited particles are frequently associates of water and gold. It is a striking case of the joint effect of the media and the gold in their action on the rays of light, and the most striking case amongst those where the medium may be changed to and fro.

When a ruby jelly is prepared with salt, and being warm is poured out in thin layers on to glass or porcelain, it first gelatinizes and then dries up; in which case the salt is excluded and crystallizes. When the dry jelly is put into cold water, the salt dissolves and can be removed. The jelly then swells to a certain amount, after which it can be left soaking in water for a week or longer, until everything soluble is separated. No change takes place in the ruby tint, no gold is removed. When the last water is poured off and the remaining jelly warmed, it melts, forming a fine ruby fluid, which can either be dissolved in more water, or regelatinized, or be dried and preserved for

any length of time. It is perfectly neutral; gives no signs of
dissolved gold by any of the tests of the metal; is not changed
by sulphuretted hydrogen, gallic acid, pyrogallic acid, dilute
caustic alkalies, or carbonated alkalies or lime-water ; or by
dilute sulphuric, hydrochloric or nitric acids, the actions being
continued for fourteen days :—being boiled with zinc filings, it
does not change; and even when dilute sulphuric or hydrochloric acid is added to evolve nascent hydrogen, still the ruby
character undergoes no alteration. Strong sulphuric, or nitric,
or hydrochloric acid does not alter it whilst cold; but when
warmed, the first causes the gold to separate as dark aggregated metallic particles, and the two latter gradually cause the
change to amethyst and blue formerly described.   Chlorine, or
a mixture of hydrochloric and nitric acids, dissolves the gold,
the ruby colour disappears, and the ordinary solution of gold
is produced.   In all these cases the ruby gold behaves exactly
as metallic gold would do with the same agents, and quite unlike
what would be expected from any possible combination of oxygen
and gold.

  In some of these jellies the ruby particles are so determinate
as to give the brown reflexion by common observation ; in others
they are so fine as to look like ruby solutions, unless a strong
sunlight and a lens be employed; and the impression again
arises, that gold may exist in particles so minute as to have
little or no power of reflecting light. Ruby particles of extreme
fineness, when present in small amount in water, appear to remain equally diffused for any length of time ; if in larger amount,
that which settles to the bottom will remain for weeks and
months as a dense ruby fluid, but without coming together : both
circumstances seem to imply an association of the particles of
gold with envelopes of water.   Many circumstances about the
ruby jellies imply a like association with that animal substance,
and many of the stains of gold upon organic substances probably include an affinity of the metal of the like kind.

Relations of Gold (and other metals) to polarized Light.

  It has been already stated, that when a ray of common light
passes through a piece of gold-leaf inclined to the ray, the light
is polarized.   When the angle between the leaf and the ray is
small, about 15°, nearly all the light that passes is polarized;

but as the leaf is really very irregular in thickness, and ill-stretched as a film, parts inclined at different angles are always present at once. The light transmitted is polarized in the same direction as that transmitted by a bundle of thin plates of glass, inclined in the same direction. The proportion of light transmitted is small, as might be expected from the high reflective power of the metal. The polarization does not seem due to any constrained condition of the beaten gold, for it is produced, as will be shortly seen, by the annealed colourless leaf-gold, and also by deposits of gold particles; but is common to it with other uncrystallized transparent substances. It would seem that a very small proportion of the gold-leaf can be occupied by apertures, since the light which passes is nearly all polarized. On subjecting thin gold-leaf, or heated gold-leaf, or films of gold, or any preparations which required the support of glass, results of polarization were obtained, but the observations were imperfect because of the interfering effect of the glass.

Proceeding to employ a polarized ray of light, it was found that a leaf of gold produced generally the same depolarizing effect as other transparent bodies. Thus, if a plate of glass be held perpendicular to the ray, or inclined to it either in the plane of polarization or at right angles to it, there is no depolarization; but if inclined in the intermediate positions, the ray is more or less depolarized. So it is with gold-leaf; the same effects are produced by it. Further, the depolarization is accompanied by a rotation of the ray, and in this respect the quadrants alternate, the rotation being to the right-hand in two opposite quadrants, and to the left in the intervening quadrants. So it is with gold-leaf; the same effects are produced by it, and the rotation is in the same direction with that produced by glass, when inclined in the same quadrant.

As further observation in this direction was stopped by the necessity of employing glass supports for the leaves, films, &c., I sought for a medium so near glass in its character, as should either reduce its effect to nothing, or render it so small as to cause its easy elimination. Either camphine or sulphide of carbon was found to answer the purpose with crown-glass; but the latter, as it possesses no sensible power of rotation under ordinary circumstances, is to be preferred. Should a medium of higher optic force be required, it would probably be supplied by the use of that dangerous fluid, phosphorus dissolved in sulphide of carbon. A rectangular glass cell being provided, which did not itself affect the polarized ray, was placed in its course and filled to a certain height with sulphide of carbon. A plate of crown-glass was then introduced perpendicularly to the ray; it did not affect it; being inclined as before described, the effect on the ray was still insensible, the glass appearing to be, for all ordinary observations such as mine, quite as the medium about it. I could now introduce gold-leaf attached to glass into the course of the polarized ray, its condition as a flat film or plane being far finer than when stretched on a wire ring as before. It proved to be so far above the sulphide of carbon, as to have powers of depolarization apparently as great as those it had in air, and being inclined, brought in the image at the analyser exceedingly well. It was indeed very striking to see, when the plate was moved parallel to itself, the darkness when mere glass intervened, and the light which sprung up when the gold-leaf came into its place; the opake metal and the transparent glass having apparently changed characters with each other. By care I was able to introduce a stretched piece of gold-leaf (without glass) into the sulphide of carbon; its effects were the same with those just described.

In all the experiments to be described, the plane of polarization and the plane of inclination had the same relation to each other: the figure shows the position of the polarizing Nicol prism, as the eye looks through it at the light, and *a*, *b* represents the vertical axis, about which the plates were inclined. Whether they were inclined in one direction or the other, or had the glass face or the metal face towards the eye, made no  difference. In all cases with gold-leaf, it was found that the ray had been rotated; that it required a little direct rotation of the analyser to regain the minimum light; that short of that red tints appeared, and beyond it blue or cold, these being necessarily affected in some degree by the green colour of the gold-leaf. Thinned gold-leaf produced the same results; but as holes appeared in those that were thinnest, the results were interfered with, because the light passing through them was affected by the analyser in a different manner, and yet mingled its result with that of the light which had passed through the gold.

The gold-leaf plates, deprived of green colour by heating in oil, were found with the glass in such good annealed condition, as not to affect the ray; but when they were moved, until the oblique colourless gold came into the course of the ray, it was depolarized; a red image appeared; direct rotation of the analyser reduced this a little in intensity and then changed the colour to blue. The reduction was not much, and both in that and the first appearance of the red image, there is a difference between the heated and the unheated gold: probably the green tint of the latter, which would tend to extinguish the red and produce a minimum, may be sufficient to account for the effect. Gold which had been re-greened by agate pressure acted in like manner on the polarized ray, but the experiments were imperfect.

A glass plate having gold-leaf on one part of it, had a second glass plate put over it and gummed at the edges. In the sulphide of carbon, therefore, it represented in one part a plate of air, and in another a compound plate of air and gold; both acted in the same direction, but the air and gold much more than the air. Gold on glass in this medium, or gold in air, or glass in air, all gave results in the same direction, *i. e.* required direct rotation of the analyser to compensate for them.

I proceeded to examine the other forms of gold; and first, the deposits on glass obtained by *electric deflagration*. These affected the ray of polarized light exactly in the manner of gold-leaf, and that even at the distant parts of the deposit. It was most striking to contrast the thinnest and faintest portion of such a film with the neighbouring parts of the glass from which it had been wiped off. It must be remembered that such a preparation is a layer of separate particles; that these particles are not like those of starch or of crystals, for they have no action whilst in a plane perpendicular to the polarized ray; nor have they a better action for being in a thick layer, as in the central parts of the deposit. The particles seem to form the equivalent of a continuous plate of transparent substance; and as in such a plate it is the two surfaces which act, so there appears to be the equivalent of these two surfaces

here ; which would seem to imply that the particles are so
small and so near, that two or more can act at once upon the
individual atoms of the vibrating ether. Their association is
such as to present as it were an optical continuity.

The gold films by phosphorus were then submitted to experiment, and gave exactly the same result. All of them depolarized, and required direct rotation of the analyser to arrive at a minimum, or to pass from the red to the blue tints.
Graduated films, of which I should judge from the depth of
tint that one place was at least twenty times as thick as another, gave the effect as well in the thinnest as the thicker or
any intermediate part ; indicating that thickness of the plate,
and therefore any quality equivalent to crystalline force of the
particles, had nothing to do with the matter. A glass beaker,
which had been employed to contain ruby fluid, had a film of
gold deposited on its inner surface so thin, as to be scarcely
perceptible either by reflexion or otherwise, except by a ruby
tint which appeared upon it in certain positions ; but being examined by a polarized ray, it gave an effect as strong and as
perfect as gold-leaf, showing how thin a film of gold was sufficient for the purpose. This thin film appeared to be almost
perfect in its continuity, for when the red image was brought
in, direct rotation of the analyser reduced it to a minimum
which was quite dark ; after which, further rotation brought
in a good blue image. The least touch of the finger removed
the film of gold and all these effects with it. These films,
though they are certainly porous to gas, and to water in some
form, for it can evaporate from beneath them through its body,
have evidently optical continuity.

In order to submit the gold fluids to experiments, cells were
made of two glass plates, separated by the thickness of a card,
and fastened at the edges by varnish internally and gum ex-
ternally. These being filled with dense ruby or blue fluid,
gave no indication of action on the ray, showing that the
diffused particles were inoperative. The same fluids, dried
on plates of glass so as to leave films, did act just as the gold
deflagrations had done ; for though the particles were very
irregularly spread, parts of the general deposit, and these not
the thickest, could be selected, which produced the effect
excellently well.

When the coloured jellies are laid upon glass plates and allowed to dry, the plates introduced obliquely into the sulphide of carbon affect the ray, but not as gold films; the light image becomes visible, but the plane of polarization is not changed; the light is coloured by the ruby or blue tint of the gold present, but a film of jelly without gold makes it visible to the same extent. In this case the gold is not in one plane, but diffused through the dry jelly, and the effect is the same as if it were diffused through water, being negative.

Such are the effects with the various preparations of divided gold. I will hastily notice what occurs with some other metals. *Platinum deflagrated* in hydrogen: it depolarized the ray, required direct rotation of the analyser to attain a minimum, therefore rotated the plane of polarization; but did not present sensible colour on either side of the minimum of light. *Palladium deflagrated* in hydrogen: it depolarized, producing a red image; direct rotation of the analyser lessened the light to a minimum, and then brought in a blue image. The films of palladium obtained by phosphorus acted well in the same manner. These films appear to be exceedingly continuous, and it could be observed in them, that though the thickest were not the best, yet films could be obtained so thin as to be distinctly inferior to other parts a little thicker; also that where the brilliancy of reflexion which indicates perfect smoothness passed in any degree into dulness, the action of the film was injured: the perfect condition of the surfaces of the films seems to be essential to their good action. *Rhodium films* by phosphorus gave good actions, like those produced by gold. *Silver deflagrations*, either in air or hydrogen, gave depolarizing results like those with gold. Silver films also gave excellent results of the like kind. A thin pale brown film was much better than a thicker one. *Copper* deflagrated in hydrogen: depolarized, bringing in a red image, which by direct rotation of the analyser was lowered a little, and then converted to blue. The copper films obtained from oil acted in the same manner; the red and blue images appeared in their order; but very little direct rotation of the analyser was required to produce the minimum of light. *Tin* deflagrated in hydrogen: depolarized and rotated the ray, as with gold; the images were only feeble in colour. *Lead* deflagrated in hydrogen: acted as tin.   *Iron* deflagrated in hydrogen: acted as tin. *Zinc* deflagrated in hydrogen: acted as tin. *Aluminium* deflagrated in hydrogen: had like action with the rest; the image brought in by it was red, which direct revolution of the analyser reduced at a little distance to a minimum, and then converted to blue. A film of *mercury* produced by sublimation, a film of *arsenic* produced in like manner, and a film of *smoke* from a candle, though all of them sufficiently pervious to light, did not produce any result of depolarization. Films of the smoke of burning *zinc*, of *antimony*, or of *oxide of iron* produced no effect.

I placed some metallic solutions in a weak atmosphere of sulphuretted hydrogen. Gold and platinum gave no films; silver so poor a film as to be of no use; and lead one so brittle as to be unserviceable. That obtained with palladium I believe to be the metal itself. The films of sulphuret of mercury, sulphuret of antimony which was orange, and sulphuret of copper which was pale brown, all acted on the light, and depolarized it. The sulphuret of copper presented a difference from the metals generally, worth recording; it depolarized the light, producing an image, which, if not blue at once, was rendered blue by a little direct rotation of the analyser; after which the same motion brought in a minimum and then produced an orange or red tint, *i. e.* with the sulphuret of copper the warm and cold tints appear on opposite sides of the minimum to those where they occur when films of the metals are employed, though the minimum in both cases is in the same direction.

Many of the results obtained in the sulphide of carbon were produced also in camphine, the analyser being in each case adjusted to the minimum of light before the metallic plate or film was introduced. I pass, however, to a very brief account of some polarizations effected by the metals themselves in the sulphide of carbon, in which case the polarizing Nicol prism was dispensed with. The results show that all the dry forms of gold accord in giving the same manifestation of action on light, whatever the state of their division, provided they be disposed in a thin regular layer, equivalent to a continuous film. It was first ascertained that a plate of crown-glass in an inclined position in sulphide of carbon gave no signs of polarity to a ray of light passing through it. When fine gold-leaf was on the glass and inclined to the ray, it polarized the light, and exactly in the same manner and direction as a bundle of glass plates in the same position in the air. More light passed than when the gold-leaf was in air, but it could not be so completely polarized; the minimum light was of a pale bluish colour. A thinned gold-leaf produced the same effect, but let more common light through. I think the difference between gold-leaf and sulphide of carbon is sensibly less than that between the metal and air. The depositions of deflagrated gold, the films of gold obtained by phosphorus, and even the heated deflagrated gold, produced polarizing effects, which, though not large, were easily recognized and distinguished from the non-action of the glass. Gold-leaf and gold films on glass produced a like effect in a camphine-bath, the results being easily distinguished from those of the glass and camphine only, in places where the glass had been cleared from gold.

Films of palladium, rhodium, silver, a plate with deposited gold particles, and a layer of deflagrated silver particles gave a like result, the effect varying in degree. The sulphuret of copper before spoken of as in contrast with the metals, gave only doubtful result, if any.

Before concluding, I may briefly describe the following negative results with the preparations of gold. I prepared a powerful electro-magnet, sent a polarized ray across the magnetic field, parallel to the magnetic axis, and then placed portions of the ruby and violet fluids, also of their deposits wet and dry, also portions of the gold films, of gold-leaf, the results of deflagrations, &c., in the course of the ray; but on exciting the magnet, could not obtain any effect beyond that due to the water or glass, which in any case accompanied the substance into the magnetic field. In some cases very dense preparations of the ruby and blue deposits were employed, the intense electric lamplight being required to penetrate them.

I passed the coloured rays of the solar beam through the various gold fluids and films that have been described. For this purpose a beam of sunlight entering a dark room through an aperture $\frac{1}{6}$th of an inch in width, was sent through two of Bontemps's flint-glass prisms, and its rays were either separated, or at once thrown on to a pure white screen; the different objects were then interposed in the course of the ray, but I could not perceive when any portion of a ray passed (and that was generally the case) that it differed sensibly in colour or quality from the ray passing into the preparation. In like manner, the objects were put into the differently coloured rays and observed by the reflected light, a lens being sometimes employed to concentrate the light; but I could not find any marked difference between the colour or character of the ray reflected and the impinging ray, except in quantity.

On the Conservation of Force*.

Various circumstances induce me at the present moment to put forth a consideration regarding the conservation of force. I do not suppose that I can utter any truth respecting it that has not already presented itself to the high and piercing intellects which move within the exalted regions of science; but the course of my own investigations and views makes me think that the consideration may be of service to those persevering labourers (amongst whom I endeavour to class myself), who, occupied in the comparison of physical ideas with fundamental principles, and continually sustaining and aiding themselves by experiment and observation, delight to labour for the advance of natural knowledge, and strive to follow it into undiscovered regions.

There is no question which lies closer to the root of all physical knowledge, than that which inquires whether force can be destroyed or not. The progress of the strict science of modern times has tended more and more to produce the conviction that " force can neither be created nor destroyed," and to render daily more manifest the value of the knowledge of that truth in experimental research. To admit, indeed, that force may be destructible or can altogether disappear, would be to admit that matter could be uncreated; for we know matter only by its forces: and though one of these is most commonly referred to, namely gravity, to prove its presence, it is not because gravity has any pretension, or any exemption amongst the forms of force, as regards the principle of *conservation*; but simply that being, as far as we perceive, inconvertible in its nature and un-

* Proceedings of the Royal Institution, Feb. 27, 1857, vol. ii. p. 352.

On the Conservation of Force.

[1857.

changeable in its manifestation, it offers an unchanging test of the matter which we recognize by it.

Agreeing with those who admit the conservation of force to be a principle in physics as large and sure as that of the indestructibility of matter, or the invariability of gravity, I think that no particular idea of force has a right to unlimited or unqualified acceptance, that does not include *assent* to it; and also, to *definite amount* and *definite disposition of the force*, either in one effect or another, for these are necessary consequences: therefore, I urge, that the conservation of force ought to be admitted as a physical principle in all our hypotheses, whether partial or general, regarding the actions of matter. I have had doubts in my own mind whether the considerations I am about to advance are not rather metaphysical than physical. I am unable to define what is metaphysical in physical science; and am exceedingly adverse to the easy and unconsidered admission of one supposition upon another, suggested as they often are by very imperfect induction from a small number of facts, or by a very imperfect observation of the facts themselves: but, on the other hand, I think the philosopher may be bold in his application of principles which have been developed by close inquiry, have stood through much investigation, and continually increase in force. For instance, *time* is growing up daily into importance as an element in the exercise of force. The earth moves in its orbit in time; the crust of the earth moves in time; light moves in time; an electro-magnet requires time for its charge by an electric current: to inquire, therefore, whether power, acting either at sensible or insensible distances, always acts in *time*, is not to be metaphysical; if it acts in time and across space, it must act by physical lines of force; and our view of the nature of the force may be affected to the extremest degree by the conclusions, which experiment and observation on time may supply; being, perhaps, finally determinable only by them. To inquire after the possible time in which gravitating, magnetic, or electric force is exerted, is no more metaphysical than to mark the times of the hands of a clock in their progress; or that of the temple of Serapis in its ascents and descents; or the periods of the occultations of Jupiter's satellites; or that in which the light from them comes to the earth. Again, in some of the known cases of action in

ERRATUM.

Page 445, line 10, *for* willingly or unwillingly *read* wittingly or unwittingly.

time, something happens whilst the *time* is passing which did not happen before, and does not continue after : it is therefore not metaphysical to expect an effect in *every* case, or to endeavour to discover its existence and determine its nature. So in regard to the principle of the conservation of force; I do not think that to admit it, and its consequences, whatever they may be, is to be metaphysical: on the contrary, if that word have any application to physics, then I think that any hypothesis, whether of heat, or electricity, or gravitation, or any other form of force, which either willingly or unwillingly dispenses with the principle of conservation, is more liable to the charge, than those which, by including it, become so far more strict and precise.

Supposing that the truth of the principle of the conservation of force is assented to, I come to *its uses.* No hypothesis should be admitted, nor any assertion of a fact credited, that denies the principle. No view should be inconsistent or incompatible with it. Many of our hypotheses in the present state of science may not comprehend it, and may be unable to suggest its consequences; but none should oppose or contradict it.

If the principle be admitted, we perceive at once, that a theory or definition, though it may not contradict the principle, cannot be accepted as sufficient or complete unless the former be contained in it; that however well or perfectly the definition may include and represent the state of things commonly considered under it, that state or result is only partial, and must not be accepted as exhausting the power or being the full equivalent, and therefore cannot be considered as representing its *whole nature*; that, indeed, it may express only a very small part of the whole, only a residual phenomenon, and hence give us but little indication of the full natural truth. Allowing the principle its force, we ought, in every hypothesis, either to account for its consequences by saying what the changes are when force of a given kind apparently disappears, as when ice thaws, or else should leave space for the idea of the conversion. If any hypothesis, more or less trustworthy on other accounts, is insufficient in expressing it or incompatible with it, the place of deficiency or opposition should be marked as the most important for examination; for there lies the hope of a discovery of new laws or a new condition of force. The deficiency should never be accepted as satisfactory, but be remembered and used as a stimulant to further inquiry; for conversions of force may here be hoped for. Suppositions may be accepted for the time, provided they are not in contradiction with the principle. Even an increased or diminished capacity is better than nothing at all; because such a supposition, if made, must be consistent with the nature of the original hypothesis, and may therefore, by the application of experiment, be converted into a further test of probable truth. The case of a force simply removed or suspended, without a transferred exertion in some other direction, appears to me to be absolutely impossible.

If the principle be accepted as true, we have a right to pursue it to its consequences, no matter what they may be. It is, indeed, a duty to do so. A theory may be perfection as far as it goes, but a consideration going beyond it, is not for that reason to be shut out. We might as well accept our limited horizon as the limits of the world. No magnitude, either of the phenomena or of the results to be dealt with, should stop our exertions to ascertain, by the use of the principle, that something remains to be discovered, and to trace in what direction that discovery may lie.

I will endeavour to illustrate some of the points which have been urged, by reference, in the first instance, to a case of power which has long had great attractions for me, because of its extreme simplicity, its promising nature, its universal presence, and its invariability under like circumstances; on which, though I have experimented* and as yet failed, I think experiment would be well bestowed: I mean the force of gravitation. I believe I represent the received idea of the gravitating force aright, in saying, that it is *a simple attractive force exerted between any two or all the particles or masses of matter, at every sensible distance, but with a strength varying inversely as the square of the distance.* The usual idea of the force implies *direct* action at a distance; and such a view appears to present little difficulty except to Newton, and a few, including myself, who in that respect may be of like mind with him†.

This idea of gravity appears to me to ignore entirely the principle of the conservation of force; and by the terms of its

* Philosophical Transactions, 1851, p. 1.   * See note, p. 451.

definition, if taken in an absolute sense, "*varying* inversely as the square of the distance," to be in direct opposition to it; and it becomes my duty, now, to point out where this contradiction occurs, and to use it in illustration of the principle of conservation. Assume two particles of matter, A and B, in free space, and a force in each or in both by which they gravitate towards each other, the force being unalterable for an unchanging distance, but varying inversely as the square of the distance when the latter varies. Then, at the distance of 10, the force may be estimated as 1; whilst at the distance of 1, *i. e.* one-tenth of the former, the force will be 100: and if we suppose an elastic spring to be introduced between the two as a measure of the attractive force, the power compressing it will be a hundred times as much in the latter case as in the former. But from whence can this enormous increase of the power come? If we say that it is the character of this force, and content ourselves with that as a sufficient answer, then it appears to me we admit a *creation* of power, and that to an enormous amount; yet by a change of condition so small and simple, as to fail in leading the least instructed mind to think that it can be a sufficient cause:—we should admit a result which would equal the highest acts our minds can appreciate of the working of infinite power upon matter; we should let loose the highest law in physical science which our faculties permit us to perceive, namely, the *conservation of force.* Suppose the two particles A and B removed back to the greater distance of 10, then the force of attraction would be only a hundredth part of that they previously possessed; this, according to the statement that the force varies inversely as the square of the distance, would double the strangeness of the above results; it would be an *annihilation* of force; an effect equal in its infinity and its consequences with *creation*, and only within the power of Him who has created.

We have a right to view gravitation under every form that either its definition or its effects can suggest to the mind; it is our privilege to do so with every force in nature; and it is only by so doing, that we have succeeded, to a large extent, in relating the various forms of power, so as to derive one from another, and thereby obtain confirmatory evidence of the great principle of the conservation of force. Then let us consider the two particles A and B as attracting each other by the force of gravitation, under another view. According to the definition, the force depends upon both particles, and if the particle A or B were by itself, it could not gravitate, *i. e.* it could have no attraction, no *force* of gravity. Supposing A to exist in that isolated state and without gravitating force, and then B placed in relation to it, gravitation comes on, as is supposed on the part of both. Now, without trying to imagine *how* I , which had no gravitating force, can raise up gravitating force in A, and how A, equally without force beforehand, can raise up force in B, still, to imagine it as a fact done, is to admit a creation of force in both particles, and so to bring ourselves within the impossible consequences which have already been referred to.

It may be said we cannot have an idea of one particle by itself, and so the reasoning fails. For my part I can comprehend a particle by itself just as easily as many particles; and though I cannot conceive the relation of a lone particle to gravitation, according to the limited view which is at present taken of that force, I can conceive its relation to something which causes gravitation, and with which, whether the particle is alone, or one of a universe of other particles, it is always related. But the reasoning upon a lone particle does not fail; for as the particles can be separated, we can easily conceive of the particle B being removed to an infinite distance from A and then the power in A will be infinitely diminished. Such removal of B will be as if it were annihilated in regard to A, and the force in A will be annihilated at the same time : so that the case of a lone particle, and that where different distances only are considered, become one, being identical with each other in their consequences. And as removal of B to an infinite distance is as regards A annihilation of B, so removal to the smallest degree is, in principle, the same thing with displacement through infinite space : the smallest increase in distance involves annihilation of power; the annihilation of the second particle, so as to have A alone, involves no other consequence in relation to gravity; there is difference in degree, but no difference in the character of the result.

It seems hardly necessary to observe, that the same line of thought grows up in the mind if we consider the mutual gra-

vitating action of one particle and many.   The particle A will
attract the particle B at the distance of a mile with a certain
degree of force; it will attract a particle C at the same distance
of a mile with a power equal to that by which it attracts B; if
myriads of like particles be placed at the given distance of a
mile, A will attract each with equal force; and if other particles be accumulated round it, within and without the sphere
of two miles diameter, it will attract them all with a force varying inversely with the square of the distance.   How are we to
conceive of this force growing up in A to a million-fold or more?
and if the surrounding particles be then removed, of its diminution in an equal degree?   Or, how are we to look upon the
power raised up in all these outer particles by the action of A
on them, or by their action one on another, without admitting,
according to the limited definition of gravitation, the facile generation and annihilation of force?

The assumption which we make for the time with regard to
the nature of a power (as gravity, heat, &c.), and the form of
words in which we express it, *i. e.* its definition, should be
consistent with the fundamental principles of force generally.
The conservation of force is a fundamental principle; hence
the assumption with regard to a particular form of force, ought
to imply what becomes of the force when its action is *increased*
or *diminished*, or its *direction changed*; or else the assumption
should admit that it is deficient on that point, being only half
competent to represent the force; and, in any case, should not
be opposed to the principle of conservation.   The usual definition of gravity as *an attractive force between the particles of
matter* VARYING *inversely as the square of the distance*, whilst
it stands as a full definition of the power, is inconsistent with
the principle of the conservation of force.   If we accept the
principle, such a definition must be an imperfect account of
the whole of the force, and is probably only a description of
one exercise of that power, whatever the nature of the force
itself may be.   If the definition be accepted as tacitly including the conservation of force, then it ought to admit that
consequences must occur during the suspended or diminished
degree of its power as gravitation, equal in importance to the
power suspended or hidden; being in fact equivalent to that
diminution.   It ought also to admit, that it is incompetent to

suggest or deal with any of the consequences of the changed part or condition of the force, and cannot tell whether they depend on, or are related to, conditions *external* or *internal* to the gravitating particle; and, as it appears to me, can say neither yes nor no to any of the arguments or probabilities belonging to the subject.

If the definition *denies* the occurrence of such contingent results, it seems to me to be unphilosophical; if it simply *ignores* them, I think it is imperfect and insufficient; if it *admits* these things, or any part of them, then it prepares the natural philosopher to look for effects and conditions as yet unknown, and is open to any degree of development of the consequences and relations of power :—by denying, it opposes a dogmatic barrier to improvement; by ignoring, it becomes in many respects an inert thing, often much in the way; by admitting, it rises to the dignity of a stimulus to investigation, a pilot to human science.

The principle of the conservation of force would lead us to assume, that when A and B attract each other less because of increasing distance, then some other exertion of power, either within or without them, is proportionately growing up; and again, that when their distance is diminished, as from 10 to 1, the power of attraction, now increased a hundred-fold, has been produced out of some other form of power which has been equivalently reduced. This enlarged assumption of the nature of gravity is not more metaphysical than the half assumption; and is, I believe, more philosophical, and more in accordance with all physical considerations. The half assumption is, in my view of the matter, more dogmatic and irrational than the whole, because it leaves it to be understood that power can be created and destroyed almost at pleasure.

When the equivalents of the various forms of force, as far as they are known, are considered, their differences appear very great; thus, a grain of water is known to have electric relations equivalent to a very powerful flash of lightning. It may therefore be supposed that a very large apparent amount of the force causing the phenomena of gravitation, may be the equivalent of a very small change in some unknown condition of the bodies, whose attraction is varying by change of distance. For my own part, many considerations urge my mind towards the idea of a cause of gravity, which is not resident in the particles of matter merely, but conjointly in them, and all space. I have already put forth considerations regarding gravity which partake of this idea*, and it seems to have been unhesitatingly accepted by Newton†.

There is one wonderful condition of matter, perhaps its only true indication, namely *inertia*; but in relation to the ordinary definition of gravity, it only adds to the difficulty. For if we consider two particles of matter at a certain distance apart, attracting each other under the power of gravity and free to approach, they will approach; and when at only half the distance, each will have had stored up in it, because of its *inertia*, a certain amount of mechanical force. This must be due to the force exerted, and, if the conservation principle be true, must have consumed an equivalent proportion of the cause of attraction; and yet, according to the definition of gravity, the attractive force is not diminished thereby, but increased fourfold, the force growing up within itself the more rapidly, the more it is occupied in producing other force. On the other hand, if mechanical force from without be used to separate the particles to twice their distance, this force is not stored up in momentum or by inertia, but disappears; and three-fourths of the attractive force at the first distance disappears with it: how can this be?

We know not the physical condition or action from which *inertia* results; but inertia is always a pure case of the conservation of force. It has a strict relation to gravity, as appears by the proportionate amount of force which gravity can communicate to the inert body; but it appears to have the same strict relation to other forces acting at a distance, as those of magnetism or electricity, when they are so applied by the tan-

* Proceedings of the Royal Institution, 1855, vol. ii. p. 10, &c.

† "That gravity should be innate, inherent, and essential to matter, so that one body may act upon another at a distance, through a *vacuum*, without the mediation of anything else, by and through which their action and force may be conveyed from one to another, is to me so great an absurdity, that I believe no man who has in philosophical matters a competent faculty of thinking, can ever fall into it. Gravity must be caused by an agent, acting constantly according to certain laws; but whether this agent be material or immaterial, I have left to the consideration of my readers."—See Newton's 'Third Letter to Bentley.'

gential balance as to act independent of the gravitating force. It has the like strict relation to force communicated by impact, pull, or in any other way. It enables a body to take up and conserve a given amount of force until that force is transferred to other bodies, or changed into an equivalent of some other form; that is all that we perceive in it: and we cannot find a more striking instance amongst natural or possible phenomena of the necessity of the conservation of force as a law of nature; or one more in contrast with the assumed variable condition of the gravitating force supposed to reside in the particles of matter.

Even gravity itself furnishes the strictest proof of the conservation of force in this, that its power is unchangeable for the same distance; and is by that in striking contrast with the variation which we assume in regard to the *cause of gravity*, to account for the *results* at different distances.

It will not be imagined for a moment that I am opposed to what may be called the *law of gravitating action*, that is, the law by which all the known effects of gravity are governed; what I am considering, is the definition of the *force* of gravitation. That the result of one exercise of a power may be inversely as the square of the distance, I believe and admit; and I know that it is so in the case of gravity, and has been verified to an extent that could hardly have been within the conception even of Newton himself when he gave utterance to the law: but that the *totality* of an inherent force can be employed according to that law I do not believe, either in relation to gravitation, or electricity, or magnetism, or any other supposed form of power.

I might have drawn reasons for urging a continual recollection of, and reference to, the principle of the conservation of force from other forms of power than that of gravitation; but I think that when founded on gravitating phenomena, they appear in their greatest simplicity; and precisely for this reason, that gravitation has not yet been connected by any degree of convertibility with the other forms of force. If I refer for a few minutes to these other forms, it is only to point in their variations, to the proofs of the value of the principle laid down, the consistency of the known phenomena with it, and the suggestions of research and discovery which arise from it*. *Heat,* for instance, is a mighty form of power, and its effects have been greatly developed; therefore, assumptions regarding its nature become useful and necessary, and philosophers try to define it. The most probable assumption is, that it is a motion of the particles of matter; but a view, at one time very popular, is, that it consists of a particular fluid of heat. Whether it be viewed in one way or the other, the principle of conservation is admitted, I believe, with all its force. When transferred from one portion to another portion of like matter, the full amount of heat appears. When transferred to matter of another kind, an apparent excess or deficiency often results; the word " capacity " is then introduced, which, whilst it acknowledges the principle of conservation, leaves space for research. When employed in changing the state of bodies, the appearance and disappearance of the heat is provided for consistently by the assumption of enlarged or diminished motion, or else space is left by the term " capacity" for the partial views which remain to be developed. When converted into mechanical force, in the steam- or air-engine, and so brought into direct contact with gravity, being then easily placed in relation to it, still the conservation of force is fully respected and wonderfully sustained. The constant amount of heat developed in the whole of a voltaic current described by M. P. A. Favre†, and the present state of the knowledge of thermo-electricity, are again fine partial or subordinate illustrations of the principle of conservation. Even when rendered radiant, and for the time giving no trace or signs of ordinary heat action, the assumptions regarding its nature have provided for the belief in the conservation of force, by admitting, either that it throws the ether into an equivalent state, in sustaining which for the time the power is engaged; or else, that the motion of the particles of heat is employed altogether in their own transit from place to place.

It is true that heat often becomes evident or insensible in a manner unknown to us; and we have a right to ask what is happening when the heat disappears in one part, as of the

---

* Helmholtz, "On the Conservation of Force." Taylor's 'Scientific Memoirs,' 2nd series, 1853, p. 114.

† Comptes Rendus, 1854, vol. xxxix. p. 1212.

thermo-voltaic current, and appears in another; or when it enlarges or changes the state of bodies; or what would happen, if the heat being presented, such changes were purposely opposed. We have a right to ask these questions, but not to ignore or deny the conservation of force; and one of the highest uses of the principle is to suggest such inquiries. Explications of similar points are continually produced, and will be most abundant from the hands of those who, not desiring to ease their labour by forgetting the principle, are ready to admit it either tacitly, or better still, effectively, being then continually guided by it. Such philosophers believe that heat must do its equivalent of work; that if in doing work it seem to disappear, it is still producing its equivalent effect, though often in a manner partially or totally unknown; and that if it give rise to another form of force (as we imperfectly express it), that force is equivalent in power to the heat which has disappeared.

What is called *chemical attraction,* affords equally instructive and suggestive considerations in relation to the principle of the conservation of force. The indestructibility of individual matter is one case, and a most important one, of the conservation of chemical force. A molecule has been endowed with powers which give rise in it to various qualities, and these never change, either in their nature or amount. A particle of oxygen is ever a particle of oxygen—nothing can in the least wear it. If it enters into combination and disappears as oxygen,—if it pass through a thousand combinations, animal, vegetable, mineral,—if it lie hid for a thousand years and then be evolved, it is oxygen with its first qualities, neither more nor less. It has all its original force, and only that; the amount of force which it disengaged when hiding itself, has again to be employed in a reverse direction when it is set at liberty; and if, hereafter, we should decompose oxygen, and find it compounded of other particles, we should only increase the strength of the proof of the conservation of force, for we should have a right to say of these particles, long as they have been hidden, all that we could say of the oxygen itself.

Again, the body of facts included in the theory of definite proportions, witnesses to the truth of the conservation of force; and though we know little of the cause of the change of properties of the acting and produced bodies, or how the forces of the former are hid amongst those of the latter, we do not for an instant doubt the conservation, but are moved to look for the manner in which the forces are, for the time, disposed, or if they have taken up another form of force, to search what that form may be.

Even chemical action at a distance, which is in such antithetical contrast with the ordinary exertion of chemical affinity, since it can produce effects miles away from the particles on which they depend, and which are effectual only by forces acting at insensible distances, still proves the same thing, the conservation of force. Preparations can be made for a chemical action in the simple voltaic circuit, but until the circuit be complete that action does not occur; yet in completing we can so arrange the circuit, that a distant chemical action, the perfect equivalent of the dominant chemical action, shall be produced; and this result, whilst it establishes the electro-chemical equivalent of power, establishes the principle of the conservation of force also, and at the same time suggests many collateral inquiries which have yet to be made and answered, before all that concerns the conservation in this case can be understood.

This and other instances of chemical action at a distance, carry our inquiring thoughts on from the facts to the physical mode of the exertion of force; for the qualities which seem located and fixed to certain particles of matter appear at a distance in connexion with particles altogether different. They also lead our thoughts to the *conversion* of one form of power into another: as for instance, in the *heat* which the elements of a voltaic pile may either show at the place where they act by their combustion or combination together; or in the distance, where the electric spark may be rendered manifest; or in the wire or fluids of the different parts of the circuit.

When we occupy ourselves with the dual forms of power, electricity and magnetism, we find great latitude of assumption; and necessarily so, for the powers become more and more complicated in their conditions. But still there is no apparent desire to loosen the force of the principle of conservation, even in those cases where the appearance and disappearance of force may seem most evident and striking. Electricity appears when there is consumption of no other force than that required for friction; we do not know how, but we search to know, not being willing to admit that the electric force can arise out of nothing. The two electricities are developed in equal proportions; and having appeared, we may dispose variously of the influence of one upon successive portions of the other, causing many changes in relation, yet never able to make the sum of the force of one kind in the least degree exceed or come short of the sum of the other. In that necessity of equality, we see another direct proof of the conservation of force, in the midst of a thousand changes that require to be developed in their principles before we can consider this part of science as even moderately known to us.

One assumption with regard to electricity is, that there is an electric fluid rendered evident by excitement in plus and minus proportions. Another assumption is, that there are two fluids of electricity, each particle of each repelling all particles like itself, and attracting all particles of the other kind always, and with a force proportionate to the inverse square of the distance, being so far analogous to the definition of gravity. This hypothesis is antagonistic to the law of the conservation of force, and open to all the objections that have been, or may be, made against the ordinary definition of gravity. Another assumption is, that each particle of the two electricities has a given amount of power, and can only attract contrary particles with the sum of that amount, acting upon each of two with only half the power it could in like circumstances exert upon one. But various as are the assumptions, the conservation of force (though wanting in the second) is, I think, intended to be included in all. I might repeat the same observations nearly in regard to magnetism,—whether it be assumed as a fluid, or two fluids or electric currents,—whether the external action be supposed to be action at a distance, or dependent on an external condition and lines of force—still all are intended to admit the conservation of power as a principle to which the phenomena are subject.

The principles of physical knowledge are now so far developed as to enable us not merely to define or describe the *known*, but to state reasonable expectations regarding the *unknown*; and I think the principle of conservation of force may greatly aid experimental philosophers in that duty to science, which consists in the enunciation of problems to be solved. It will lead us, in any case where the force remaining unchanged in form is altered in direction only, to look for the new disposition of the force; as in the cases of magnetism, static electricity, and perhaps gravity, and to ascertain that, as a whole, it remains unchanged in amount :—or, if the original force disappear, either altogether or in part, it will lead us to look for the new condition or form of force which should result, and to develope its equivalency to the force that has disappeared. Likewise, when force is developed, it will cause us to consider the previously existing equivalent to the force so appearing; and many such cases there are in chemical action.   When force disappears, as in the electric or magnetic induction after more or less discharge, or that of gravity with an increasing distance, it will suggest a research as to whether the equivalent change is one within the apparently acting bodies, or one *external* (in part) to them. It will also raise up inquiry as to the nature of the internal or external state, both before the change and after. If supposed to be external, it will suggest the necessity of a physical process, by which the power is communicated from body to body; and in the case of external action, will lead to the inquiry, whether, in any case, there can be truly action at a distance, or whether the ether, or some other medium, is not necessarily present.

We are not permitted as yet to see the nature of the source of physical power, but we are allowed to see much of the consistency existing amongst the various forms in which it is presented to us. Thus if, in static electricity, we consider an act of induction, we can perceive the consistency of all other like acts of induction with it. If we then take an electric current, and compare it with this inductive effect, we see their relation and consistency. In the same manner we have arrived at a knowledge of the consistency of magnetism with electricity; and also of chemical action and of heat with all the former; and if we see not the consistency between gravitation with any of these forms of force, I am strongly of the mind that it is because of our ignorance only. How imperfect would our idea of an electric current now be, if we were to leave out of sight its origin, its state and dynamic induction, its magnetic influence, its chemical and heating effects !—or our idea of any one of these results, if we left any of the others unregarded! That there should be a power of gravitation existing by itself, having *no relation to the other natural powers, and no respect to the law of the conservation of force,* is as little likely as that there should be a principle of levity as well as of gravity. Gravity may be only the residual part of the other forces of nature, as Mossotti has tried to show; but that it should fall out from the law of all other force, and should be outside the reach either of further experiment or philosophical conclusions, is not probable. So we must strive to learn more of this outstanding power, and endeavour to avoid any definition of it which is incompatible with the principles of force generally, for all the phenomena of nature lead us to believe that the great and governing law is one. I would much rather incline to believe that bodies affecting each other by gravitation act by lines of force of definite amount (somewhat in the manner of magnetic or electric induction, though without polarity), or by an ether pervading all parts of space, than admit that the conservation of force could be dispensed with.

It may be supposed, that one who has little or no mathematical knowledge should hardly assume a right to judge of the generality and force of a principle such as that which forms the subject of these remarks. My apology is this: I do not perceive that a mathematical mind, simply as such, has any advantage over an equally acute mind not mathematical, in perceiving the nature and power of a natural principle of action. It cannot of itself introduce the knowledge of any new principle. Dealing with any and every amount of static electricity, the mathematical mind can, and has balanced and adjusted them with wonderful advantage, and has foretold results which the experimentalist can do no more than verify. But it could not discover dynamic electricity, nor electro-magnetism, nor magneto-electricity, or even suggest them; though when once discovered by the experimentalist, it can take them up with extreme facility. So in respect of the force of gravitation, it has calculated the results of the power in such a wonderful manner as to trace the known planets through their courses and perturbations, and in so doing has *discovered* a planet before unknown; but there may be results of the gravitating force of other kinds than attraction inversely as the square of the distance, of which it knows nothing, can discover nothing, and can neither assert nor deny their possibility or occurrence. Under these circumstances, a principle, which may be accepted as equally strict with mathematical knowledge, comprehensible without it, applicable by all in their philosophical logic whatever form that may take, and above all, suggestive, encouraging, and instructive to the mind of the experimentalist, should be the more earnestly employed and the more frequently resorted to when we are labouring either to discover new regions of science, or to map out and develope those which are known into one harmonious whole; and if in such strivings, we, whilst applying the principle of conservation, see but imperfectly, still we should endeavour to see, for even an obscure and distorted vision is better than none. Let us, if we can, discover a new thing in *any shape*; the true appearance and character will be easily developed afterwards.

Some are much surprised that I should, as they think, venture to oppose the conclusions of Newton: but here there is a mistake. I do not oppose Newton on any point; it is rather those who sustain the idea of action at a distance that contradict him. Doubtful as I ought to be of myself, I am certainly very glad to feel that my convictions are in accordance with his conclusions. At the same time, those who occupy themselves with such matters ought not to depend altogether upon authority, but should find reason within themselves, after careful thought and consideration, to use and abide by their own judgment. Newton himself, whilst referring to those who were judging his views, speaks of such as are competent to form an opinion in such matters, and makes a strong distinction between them and those who were incompetent for the case.

But, after all, the principle of the conservation of force may by some be denied. Well, then, if it be unfounded even in its application to the smallest part of the science of force, the proof must be within our reach, for all physical science is so. In that case, discoveries as large or larger than any yet made may be anticipated. I do not resist the search for them, for no one can do harm, but only good, who works with an earnest and truthful spirit in such a direction. But let us not admit the destruction or creation of force without clear and constant proof. Just as the chemist owes all the perfection of his science to his dependence on the certainty of gravitation applied by the balance, so may the physical philosopher expect to find the greatest security and the utmost aid in the principle of the conservation of force. All that we have that is good and safe, as the steam-engine, the electric telegraph, &c., witness to that principle,—it would require a perpetual motion, a fire without heat, heat without a source, action without reaction, cause without effect, or effect without a cause, to displace it from its rank as a law of nature.

During the year that has passed since the publication of the preceding views regarding gravitation, &c., I have come to the knowledge of various observations upon them, some adverse, others favourable; these have given me no reason to change my own mode of viewing the subject, but some of them make me think that I have not stated the matter with sufficient precision. The word "force" is understood by many to mean simply "the tendency of a body to pass from one place to another," which is equivalent, I suppose, to the phrase "mechanical force;" those who so restrain its meaning must have found my argument very obscure. What I mean by the word "force," is the *cause* of a physical action; the source or sources of all possible changes amongst the particles or materials of the universe.

It seems to me that the idea of the conservation of force is absolutely independent of any notion we may form of the nature of force or its varieties, and is as sure and may be as firmly held in the mind, as if we, instead of being very ignorant, understood perfectly every point about the cause of force and the varied effects it can produce. There may be perfectly distinct and separate causes of what are called chemical actions, or electrical actions, or gravitating actions, constituting so many forces; but if the "conservation of force" is a good and true principle, each of these forces must be subject to it: none can vary in its absolute amount; each must be definite at all times, whether for a particle, or for all the particles in the universe; and the sum also of the three forces must be equally unchangeable. Or, there may be but one cause for these three sets of actions, and in place of three forces we may really have but one, convertible in its manifestations; then the proportions between one set of actions and another, as the chemical and the electrical,

may become very variable, so as to be utterly inconsistent with
the idea of the conservation of two separate forces (the electrical and the chemical), but perfectly consistent with the conservation of a force being the common cause of the two or
more sets of action.

It is perfectly true that we cannot always trace a force by its
actions, though we admit its conservation. Oxygen and hydrogen may remain mixed for years without showing any signs
of chemical activity; they may be made at any given instant to
exhibit active results, and then assume a new state, in which again
they appear as passive bodies. Now, though we cannot clearly
explain what the chemical force is doing, that is to say, what
are its effects during the three periods before, at, and after
the active combination, and only by very vague assumption can
approach to a feeble conception of its respective states, yet we
do not suppose the creation of a new portion of force for the
active moment of time, or the less believe that the forces
belonging to the oxygen and hydrogen exist unchanged in
their amount at all these periods, though varying in their
results. A part may at the active moment be thrown off as
mechanical force, a part as radiant force, a part disposed of we
know not how; but believing, by the principle of conservation,
that it is not increased or destroyed, our thoughts are directed
to search out what at all and every period it is doing, and how
it is to be recognized and measured. A problem, founded on
the physical truth of nature, *is stated*, and, being stated, is
on the way to its solution.

Those who admit the possibility of the common origin of all
physical force, and also acknowledge the principle of conservation, apply that principle to the sum total of the force. Though
the amount of mechanical force (using habitual language for
convenience sake) may remain unchanged and definite in its character for a long time, yet when, as in the collision of two equal
inelastic bodies, it appears to be lost, they find it in the form
of heat and whether they admit that heat to be a continued
mechanical action (as is most probable), or assume some other
idea, as that of electricity, or action of a heat-fluid, still they
hold to the principle of conservation by admitting that the sum
of force, *i. e.* of the "cause of action," is the same, whatever
character the effects assume. With them the convertibility of

heat, electricity, magnetism, chemical action and motion is a familiar thought; neither can I perceive any reason why they should be led to exclude, *a priori*, the cause of gravitation from association with the cause of these other phenomena respectively. All that they are limited by in their various investigations, whatever directions they may take, is the necessity of making no assumption directly contradictory of the conservation of force applied to the sum of all the forces concerned, and to endeavour to discover the different directions in which the various parts of the total force have been exerted.

Those who admit separate forces inter-unchangeable, have to show that each of these forces is separately subject to the principle of conservation. If gravitation be such a separate force, and yet its power in the action of two particles be supposed to be diminished fourfold by doubling the distance, surely some new action, having true gravitation character, and that alone, ought to appear; for how else can the totality of the force remain unchanged? To define the force as " a simple attractive force exerted between any two or all the particles of matter, with a strength varying inversely as the square of the distance," is not to answer the question; nor does it indicate or even assume what are the other complementary results which occur; or allow the supposition that such are necessary: it is simply, as it appears to me, to *deny* the conservation of force.

As to the gravitating force, I do not presume to say that I have the least idea of what occurs in two particles when their power of mutually approaching each other is changed by their being placed at different distances; but I have a strong conviction, through the influence on my mind of the doctrine of conservation, that there is a change; and that the phenomena resulting from the change will probably appear some day as the result of careful research. If it be said that "'twere to consider too curiously to consider so," then I must dissent. To refrain to consider, would be to ignore the principle of the conservation of force, and to stop the inquiry which it suggests :—whereas to admit the proper logical force of the principle in our hypotheses and considerations, and to permit its guidance in a cautious yet courageous course of investigation, may give us power to enlarge the generalities we already possess in respect of heat, motion, electricity, magnetism, &c. ;

to associate gravity with them ; and perhaps enable us to know whether the essential force of gravitation (and other attractions) is internal or external as respects the attracted bodies [*].

Returning once more to the definition of the gravitating power as " *a simple attractive force exerted between any two or all the particles or masses of matter at every sensible distance, but with a* STRENGTH VARYING *inversely as the square of the distance,*" I ought perhaps to suppose there are many who accept this as a true and sufficient description of the force, and who therefore, in relation to it, deny the principle of conservation. If both are accepted and are thought to be consistent with each other, it cannot be difficult to add words which shall make " varying strength " and " conservation " agree together. It cannot be said that the definition merely applies to the *effects* of gravitation as far as we know them. So understood, it would form no barrier to progress ; for, that particles at different distances are urged towards each other with a power varying inversely as the square of the distance, is a truth ; but the definition has not that meaning ; and what I object to is the pretence of knowledge which the definition sets up, when it assumes to describe, not the partial effects of the force, but the nature of the force as a whole.

June, 1858.   M. F.

Observations on Mental Education [†].

[These observations were delivered as a lecture before His Royal Highness The Prince Consort and the Members of the Royal Institution on the 6th of May, 1854. They are so immediately connected in their nature and origin with my own experimental life, considered either as cause or consequence, that I have thought the close of this volume not an unfit place for their reproduction.]

I TAKE courage, Sir, from your presence here this day, to speak boldly that which is upon my mind. I feared that it

---

[*] Dr. Winslow, of West Newton (Mass.), U.S., states, that from the examination of a record of 850 earthquakes and volcanic eruptions, it appears that the greater number occur in the winter months, when the sun is nearest to the earth, and the attraction of gravity greatest. Their occurrence is more rare as the distance is greater, the number being for December 102, which in the intervening months gradually decreases to and increases from 44 for June. Hence he draws conclusions regarding other exhibitions of the gravitating force than mere attraction, when that attraction is varied by change of distance.—*Annual of Scientific Discovery.*

[†] Lectures on Education, 1855. Parker and Son.

might be unpleasant to some of my audience, but as I know that your Royal Highness is a champion for and desires the truth, I will believe that all here are united in the same cause, and therefore will give utterance, without hesitation, to what I have to say regarding the present condition of Mental Education.

If the term education may be understood in so large a sense as to include all that belongs to the improvement of the mind, either by the acquisition of the knowledge of others, or by increase of it through its own exertions, then I may hope to be justified for bringing forward a few desultory observations respecting the exercise of the mental powers in a particular direction, which otherwise might seem out of place. The points I have in view are general, but they are manifest in a striking manner, among the physical matters which have occupied my life; and as the latter afford a field for exercise in which cogitations and conclusions can be subjected to the rigid tests of fact and experiment—as all classes employ themselves more or less in the consideration of physical matters, and may do so with great advantage, if inclined in the least degree to profit by educational practices—so I hope that what I may say will find its application in every condition of life.

Before entering upon the subject, I must take one distinction which, however it may appear to others, is to me of the utmost importance. High as man is placed above the creatures around him, there is a higher and far more exalted position within his view; and the ways are infinite in which he occupies his thoughts about the fears, or hopes, or expectations of a future life. I believe that the truth of that future cannot be brought to his knowledge by any exertion of his mental powers, however exalted they may be; that it is made known to him by other teaching than his own, and is received through simple belief of the testimony given. Let no one suppose for a moment that the self-education I am about to commend in respect of the things of this life, extends to any considerations of the hope set before us, as if man by reasoning could find out God. It would be improper here to enter upon this subject further than to claim an absolute distinction between religious and ordinary belief. I shall be reproached with the weakness of refusing to apply those mental operations which I think good in respect of high things to the very highest. I am content to bear the reproach. Yet, even in earthly matters, I believe that the invisible things of HIM from the creation of the world are clearly seen, being understood by the things that are made, even His eternal power and Godhead; and I have never seen anything incompatible between those things of man which can be known by the spirit of man which is within him, and those higher things concerning his future, which he cannot know by that spirit.

Claiming, then, the use of the ordinary faculties of the mind in ordinary things, let me next endeavour to point out what appears to me to be a great deficiency in the exercise of the mental powers in every direction; three words will express this great want, *deficiency of judgment*. I do not wish to make any startling assertion, but I know that in physical matters multitudes are ready to draw conclusions who have little or no power of judgment in the cases; that the same is true of other departments of knowledge; and that, generally, mankind is willing to leave the faculties which relate to judgment almost entirely uneducated, and their decisions at the mercy of ignorance, prepossessions, the passions, or even accident.

Do not suppose, because I stand here and speak thus, making no exceptions, that I except myself. I have learned to know that I fall infinitely short of that efficacious exercise of the judgment which may be attained. There are exceptions to my general conclusion, numerous and high; but if we desire to know how far education is required, we do not consider the few who need it not, but the many who have it not; and in respect of judgment, the number of the latter is almost infinite. I am moreover persuaded, that the clear and powerful minds which have realized in some degree the intellectual preparation I am about to refer to, will admit its importance, and indeed its necessity; and that they will not except themselves, nor think that I have made my statement too extensive.

As I believe that a very large proportion of the errors we make in judgment is a simple and direct result of our perfectly unconscious state, and think that a demonstration of the liabilities we are subject to would aid greatly in providing a remedy, I will proceed first to a few illustrations of a physical

2 H

nature. Nothing can better supply them than the intimations we derive from our senses; to them we trust directly; by them we become acquainted with external things, and gain the power of increasing and varying facts, upon which we entirely depend. Our sense-perceptions are wonderful. Even in the observant, but unreflective infant, they soon produce a result which looks like intuition, because of its perfection. Coming to the mind as so many data, they are stored up, and without our being conscious, are ever after used in like circumstances in forming our judgment; and it is not wonderful that man should be accustomed to trust them without examination. Nevertheless, the result is the effect of education: the mind has to be instructed with regard to the senses and their intimations through every step of life; and where the instruction is imperfect, it is astonishing how soon and how much their evidence fails us. Yet, in the latter years of life, we do not consider this matter, but, having obtained the ordinary teaching sufficient for ordinary purposes, we venture to judge of things which are extraordinary for the time, and almost always with the more assurance as our powers of observation are less educated. Consider the following case of a physical impression, derived from the sense of touch, which can be examined and verified at pleasure:—If the hands be brought towards each other so that the tips of the corresponding fingers touch, the end of any finger may be considered as an object to be felt by the opposed finger; thus the two middle fingers may for the present be so viewed. If the attention be directed to them, no difficulty will be experienced in moving each lightly in a circle round the tip of the other, so that they shall each feel the opposite, and the motion may be either in one direction or the other—looking at the fingers, or with eyes employed elsewhere—or with the remaining fingers touching quiescently, or moving in a like direction; all is easy, because each finger is employed in the ordinary or educated manner whilst obeying the will, and whilst communicating through the sentient organ with the brain. But turn the hands half-way round, so that their backs shall be towards each other, and then, crossing them at the wrists, again bring the like fingers into contact at the tips. If it be now desired to move the extremities of the middle fingers round each other, or to follow the contour of one finger by the tip of the opposed one, all sorts of confusion in the motion will ensue; and as the finger of one hand tries, under the instruction of the will, to move in one course, the touched finger will convey an intimation that it is moving in another. If all the fingers move at once, all will be in confusion, the ease and simplicity of the first case having entirely disappeared. If, after some considerable trial, familiarity with the new circumstances have removed part of the uncertainty, then, crossing the hands at the opposite sides of the wrists will renew it. These contrary results are dependent not on any change in the nature of the sentient indication, or of the surfaces or substances which the sense has to deal with; but upon the trifling circumstance of a little variation from the direction in which the sentient organs of these parts are usually exerted; and they show to what an extraordinary extent our interpretations of the sense impressions depend upon the experience, *i. e.* the education which they have previously received, and their great inability to aid us at once in circumstances which are entirely new.

At other times they fail us because we cannot keep a true remembrance of former impressions. Thus, on the evening of the 11th of March last, I and many others were persuaded that at one period the moon had a real green colour, and though I knew that the prevailing red tints of the general sky were competent to produce an effect of such a kind, yet there was so little of that in the neighbourhood of the planet, that I was doubtful whether the green tint was not produced on the moon by some aërial medium spread before it; until by holding up white cards in a proper position, and comparing them with our satellite, I had determined experimentally that the effect was only one of contrast. In the midst of the surrounding tints, my memory could not recall the true sentient impression which the white of the moon most surely had before made upon the eye.

At other times the failure is because one impression is overpowered by another; for as the morning star disappears when the sun is risen, though still above the horizon and shining brightly as ever, so do stronger phenomena obscure weaker, even when both are of the same kind; till an uninstructed person is apt to pass the weaker unobserved, and even deny their existence.

So, error results occasionally from *believing* our senses : it ought to be considered, rather, as an *error of the judgment* than of the sense, for the latter has performed its duty; the indication is always correct, and in harmony with the great truth of nature. Where, then, is the mistake?—almost entirely with our judgment. We have not had that sufficient instruction by the senses which would justify our making a conclusion; we have to contrive extra and special means, by which their first impressions shall be corrected, or rather enlarged; and it is because our procedure was hasty, our data too few, and our judgment untaught, that we fell into mistake; not because the data were wrong. How frequently may each one of us perceive, in our neighbours, at least, that a result like this, derived from the observation of physical things, happens in the ordinary affairs of common life !

When I become convicted of such haste, which is not unfrequently the case, I look back upon the error as one of ‘ presumptuous judgment.’ Under that form it is easily presentable to the mind, and has a useful corrective action. I do not think the expression too strong; for if we are led, either by simplicity or vanity, to give an opinion upon matters respecting which we are not instructed, either by the knowledge of others, or our own intimate observation; if we are induced to ascribe an effect to one force, or deny its relation to another, knowing little or nothing of the laws of the forces, or the necessary conditions of the effect to be considered; surely our judgment must be qualified as ‘ presumptuous.’

There are multitudes who think themselves competent to decide, after the most cursory observation, upon the cause of this or that event (and they may be really very acute and correct in things familiar to them) :—a not unusual phrase with them is, that ‘ it stands to reason’ that the effect they expect should result from the cause they assign to it, and yet it is *very difficult*, in numerous cases that appear plain, to show this reason, or to deduce the true and only rational relation of cause and effect. In matters connected with natural philosophy, we have wonderful aid in the progress and assurance in the character, of our final judgment, afforded us by the facts which supply our data, and the experience which multiplies their number and varies their testimony. A fundamental fact, like an elementary principle, never fails us, its evidence is

always true ; but, on the other hand, we frequently have to
ask what is the fact ?—often fail in distinguishing it,—often fail
in the very statement of it,—and mostly overpass or come short
of its true recognition.

If we are subject to mistake in the interpretation of our
mere sense impressions, we are much more liable to error
when we proceed to deduce from these impressions (as supplied
to us by our ordinary experience) the relation of cause and
effect ; and the accuracy of our judgment, consequently, is
more endangered. Then our dependence should be upon
carefully observed facts, and the laws of nature ; and I shall
proceed to a further illustration of the mental deficiency I
speak of, by a brief reference to one of these.

The *laws of nature,* as we understand them, are the foundation of our knowledge in natural things. So much as we
know of them has been developed by the successive energies
of the highest intellects, exerted through many ages. After a
most rigid and scrutinizing examination upon principle and
trial, a definite expression has been given to them ; they have
become, as it were, our belief or trust. From day to day we
still examine and test our expressions of them. We have no
interest in their retention if erroneous ; on the contrary, the
greatest discovery a man could make would be to prove that
one of these accepted laws was erroneous, and his greatest
honour would be the discovery. Neither should there be any
desire to retain the former expression :—for we know that the
new or the amended law would be far more productive in results,
would greatly increase our intellectual acquisitions, and would
prove an abundant source of fresh delight to the mind.

These laws are numerous, and are more or less comprehen-
sive. They are also precise ; for a law may present an apparent
exception, and yet not be less a law to us, when the exception
is included in the expression. Thus, that elevation of temperature expands all bodies is a well-defined law, though there
be an exception in water for a limited temperature ; because
we are careful, whilst stating the law, to state the exception
and its limits. Pre-eminent among these laws, because of its
simplicity, its universality, and its undeviating truth, stands
that enunciated by Newton (commonly called the *law of gravitation*), that matter attracts matter with a force inversely as

the square of the distance. Newton showed, that by this law, the general condition of things on the surface of the earth is governed; and the globe itself, with all upon it, kept together as a whole. He demonstrated that the motions of the planets round the sun, and of the satellites about the planets, were subject to it. During and since his time, certain variations in the movements of the planets, which were called irregularities, and might, for aught that was then known, be due to some cause other than the attraction of gravitation, were found to be its necessary consequences. By the close and scrutinizing attention of minds the most persevering and careful, it was ascertained that even the distant stars were subject to this law; and at last, to place as it were the seal of assurance to its never-failing truth, it became, in the minds of Leverrier and Adams (1845), the foreteller and the discoverer of an orb rolling in the depths of space, so large as to equal nearly sixty earths, yet so far away as to be invisible to the unassisted eye. What truth, beneath that of revelation, can have an assurance stronger than this?

Yet this law is often cast aside as of no value or authority, because of the unconscious ignorance amidst which we dwell. You hear at the present day, that some persons can place their fingers on a table, and then elevating their hands, the table will rise up and follow them; that the piece of furniture, though heavy, will ascend, and that their hands bear no weight, or are not drawn down to the wood; you do not hear of this as a conjuring manœuvre, to be shown for your amusement; but are expected seriously to believe it, and are told that it is an important fact, a great discovery amongst the truths of nature. Your neighbour, a well-meaning, conscientious person, believes it; and the assertion finds acceptance in every rank of society, and amongst classes which are esteemed to be educated. Now, what can this imply but that society, speaking generally, is not only ignorant as respects education of the judgment, but is also ignorant of its ignorance. The parties who are thus persuaded, and those who are inclined to think and to hope that they are right, throw up Newton's law at once, and *that* in a case which of all others is fitted to be tested by it; or if the law be erroneous, to test the law. I will not say they oppose the law, though I *have* heard the supposed fact quoted triumphantly against it; but as far as my observation has gone, they will not apply it. The law affords the simplest means of testing the fact; and if there be, indeed, anything in the latter new to our knowledge (and who shall say that new matter is not presented to us daily, passing away unrecognized?), it also affords the means of placing *that* before us separately in its simplicity and truth. Then why not consent to apply the knowledge we have to that which is under development? Shall we educate ourselves in what is known, and then casting away all we have acquired, turn to our ignorance for aid to guide us among the unknown? If so, instruct a man to write, but employ one who is unacquainted with letters to read that which is written; the end will be just as unsatisfactory, though not so injurious; for the book of nature, which we have to read, is written by the finger of God. Why should not one who can thus lift a table, proceed to verify and simplify his fact, and bring it into relation with the law of Newton? Why should he not take the top of his table (it may be a small one), and placing it in a balance, or on a lever, proceed to ascertain how much weight he can raise by the draught of his fingers upwards; and of this weight, so ascertained, how much is unrepresented by any pull upon the fingers downward? He will then be able to investigate the further question, whether electricity, or any new force of matter, is made manifest in his operations; or whether action and reaction being unequal, he has at his command the source of a perpetual motion. Such a man, furnished with a nicely constructed carriage on a railway, ought to travel by the mere draught of his own fingers. A far less prize than this would gain him the attention of the whole scientific and commercial world; and he may rest assured, that if he can make the most delicate balance incline or decline by attraction, though it be only with the fourth of an ounce, or even a grain, he will not fail to gain universal respect and most honourable reward.

When we think of the laws of nature (which by continued observation have become known to us) as the proper tests to which any new fact or our theoretical representation of it should in the first place be subjected, let us contemplate their assured and large character. Let us go out into the field and look at the heavens with their solar, starry, and planetary glories; the sky with its clouds; the waters descending from above or wandering at our feet; the animals, the trees, the plants; and consider the permanency of their actions and conditions under the government of these laws. The most delicate flower, the tenderest insect, continues in its species through countless years; always varying, yet ever the same. When we think we have discovered a departure, as in the *Aphides, Medusæ, Distomæ,* &c.*, the law concerned is itself the best means of instituting an investigation, and hitherto we have always found the witness to return to its original testimony. These frail things are never-ceasing, never-changing, evidence of the law's immutability. It would be well for a man who has an anomalous case before him, to contemplate a blade of grass, and when he has considered the numerous ceaseless, yet certain actions there located, and his inability to change the character of the least among them, to recur to his new subject; and, in place of accepting unwatched and unchecked results, to search for a like certainty and recurrence in the appearances and actions which belong to it.

Perhaps it may be said, the delusion of table-moving is past, and need not be recalled before an audience like the present†; —even granting this, let us endeavour to make the subject leave one useful result; let it serve for an example, not to pass into forgetfulness. It is so recent, and was received by the public in a manner so strange, as to justify a reference to it, in proof of the uneducated condition of the general mind. I do not object to table-moving, for *itself*; for being once stated, it becomes a fit, though a very unpromising subject for experiment; but I am opposed to the unwillingness of its advocates to investigate; their boldness to assert; the credulity of the

---

* See Claparède's Account of Alternating Generation and the Metamorphoses of Inferior Animals.—*Bibl. Univ.* Mar. 1854, p. 229.

† As an illustration of the present state of the subject, I will quote one letter from among many like it which I have received.

"———— *April* 5, 1854.

"Sir,—I am one of the clergymen of this parish, and have had the subject of table-turning brought under my notice by some of my young parishioners; I gave your solution of it as a sufficient answer to the mystery. The reply was made, that you had since seen reason to alter your opinion. Would you have the politeness to inform me if you have done so? With many apologies for troubling you,

"I am, your obedient servant,
"————— "

lookers-on; their desire that the reserved and cautious objector should be in error; and I wish, by calling attention to these things, to make the general want of mental discipline and education manifest.

Having endeavoured to point out this great deficiency in the exercise of the intellect, I will offer a few remarks upon the means of subjecting it to the improving processes of instruction. Perhaps many who watch over the interests of the community, and are anxious for its welfare, will conclude that the development of the judgment cannot properly be included in the general idea of education; that as the education proposed must, to a very large degree, be of *self*, it is so far incommunicable; that the master and the scholar merge into one, and both disappear; that the instructor is no wiser than the one to be instructed, and thus the usual relations of the two lose their power. Still, I believe that the judgment may be educated to a very large extent, and might refer to the fine arts, as giving proof in the affirmative; and though, as repects the community and its improvement in relation to common things, any useful education must be of *self*, I think that society, as a body, may act powerfully in the cause. Or it may still be objected that my experience is imperfect, is chiefly derived from exercise of the mind within the precincts of natural philosophy, and has not that generality of application which can make it of any value to society at large. I can only repeat my conviction, that society occupies itself now-a-days about physical matters and judges them as common things. Failing in relation to them, it is equally liable to carry such failures into other matters of life. The proof of deficient judgment in one department shows the habit of mind, and the general want, in relation to others. I am persuaded that all persons may find in natural things an admirable school for self-instruction, and a field for the necessary mental exercise; that they may easily apply their habits of thought, thus formed, to a social use; and that they ought to do this, as a duty to themselves and their generation.

Let me try to illustrate the former part of the case, and at the same time state what I think a man may and ought to do for himself.

The *self-education* to which he should be stimulated by the desire to improve his judgment, requires no blind dependence upon the dogmas of others, but is commended to him by the suggestions and dictates of his own common sense. The first part of it is founded in mental discipline: happily it requires no unpleasant avowals; appearances are preserved, and vanity remains unhurt; but it is necessary that a man *examine himself*, and *that* not carelessly. On the contrary, as he advances, he should become more and more strict, till he ultimately prove a sharper critic to himself than any one else can be; and he ought to intend this, for, so far as he consciously falls short of it, he acknowledges that others may have reason on their side when they criticise him. A first result of this habit of mind will be an internal conviction of *ignorance in many things respecting which his neighbours are taught*, and that his opinions and conclusions on such matters ought to be advanced with reservation. A mind so disciplined will be *open to correction upon good grounds in all things*, even in those it is best acquainted with, and should familiarize itself with the idea of such being the case; for though it sees no reason to suppose itself in error, yet the possibility exists. The mind is not enfeebled by this internal admission, but strengthened; for if it cannot distinguish proportionately between the probable right and wrong of things known imperfectly, it will tend either to be rash or to hesitate; whilst that which admits the due amount of probability is likely to be justified in the end. It is right that we should stand by and act on our principles; but not right to hold them in obstinate blindness, or retain them when proved to be erroneous. I remember the time when I believed a spark was produced between voltaic metals as they approached to contact (and the reasons why it might be possible yet remain); but others doubted the fact and denied the proofs, and on re-examination I found reason to admit their corrections were well-founded. Years ago I believed that electrolytes could conduct electricity by a conduction proper; that has also been denied by many through long time: though I believed myself right, yet circumstances have induced me to pay that respect to criticism as to reinvestigate the subject, and I have the pleasure of thinking that nature confirms my original conclusions. So though evidence may appear to preponderate extremely in favour of a certain decision, it is wise and proper to hear a counter-statement. You can have no idea how often and how much, under such an impression, I have desired that the marvellous descriptions which have reached me might prove, in some points, correct; and how frequently I have submitted myself to hot fires, to friction with magnets, to the passes of hands, &c., lest I should be shutting out discovery;—encouraging the strong desire that something might be true, and that I might aid in the development of a new force of nature.

Among those points of self-education which take up the form of *mental discipline*, there is one of great importance, and, moreover, difficult to deal with, because it involves an internal conflict, and equally touches our vanity and our ease. It consists in the *tendency to deceive ourselves* regarding all we wish for, and the necessity of *resistance to these desires*. It is impossible for any one who has not been constrained, by the course of his occupation and thoughts, to a habit of continual self-correction, to be aware of the amount of error in relation to judgment arising from this tendency. The force of the temptation which urges us to seek for such evidence and appearances as are in favour of our desires, and to disregard those which oppose them, is wonderfully great. In this respect we are all, more or less, active promoters of error. In place of practising wholesome self-abnegation, we ever make the wish the father to the thought: we receive as friendly that which agrees with, we resist with dislike that which opposes us; whereas the very reverse is required by every dictate of common sense. Let me illustrate my meaning by a case where the proof being easy, the rejection of it under the temptation is the more striking. In old times, a ring or a button would be tied by a boy to one end of a long piece of thread, which he would then hold at the other end, letting the button hang within a glass, or over a piece of slate-pencil, or sealing-wax, or a nail; he would wait and observe whether the button swung, and whether, in swinging, it tapped the glass as many times as the clock struck last, or moved along or across the slate-pencil, or in a circle or oval. In late times, parties in all ranks of life have renewed and repeated the boy's experiment. They have sought to ascertain a very simple fact, namely, whether the effect was as reported; but how many were unable to do this? They were sure they could keep their hands immoveable,—were sure they could do so whilst watching the result,—were sure that accordance of swing with an expected direction was *not* the result of their desires or involuntary motions. How easily all these points could be put to the proof by *not looking at the objects*, yet how difficult for the experimenter to deny himself that privilege! I have rarely found one who would freely permit the substance experimented with to be screened from his sight, and then have its position changed.

When engaged in the investigation of table-turning, I constructed a very simple apparatus*, serving as an index, to show the unconscious motions of the hands upon the table. The results were either that the index moved before the table, or that neither index nor table moved; and in numerous cases all moving power was annihilated. A universal objection was made to it by the table-turners;—it was said to paralyse the powers of the mind. But the experimenters need not see the index; they may leave their friends to watch that, and their minds may revel in any power that their expectation or their imagination can confer. So restrained, however, a *dislike* to the trial arises; but what is that except a proof, that whilst they trust themselves they doubt themselves, and are not willing to proceed to the decision, lest the trust which they like should fail them, and the doubt which they dislike rise to the authority of truth?

Again, in respect of the action of magnets on the body, it is almost impossible for an uninstructed person to enter profitably upon such an inquiry. He may observe *any* symptom which his expectation has been accidentally directed to:—yet be unconscious of any, if unaware of his subjection to the magnetic force, or of the conditions and manner of its application.

As a proof of the extent of this influence, even on the minds of those well-aware of its power, and desirous under every circumstance to escape from it, I will mention the practice of the chemist; who, dealing with the balance, that impartial decider which never fails in its indication, but offers its evidence with all simplicity, durability, and truth, still remembers

* P. 387, or Athenæum, July 2, 1853.

he should doubt himself; and, with the desire of rendering himself inaccessible to temptation, takes a counterpoised but unknown quantity of the substance for analysis, that he may remain ignorant of the proportions which he ought to obtain, and only at last compares the sum of his products with his counterpoise.

The *inclination* we exhibit in respect of any report or opinion that harmonizes with our preconceived notions, can only be compared in degree with the *incredulity* we entertain towards everything that opposes them; and these opposite and apparently incompatible, or at least inconsistent conditions, are accepted simultaneously in the most extraordinary manner. At one moment a departure from the laws of nature is admitted without the pretence of a careful examination of the proof; and at the next, the whole force of these laws, acting undeviatingly through all time, is denied, because the testimony they give is disliked.

It is my firm persuasion that no man can examine himself in the most common things, having any reference to him personally, or to any person, thought or matter related to him, without being soon made aware of *the temptation* and the difficulty of opposing it. I could give you many illustrations personal to myself, about atmospheric magnetism, lines of force, attraction, repulsion, unity of power, nature of matter, &c.; or in things more general to our common nature, about likes and dislikes, wishes, hopes, and fears; but it would be unsuitable and also unnecessary, for each must be conscious of a large field sadly uncultivated in this respect. *I will simply express my strong belief, that that point of self-education which consists in teaching the mind to resist its desires and inclinations, until they are proved to be right, is the most important of all, not only in things of natural philosophy, but in every department of daily life.*

There are numerous precepts resulting more or less from the principles of mental discipline already insisted on as essential, which are very useful in forming a judgment about matters of fact, whether among natural things or between man and man. Such a precept, and one that should recur to the mind early in every new case, is, to *know the conditions* of the matter respecting which we are called upon to make a judgement. To suppose that any would judge before they professed to know the conditions would seem to be absurd; on the other hand, to assume that the community *does wait* to know the conditions before it judges, is an assumption so large that I cannot accept it. Very few search out the conditions; most are anxious to sink those which oppose their preconceptions; yet none can be left out if a right judgment is to be formed. It is true, that many conditions must ever remain unknown to us, even in regard to the simplest things in nature : thus as to the wonderful action of gravity, whose law never fails us, we cannot say whether the bodies are acting truly at a distance, or by a physical line of force as a connecting link between them*. The great majority think the former is the case; Newton's judgment is for the latter†. But of the conditions which are within our reach, we should search out all; for in relation to those which remain unknown or unsuspected, we are in that very ignorance (regarding judgment) which it is our present object, first to make manifest, and then to remove.

One exercise of the mind, which largely influences the power and character of the judgment, is the habit of forming *clear and precise ideas*. If, after considering a subject in our ordinary manner, we return upon it with the special purpose of noticing the condition of our thoughts, we shall be astonished to find how little precise they remain. On recalling the phenomena relating to a matter of fact, the circumstances modifying them, the kind and amount of action presented, the real or probable result, we shall find that the first impressions are scarcely fit for the foundation of a judgment, and that the second thoughts will be best. For the acquirement of a good condition of mind in this respect, the thoughts should be trained to a habit of clear and precise formation, so that vivid and distinct impressions of the matter in hand, its circumstances and consequences, may remain.

Before we proceed to consider any question involving physical principles, we should set out with *clear ideas* of the naturally possible and impossible. There are many subjects uniting more or less of the most sure and valuable investiga-

* See pp. 446, 460.
† Newton's Works. Horsley's Edition, 1783, iv. p. 438; or the Third Letter to Bentley.

tions of science with the most imaginary and unprofitable speculation, that are continually passing through their various phases of intellectual, experimental, or commercial development: some to be established, some to disappear, and some to recur again and again, like ill weeds that cannot be extirpated, yet can be cultivated to no result as wholesome food for the mind. Such, for instance, in different degrees, are the caloric engine, the electric light, the Pasilalinic sympathetic compass*, mesmerism, homœopathy, odylism, the magneto-electric engine, the perpetual motion, &c.: all hear and talk of these things; all use their judgment more or less upon them, and all might do that effectively, if they were to instruct themselves to the extent which is within their reach. I am persuaded that natural things offer an admirable school for self-instruction, a most varied field for the necessary mental practice, and that those who exercise themselves therein may easily apply the habits of thought thus formed to a social use: but as a first step in such practice, clear ideas should be obtained of what is possible and what is impossible. Thus, it is impossible to *create* force. We may employ it; we may evoke it in one form by its consumption in another; we may hide it for a period; but we can neither *create* nor *destroy* it. We may cast it away; but where we dismiss it, there it will do its work. If, therefore, we desire to consider a proposition respecting the employment or evolution of power, let us carry our judgment, educated on this point, with us. If the proposal include the double use of a force with only one excitement, it implies a creation of power, and that *cannot be.* If we could by the fingers draw a heavy piece of wood or stone upward without effort, and then, letting it sink, could produce by its gravity an effort equal to its weight, that would be a creation of power, and *cannot be.*

So, again, we cannot *annihilate* matter, nor can we *create* it. But if we are satisfied to rest upon that dogma, what are we to think of table-lifting? If we could make the table to cease from acting by gravity upon the earth beneath it, or by reaction upon the hand supposed to draw it upwards, we *should annihilate it* in respect of that very property which characterizes it as matter.

* See Chambers's Journal, 1851, Feb. 15, p. 105.

Considerations of this nature are very important aids to the judgment; and when a statement is made claiming our assent, we should endeavour to reduce it to some consequence which can be immediately compared with, and tried by, these or like compact and never-failing truths. If incompatibility appears, then we have reason to suspend our conclusion, however attractive to the imagination the proposition may be, and pursue the inquiry further, until accordance is obtained; it must be a most uneducated and presumptuous mind that can at once consent to cast off the tried truth and accept in its place the mere loud assertion. We should endeavour to separate the points before us, and concentrate each, so as to evolve a clear type idea of the ruling fact and its consequences; looking at the matter on every side, with the great purpose of distinguishing the constituent reality, and recognizing it under every variety of aspect.

In like manner we should accustom ourselves to clear and definite language, especially in physical matters; giving to a word its true and full, but measured meaning, that we may be able to convey our ideas clearly to the minds of others. Two persons cannot mutually impart their knowledge, or compare and rectify their conclusions, unless both attend to the true intent and force of language. If by such words as attraction, electricity, polarity, atom, they imply different things, they may discuss facts, deny results, and doubt consequences for an indefinite time without any advantageous progress. I hold it as a great point in self-education that the student should be continually engaged in forming exact ideas, and in expressing them clearly by language. Such practice insensibly opposes any tendency to exaggeration or mistake, and increases the sense and love of truth in every part of life.

I should be sorry, however, if what I have said were understood as meaning that education for the improvement and strengthening of the judgment is to be altogether repressive of the imagination, or confine the exercise of the mind to processes of a mathematical or mechanical character. I believe that, in the pursuit of physical science, the imagination should be taught to present the subject investigated in all possible, and even in impossible views; to search for analogies of likeness and (if I may say so) of opposition—inverse or contrasted analogies; to present the fundamental idea in every form, proportion, and condition; to clothe it with suppositions and probabilities,—that all cases may pass in review, and be touched, if needful, by the Ithuriel spear of experiment. But all this must be *under government,* and the result must not be given to society until the judgment, educated by the process itself, has been exercised upon it. Let us construct our hypotheses for an hour, or a day, or for years; they are of the utmost value in the elimination of truth, ' which is evolved more freely from error than from confusion;' but, above all things, let us not cease to be aware of the temptation they offer; or, because they gradually become familiar to us, accept them as established. We could not reason about electricity without thinking of it as a fluid, or a vibration, or some other existent state or form. We should give up half our advantage in the consideration of heat if we refused to consider it as a principle, or a state of motion. We could scarcely touch such subjects by experiment, and we should make no progress in their practical application without hypothesis; still it is absolutely necessary that we should learn to doubt the conditions we assume, and acknowledge we are uncertain, whether heat and electricity are vibrations or substances, or either.

When the different data required are in our possession, and we have succeeded in forming a clear idea of each, the mind should be instructed to *balance them* one against another, and not suffered carelessly to hasten to a conclusion. This reserve is most essential; and it is especially needful that the reasons which are adverse to our expectations or our desires should be carefully attended to. We often receive truth from unpleasant sources; we often have reason to accept unpalatable truths. We are never freely willing to admit information having this unpleasant character, and it requires much self-control in this respect, to preserve us even in a moderate degree from errors. I suppose there is scarcely one investigator in original research who has not felt the temptation to disregard the reasons and results which are against his views. I acknowledge that I have experienced it very often, and will not pretend to say that I have yet learned on all occasions to avoid the error. When a bar of bismuth or phosphorus is placed between the poles of a powerful magnet, it is drawn into a position across the line joining the poles; when only one pole is near the bar, the latter recedes; this and the former effect are due to repulsion, and are strikingly in contrast with the attraction shown by iron. To account for it, I at one time suggested the idea that a polarity was induced in the phosphorus or bismuth the reverse of the polarity induced in iron, and that opinion is still sustained by eminent philosophers. But observe a necessary result of such a supposition, which appears to follow when the phenomena are referred to elementary principles. *Time* is shown, by every result bearing on the subject, to be concerned in the coming on and passing away of the inductive condition produced by magnetic force, and the consequence, as Thomson pointed out, is, that if a ball of bismuth could be suspended between the poles of a magnet, so as to encounter no resistance from the surrounding medium, or from friction or torsion, and were once put in motion round a vertical axis, it would, because of the assumed polar state, go on for ever revolving, the parts which at any moment are axial moving like the bar, so as to become the next moment equatorial. Now, as we believe the mechanical forces of nature tend to bring things into a stable, and not into an unstable condition; as we believe that a perpetual motion is impossible; so because both these points are involved in the notion of the reverse polarity, which itself is not supposed to be dependent on any consumption of power, I feel bound to hold the judgment balanced, and therefore hesitate to accept a conclusion founded on such a notion of the physical action; the more especially as the peculiar test facts* which prove the polarity of iron are not reproduced in the case of diamagnetic bodies.

As a result of this wholesome mental condition, we should be able to form a *proportionate judgment*. The mind naturally desires to settle upon one thing or another; to rest upon an affirmative or a negative; and that with a degree of absolutism which is irrational and improper. In drawing a conclusion, it is very difficult, but not the less necessary, to make it *proportionate* to the evidence: except where certainty exists (a case of rare occurrence), we should consider our decisions as probable only. The probability may appear very great, so that in affairs of the world we often accept such as certainty, and trust

* Experimental Researches in Electricity, paragraphs 2657–2681.

our welfare or our lives upon it. Still, only an uneducated mind will confound probability with certainty, especially when it encounters a contrary conclusion drawn by another from like data. This suspension in degree of judgment will not make a man less active in life, or his conclusions less certain as truths; on the contrary, I believe him to be the more ready for the right amount and direction of action on any emergency; and am sure his conclusions and statements will carry more weight in the world than those of the incautious man.

When I was young, I received from one well able to aid a learner in his endeavours toward self-improvement, a curious lesson in the mode of estimating the amount of belief we might be induced to attach to our conclusions. The person was Dr. Wollaston, who, upon a given point, was induced to offer me a wager of two to one on the affirmative. I rather impertinently quoted Butler's well-known lines* about the kind of persons who use wagers for argument, and he gently explained to me, that he considered such a wager not as a thoughtless thing, but as an expression of the amount of belief in the mind of the person offering it; combining this curious application of the wager, as a *meter*, with the necessity that ever exists of drawing conclusions, not absolute but proportionate to the evidence.

Occasionally and frequently the exercise of the judgment ought to end in *absolute reservation*. It may be very distasteful, and great fatigue, to suspend a conclusion; but as we are not infallible, so we ought to be cautious; we shall eventually find our advantage, for the man who rests in his position is not so far from right as he who, proceeding in a wrong direction, is ever increasing his distance. In the year 1824, Arago discovered† that copper and other bodies placed in the vicinity of a magnet, and having no direct action of attraction or repulsion upon it, did affect it when moved, and was affected by it. A copper plate revolving near a magnet carried the magnet with it; or if the magnet revolved, and not the copper, it carried the copper with it. A magnetic needle vibrating freely over a disc of glass or wood, was exceedingly retarded in its motion when these were replaced by a disc of copper. Arago

* " Quoth she, I 've heard old cunning stagers
  Say fools for arguments use wagers."
† Annales de Chimie, xxviii. 325.

stated most clearly all the conditions, and resolved the forces into three directions; but not perceiving the physical cause of the action, exercised a most wise and instructive reservation as to his conclusion. Others, as Haldat, considered it as the proof of the universality of a magnetism of the ordinary kind, and held to that notion though it was contradicted by the further facts; and it was only at a future period that the true physical cause, namely, magneto-electric currents induced in the copper, became known to us*. What an education Arago's mind must have received in relation to philosophical reservation; what an antithesis he forms with the mass of table-turners; and what a fine example he has left us of that condition of judgment to which we should strive to attain!

If I may give another illustration of the needful reservation of judgment, I will quote the case of oxygen and hydrogen gases, which, being mixed, will remain together uncombined for years in contact with glass, but in contact with spongy platinum combine at once. We have the same fact in many forms, and many suggestions have been made as to the mode of action; but as yet we do not know *clearly* how the result comes to pass. We cannot tell whether electricity acts or not. Then we should suspend our conclusions. Our knowledge of the fact itself, and the many varieties of it, is not the less abundant or sure; and when the truth shall hereafter emerge from the mist, we ought to have no opposing prejudice, but be prepared to receive it.

The education which I advocate will require *patience* and *labour of thought* in every exercise tending to improve the judgment. It matters not on what subject a person's mind is occupied, he should engage in it with the conviction that it will require mental labour. A powerful mind will be able to draw a conclusion more readily and more correctly than one of moderate character; but both will surpass themselves if they make an earnest, careful investigation, instead of a careless or prejudiced one; and education for this purpose is the more necessary for the latter, because the man of less ability may, through it, raise his rank and amend his position. I earnestly urge this point of self-education, for I believe it to be more or less in the power of every man greatly to improve his judgment. I do not think

* Philosophical Transactions, 1832, p. 146.

that one has the complete capacity for judgment which another is naturally without. I am of opinion that all may judge, and that we only need to declare on every side the conviction that mental education is wanting, and lead men to see that through it they hold, in a large degree, their welfare and their character in their own hands, to cause in future years an abundant development of right judgment in every class.

This education has for its first and its last step *humility.* It can commence only because of a conviction of deficiency; and if we are not disheartened under the growing revelations which it will make, that conviction will become stronger unto the end. But the humility will be founded, not on comparison of ourselves with the imperfect standards around us, but on the increase of that internal knowledge which alone can make us aware of our internal wants. The first step in correction is to learn our deficiencies, and having learned them, the next step is almost complete: for no man who has discovered that his judgment is hasty, or illogical, or imperfect, would go on with the same degree of haste, or irrationality, or presumption as before. I do not mean that all would at once be cured of bad mental habits, but I think better of human nature than to believe, that a man in any rank of life, who has arrived at the consciousness of such a condition, would deny his common sense, and still judge and act as before. And though such self-schooling must continue to the end of life to supply an experience of deficiency rather than of attainment, still there is abundant stimulus to excite any man to perseverance. What he has lost are things imaginary, not real; what he gains are riches before unknown to him, yet invaluable; and though he may think more humbly of his own character, he will find himself at every step of his progress more sought for than before, more trusted with responsibility and held in pre-eminence by his equals, and more highly valued by those whom he himself will esteem worthy of approbation.

And now a few words upon the mutual relation of two classes, namely, *those* who decline to educate their judgments in regard to the matters on which they decide, and those who, by self-education, have endeavoured to improve themselves; and upon the remarkable and somewhat unreasonable manner in which the latter are called upon, and occasionally taunted, by the former. A man who makes assertions, or draws conclusions, regarding any given case, ought to be competent to investigate it. He has no right to throw the onus on others, declaring it their duty to prove him right or wrong. His duty is to demonstrate the truth of that which he asserts, or to cease from asserting. The men he calls upon to consider and judge have enough to do with themselves, in the examination, correction, or verification of their own views. The world little knows how many of the thoughts and theories which have passed through the mind of a scientific investigator have been crushed in silence and secrecy by his own severe criticism and adverse examination; that in the most successful instances not a tenth of the suggestions, the hopes, the wishes, the preliminary conclusions have been realized. And is a man so occupied to be taken from his search after truth in the path he hopes may lead to its attainment, and occupied in vain upon nothing but a broad assertion?

Neither has the assertor of any thing new a right to claim an answer in the form of *Yes* or *No*; or think, because none is forthcoming, that he is to be considered as having established his assertion. So much is unknown to the wisest man, that he may often be without an answer: as frequently he is so, because the subject is in the region of hypothesis, and not of facts. In either case he has the right to refuse to speak. I cannot tell whether there are two fluids of electricity or any fluid at all. I am not bound to explain how a table tilts any more than to indicate how, under the conjurer's hands, a pudding appears in a hat. The means are not known to me. I am persuaded that the results, however strange they may appear, are in accordance with that which is truly known, and if carefully investigated would justify the well-tried laws of nature; but, as life is limited, I am not disposed to occupy the time it is made of, in the investigation of matters which, in what is known to me of them, offer no reasonable prospect of any useful progress, or anything but negative results. We deny the right of those who call upon us to answer their speculations ' *if we can,*' whilst we have so many of our own to develope and correct; and claim the right for ourselves of withholding either our conclusions or the reasons for them, without in the least degree admitting that their affirmations are unanswerable. We are not even called upon to give an answer to the best of our belief; nor bound to admit a bold assertion because we do not *know* to the contrary. No one is justified in claiming our assent to the spontaneous generation of insects, because we cannot circumstantially explain how a mite or the egg of a mite has entered into a particular bottle. Let those who affirm the exception to the general law of nature, or those others who upon the affirmation accept the result, work out the experimental proof. It has been done in this case by Schulze*, and is in the negative; but how few among the many who make or repeat the assertion, would have the requisite self-abnegation, the subjected judgment, the perseverance, and the precision, which has been displayed in that research!

When men, more or less marked by their advance, are led by circumstances to give an opinion adverse to any popular notion, or to the assertions of any sanguine inventor, nothing is more usual than the attempt to neutralize the force of such an opinion by reference to the mistakes which like educated men have made; and their occasional misjudgments and erroneous conclusions are quoted, as if they were less competent than others to give an opinion, being even disabled from judging like matters to those which are included in their pursuits by the very exercise of their minds upon them. How frequently has the reported judgment of Davy, upon the impossibility of gas-lighting on a large scale, been quoted by speculators engaged in tempting moneyed men into companies, or in the pages of journals occupied with the popular fancies of the day ; as if an argument were derivable from that in favour of some special object to be commended ! Why should not men taught in the matter of judgment far beyond their neighbours, be expected to err sometimes, since the very education in which they are advanced can only terminate with their lives? What is there about them, derived from *this education*, which sets up the shadow of a pretence to perfection? Such men cannot learn all things, and may often be ignorant. The very progress which science makes amongst them as a body is a continual correction of ignorance, *i. e.* of a state which is ignorance in relation to the future, though wisdom and knowledge in relation

* Müller's Physiology, or Poggendorff's Annalen, 1836, xxxix. p. 487.

to the past. In 1823, Wollaston discovered that beautiful substance which he called Titanium, believing it to be a simple metal; and it was so accepted by all philosophers. Yet this was a mistake, for Wohler*, in 1850, showed the substance was a very compound body. This is no reproach to Wollaston or to those who trusted in him; he made a step in metallurgy which advanced knowledge, and perhaps we may hereafter, through it, learn to know that metals are compound bodies. Who, then, has a right to quote his mistake as a reproach against him? Who could correct him but men intellectually educated as he himself was? Who does not feel that the investigation remains a bright gem in the circlet that memory offers to his honour?

If we are to estimate the utility of an educated judgment, do not let us hear merely of the errors of scientific men, which have been corrected by others taught in the same careful school; but let us see what, as a body, they have produced, compared with that supplied by their reproachers. Where are the established truths and triumphs of ring-swingers, table-turners, table-speakers? What one result in the numerous divisions of science or its applications can be traced to their exertions? Where is the investigation completed, so that, as in gas-lighting, all may admit that the principles are established and a good end obtained, without the shadow of a doubt?

If we look to electricity, it, in the hands of the careful investigator, has advanced to the most extraordinary results: it approaches at the motion of his hand; bursts from the metal; descends from the atmosphere; surrounds the globe: it talks, it writes, it records, it appears to him (cautious as he has learned to become) as a universal spirit in nature. If we look to photography, whose origin is of our own day, and see what it has become in the hands of its discoverers and their successors, how wonderful are the results! The light is made to yield impressions upon the dead silver or the coarse paper, beautiful as those it produces upon the living and sentient retina: its most transient impression is rendered durable for years; it is made to leave a visible or an invisible trace; to give a result to be seen now or a year hence; made to paint all natural forms and even colours; it serves the offices of war, of peace,

* Annales de Chimie, xxix. p. 166.

of art, science, and economy : it replaces even the mind of the human being in some of its lower services ; for a little camphine lamp is set down and left to itself, to perform the duty of watching the changes of magnetism, heat, and other forces of nature, and to record the results, in pictorial curves, which supply an enduring record of their most transitory actions.

What has clairvoyance, or mesmerism, or table-rapping done in comparison with results like these ? What have the snails at Paris told us from the snails at New York ? What have any of these intelligences done in *aiding* such developments ? Why did they not inform us of the possibility of photography ? or when that became known, why did they not favour us with some instructions for its improvement ? They all profess to deal with agencies far more exalted in character than an electric current or a ray of light : they also deal with mechanical forces ; they employ both the bodily organs and the mental ; they profess to lift a table, to turn a hat, to see into a box, or into the next room, or a town :—why should they not move a balance, and so give us the element of a new mechanical power ? take cognizance of a bottle and its contents, and tell us how they will act upon those of a neighbouring bottle ? either see or feel into a crystal, and inform us of what it is composed ? Why have they not added one metal to the fifty known to mankind, or one planet to the number daily increasing under the observant eye of the astronomer ? Why have they not corrected one of the *mistakes* of the philosophers ? There are no doubt very many that require it. There has been plenty of time for the development and maturation of some of the numerous public pretences that have risen up in connexion with these supposed agencies ; how is it that not one new power has been added to the means of investigation employed by the philosophers, or one valuable utilitarian application presented to society ?

In conclusion, I will freely acknowledge that all I have said regarding the great want of judgment manifested by society as a body, and the high value of any means which would tend to supply the deficiency, have been developed and declared on numerous occasions, by authority far above any I possess. The deficiency is known hypothetically, but I doubt if in reality ; the individual acknowledges the state in respect of others, but is unconscious of it in regard to himself. As to the world at large, the condition is accepted as a necessary fact ; and so it is left untouched, almost ignored. I think that education in a large sense should be applied to this state of the subject, and that society, though it can do little in the way of communicated experience, can do much, by a declaration of the evil that exists and of its remediable character, by keeping alive a sense of the deficiency to be supplied, and by directing the minds of men to the practice and enlargement of that self-education which every one pursues more or less, but which under conviction and method would produce a tenfold amount of good. I know that the multitude will always be behindhand in this education, and to a far greàter extent than in respect of the education which is founded on book learning. Whatever advance books make, they retain; but each new being comes on to the stage of life, with the same average amount of conceit, desires, and passions, as his predecessors, and in respect of self-education has all to learn. Does the circumstance that we can do little more than proclaim the necessity of instruction, justify the ignorance, or our silence, or make the plea for this education less strong? Should it not, on the contrary, gain its strength from the fact that all are wanting more or less? I desire we should admit that, as a body, we are universally deficient in judgment. I do not mean that we are utterly ignorant, but that we have advanced only a little way in the requisite education, compared with what is within our power.

If the necessity of the education of the judgment were a familiar and habitual idea with the public, it would often afford a sufficient answer to the statement of an ill-informed or incompetent person; if quoted to recall to his remembrance the necessity of a mind instructed in a matter, and accustomed to balance evidence, it might frequently be an answer to the individual himself. Adverse influence might, and would, arise from the careless, the confident, the presumptuous, the hasty, and the dilatory man, perhaps extreme opposition; but I believe that the mere acknowledgment and proclamation of the ignorance, by society at large, would, through its moral influence, destroy the opposition, and be a great means to the attainment of the good end desired: for if no more be done than to lead such to turn their thoughts inwards, a step in education is gained: if they are *convinced* in any degree, an important advance is made; if they learn only to *suspend* their judgment, the improvement will be one above price.

It is an extraordinary thing, that man, with a mind so wonderful that there is nothing to compare with it elsewhere in the known creation, should leave it to run wild in respect of its highest elements and qualities. He has powers of comparison and judgment, by which his final resolves, and all those acts of his material system which distinguish him from the brutes, are guided :—shall he omit to educate and improve them when education can do much? Is it towards the very principles and privileges that distinguish him above other creatures, he should feel indifference? Because the education is internal, it is not the less needful; nor is it more the duty of a man that he should cause his child to be taught than that he should teach himself. Indolence may tempt him to neglect the self-examination and experience which form his school, and weariness may induce the evasion of the necessary practices; but surely a thought of the prize should suffice to stimulate him to the requisite exertion: and to those who reflect upon the many hours and days, devoted by a lover of sweet sounds, to gain a moderate facility upon a mere mechanical instrument, it ought to bring a correcting blush of shame, if they feel convicted of neglecting the beautiful living instrument, wherein play all the powers of the mind.

I will conclude this subject :—believe me when I say I have been speaking from self-conviction. I did not think this an occasion on which I ought to seek for flattering words regarding our common nature ; if so, I should have felt unfaithful to the trust I had taken up ; so I have spoken from experience In thought I hear the voice, which judges me by the prece~ s I have uttered. I know that I fail frequently in that very exercise of judgment to which I call others ; and have abundant reason to believe that much more frequently I stand manifest to those around me, as one who errs, without being corrected by knowing it. I would willingly have evaded appearing before you on this subject, for I shall probably do but little good, and may well think it was an error of judgment to consent: having consented, my thoughts would flow back amongst the events and reflections of my past life, until I found nothing present itself but an open declaration, almost a confession, as the means of performing the duty due to the subject and to you.

INDEX.

THE END.

PRINTED BY TAYLOR AND FRANCIS, RED LION COURT, FLEET STREET.

Plate I.

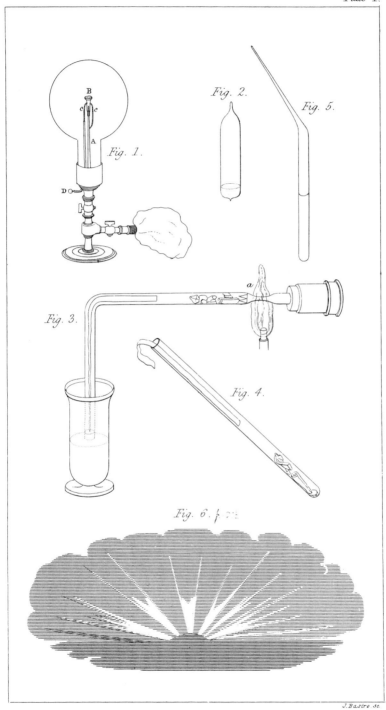

Fig. 1.

Fig. 2.

Fig. 5.

Fig. 3.

Fig. 4.

Fig. 6.

J. Basire sc.

Taylor and Francis_1859.

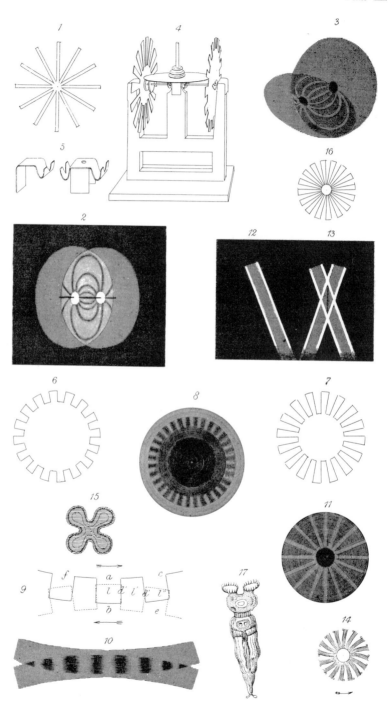

Plate II.

CHEMISTRY AND PHYSICS

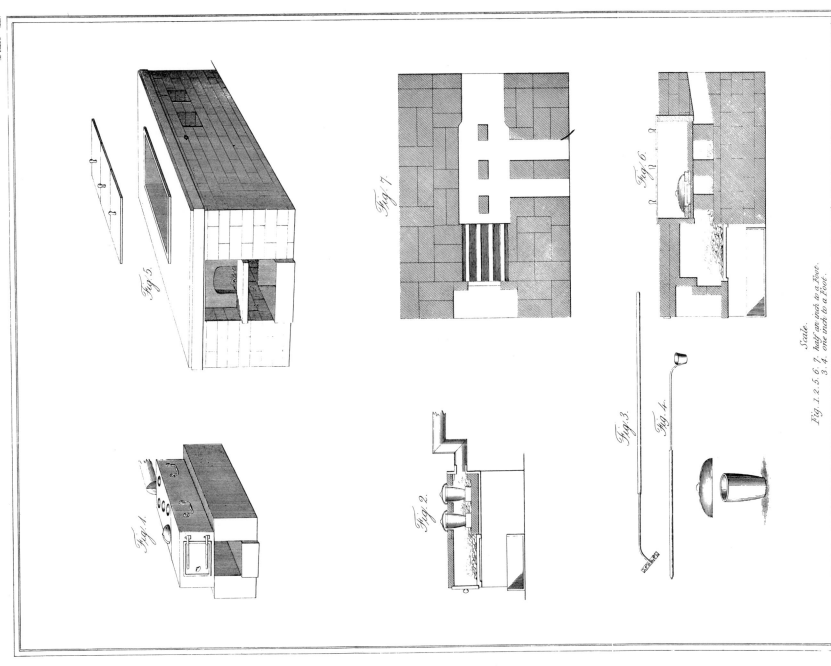

Fig. 5.

Fig. 7.

Fig. 6.

Fig. 1.

Fig. 2.

Fig. 3.

Fig. 4.

Scale.

Fig. 1. 2. 5. 6. 7. half an inch to a Foot.
3. 4. one inch to a Foot.

J. Basire, sc.

Taylor and Francis. 1859.